FALSEHOODS FLY

FALSEHOODS FLY

Why Misinformation Spreads and How to Stop It

PAUL THAGARD

Columbia University Press
New York

Columbia University Press
Publishers Since 1893
New York Chichester, West Sussex
cup.columbia.edu

Library of Congress Cataloging-in-Publication Data
Names: Thagard, Paul, author.
Title: Falsehoods fly : why misinformation spreads and how to stop it / Paul Thagard.
Description: New York : Columbia University Press, [2024] | Includes bibliographical references and index.
Identifiers: LCCN 2023028323 | ISBN 9780231213943 (hardback) | ISBN 9780231213950 (trade paperback) | ISBN 9780231560115 (ebook)
Subjects: LCSH: Misinformation—Social aspects. | Information science—Sociological aspects. | Disinformation—Social aspects. | Information integrity—Social aspects. | Truthfulness and falsehood—Social aspects.
Classification: LCC HM851 .T496 2024 | DDC 303.48/33—dc23/eng/20230731
LC record available at https://lccn.loc.gov/2023028323

Printed and bound by CPI Group (UK) Ltd, Croydon, CR0 4YY

Cover design: Henry Sene Yee

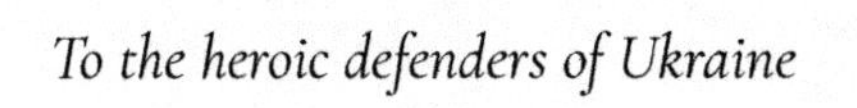

To the heroic defenders of Ukraine

CONTENTS

PREFACE

In his 1710 essay on political lying, Jonathan Swift wrote that "falsehood flies and truth comes limping after it."[1] Four hundred years later, the flight of falsehoods has been appallingly accelerated by the interaction of politics and technology, with social media making it easy for lies to spread to millions of people almost instantly. Dealing with humanity's major problems requires dealing with misinformation.

Misinformation is threatening medicine, science, politics, social justice, and international relations, affecting problems such as vaccine hesitancy, climate change denial, conspiracy theories, claims of racial inferiority, and the Russian invasion of Ukraine. Dealing with misinformation requires explanation of how information is generated and spread and how it breaks down but can be repaired. This book offers a deep account of information and misinformation and provides concrete advice on how improved thinking and communication can benefit individuals and societies.

Concern with misinformation is widespread, but this book is unique in both depth and breadth. The depth comes from an original theory of information as deriving from four processes: acquisition, inference, memory, and spread. Each of these processes is spelled out in terms of concrete mechanisms that generate real information when done well but can easily break down and produce misinformation. Fortunately, misinformation can be transformed into real information using the same mechanisms.

The breadth of the book comes from applying the theory of information and misinformation to five different domains: COVID-19, climate change, conspiracy theories, inequality, and the Russia-Ukraine war. The practical benefit of the book comes from its provision of tools for converting misinformation into real information that can help solve major social problems.

The new theories of information and misinformation developed here grew out of a concern with the impact of energy on information processing in animals and computers.[2] In November 2019, I happened to hear a biology talk by Mary O'Connor and a theoretical neuroscience talk by Chris Eliasmith, which struck me as converging on important conclusions about the nature of mind. While working out the conclusions, I did a review of recent work on information and was struck by the lack of progress toward a theory rich enough to explain the kinds of meaningful information employed by people, other animals, and advanced computers.

Along with other contemporary philosophers of science, I have long advocated the view of theories as descriptions of mechanisms, so I asked, What are the mechanisms that govern the generation and use of information? I quickly produced a list of eight plausible mechanisms that I later realized fall under the more manageable headings of acquisition, inference, memory, and spread, all captured by the acronym AIMS.

In March 2020, the COVID-19 pandemic struck and the world shut down. Scientific information exploded about the new coronavirus, but so did misinformation about its origin and treatment. It struck me that misinformation could be explained by breakdowns in the mechanisms of information that I had identified, just as disease is explained by breakdowns in the biological mechanisms that support health. I later expanded this analogy to look for ways of transforming misinformation into real information, just as effective medical treatments help to transform disease into health. I hope that the tools developed in this book for converting misinformation into real information will help people deal with medical issues along with the many other threats we now face.

Because of the practical importance of misinformation, I have tried to make this book accessible to a wide audience. A glossary summarizes key concepts explained more thoroughly in the text. I reserve more technical, academic matters for an online supplement available at paulthagard.com; this online supplement will also provide live links to all the web references in the notes.

After the theoretical basics are introduced in chapters 1 to 3, the application chapters 4 to 8 can be read in any order. Readers wanting an executive summary could first look at chapters 1 and 8.

ACKNOWLEDGMENTS

For helpful suggestions, I am grateful to Jan Angus, Adam Thagard, Dan Thagard, and Joanne Wood. For valuable comments on an earlier draft, I thank N. Emrah Aydiononat, Steve Bank, Alessandra Basso, Jim Bowers, Gerry Harnett, Teemu Lari, Uskali Mäki, Carlo Martini, Carl Pado, and Lauren Talalay.

I have recycled some material from my *Psychology Today* blog, *Hot Thought*, for which I hold copyright.

Thanks to Miranda Martin and Ben Kolstad for editorial support, Marianne L'Abbate for copyediting, and ARC Indexing for the index.

FALSEHOODS FLY

CHAPTER 1

LIES KILL

The Perils of Misinformation

Misinformation kills.

Phil Valentine was an American radio host who told his audience that they were at low risk of getting COVID-19 and should not bother getting vaccinated. In August 2021, he died of the disease.[1] By 2022, more than 1 million Americans had died of COVID-19 despite wide availability of effective vaccines, which many people declined because of misinformation about their risks and benefits.

In July 2021, Lytton, British Columbia, recorded Canada's highest ever temperature of 49.7°C (121.3°F), a heat wave that experts said was a sign of increasing global warming.[2] The perils of climate change have been evident since the 1990s, but misinformation from oil companies and their political allies has discouraged actions to slow global warming. The heat wave in the Pacific Northwest led to hundreds of deaths,[3] and the Lytton wildfire killed two people in their sixties.[4]

A mob attacked the United States Capitol Building in Washington, DC, on January 6, 2021. They were motivated by misinformation that the 2020 election had been stolen by Joseph Biden's allegedly rigged victory over Donald Trump. The riot produced many injuries, including more than one hundred police officers, and one rioter was killed by police.[5] Several subsequent deaths of officers were attributed to the stress of the traumatic confrontation.

In May 2022, Payton Gendron shot and killed ten people in a Black neighborhood of Buffalo, New York. He had posted online a document endorsing the conspiracy theory of a plot among elites and racial minorities to replace the white majority.[6]

Modern democracies proclaim that all people are equal, but even the best perpetuate inequalities that harm the disadvantaged. For example, Canada had higher rates of COVID-19 deaths for people who were poor, minorities, or recent immigrants.[7] In the United States, the decline in life expectancy brought by the pandemic was greater for Black Americans than for whites.[8]

In February 2022, Russia invaded Ukraine with casualties that quickly included thousands of deaths and millions of refugees. Russian President Vladimir Putin justified the invasion by spreading misinformation that Ukraine is dominated by neo-Nazis.[9]

In all these cases, misinformation caused deaths. Even when misinformation is not lethal, it can be dangerous. People who get bad medical advice can suffer needlessly from disease. Increased global warming will lead to food shortages and panicked migrations as well as millions of deaths. Political conspiracy theories undermine confidence in democratic governments and encourage authoritarian alternatives, with governments that imprison and torture citizens. Misinformation is used to justify social inequality that causes suffering for billions of people. Thriving in modern societies requires getting a good education, but misinformation encourages defective teaching practices. Misinformation promotes wars based on lies concerning the perceived enemy.

The harms of misinformation have been addressed using different terminology by critics of myths, misconceptions, humbug, baloney, malarkey, and bullshit. I could add "fake news" to that list but the term has been debased by purveyors of misinformation who use it to dismiss reports that contradict their own lies.

A 2022 survey found that 70 percent of people across nineteen countries viewed the spread of false information online as a major threat, second only to climate change.[10] Dealing with misinformation is an enormous medical, scientific, and social problem whose solution requires an understanding of how information is generated, understood, and spread. This book offers a systematic explanation of information and misinformation, along with concrete advice on how improved thinking and communication can benefit individuals and societies.

WHAT IS MISINFORMATION?

In 2021, the U.S. Surgeon General Vivek Murthy released an excellent *report on confronting health misinformation*. It said: "Health misinformation is a serious threat to public health. It can cause confusion, sow *mistrust*, harm people's health, and undermine public health efforts. Limiting the spread of health misinformation is a moral and civic imperative that will require a whole-of-society effort."[11]

The report defines misinformation as "information that is false, inaccurate, or misleading according to the best available evidence at the time."[12] This definition includes information that is misleading as well as false because even true reports can be harmful. For example, someone who posts on social media that a friend got a blood clot after getting a COVID-19 vaccination may be reporting the truth, but the anecdote is misleading if it suggests that vaccines are dangerous despite ample evidence that their risks are minuscule compared with the dangers of COVID-19. The standard of evaluating information according to the best available evidence recognizes that evidence can change over time but can still be used wisely. In contrast to misinformation, we can recognize *real* information as true, accurate, and trustworthy in its descriptions, explanations, and practical advice.

The major problem with the surgeon general's definition is that it depends on the meaning of information, which dictionaries unhelpfully define as knowledge, facts, or data, which in turn are often defined as information. Information and misinformation can best be characterized together by developing a joint theory that applies to both. The difference between information and misinformation can be captured by identifying the psychological and social mechanisms that produce information and by describing how breakdowns in these mechanisms promote misinformation.[13] At the end of this chapter, I characterize information by specifying examples, typical features, and explanations.

Disinformation is misinformation that is spread deliberately by people who know it is false. Disinformation consists of intentional lies, whereas misinformation can be unwittingly spread by people who actually believe it. Former president Barack Obama has described disinformation as "the single biggest threat to our democracy."[14]

In biology, life is understood as resulting from mechanisms that include metabolism, reproduction, genetics, locomotion, and natural selection. These mechanisms specify parts such as cells, connections such as cell adhesion, and interactions such as cell signaling; together, they explain the life of organisms. Failures of these mechanisms then explain the end of life, for example, when starvation leads to death. Everyone is familiar with mechanisms such as scissors, toasters, and bicycles, which are combinations of connected parts whose interactions produce regular changes. Scissors have just three parts (two blades and

"I just feel fortunate to live in a world with so much disinformation at my fingertips."

1.1 Disinformation.

Source: www.CartoonStock.com. Reprinted by permission.

a bolt that holds them together) but people can use them to cut paper or cloth. Mechanisms break when problems with the parts, connections, or interactions interfere with the functional changes, for example, when a flat tire on a bicycle prevents it from moving easily.

Similarly, human health is the proper functioning of an organism that depends on mechanisms that include digestion, respiration, blood circulation, the immune system, and cell division.[15] Diseases are breakdowns in these mechanisms, for example, when a blocked artery causes a heart attack or when faulty cell division produces a cancerous tumor. Medical treatments attempt to cure diseases or reduce their symptoms by fixing the broken mechanisms, for example, by using a stent to unblock an artery or a drug to kill tumor cells.

Health requires functioning mechanisms that are damaged in diseases but fixed or improved by medical treatments. In a similar way, information requires

TABLE 1.1 Medical analogy for information

	Medicine	**Information**
Causal mechanisms	Health	Real information
Breakdowns in mechanisms	Disease	Misinformation
Fix breakdowns	Treatment	Reinformation
Avoid breakdowns	Prevention	Preinformation

psychological and social mechanisms that can break to produce misinformation, which can be mended by identifiable strategies. I will describe how these strategies work in domains that include medicine, science, politics, society, and war. Together they comprise a tool kit for fostering the spread of valuable information and inhibiting the generation and propagation of harmful misinformation.

I propose a new word, "**re**information," to describe the process of correcting misinformation by *re*pairing or *re*medying it to *re*store, *re*claim, or *re*cover real information. The goal is to explain how information often works well, sometimes breaks into misinformation, but can be mended by reinformation. Tools for reinformation are the keys to transforming misinformation into real information. Reinformation is to misinformation and real information what medical treatment is to disease and health.

My medical analogy is spelled out in table 1.1, which summarizes the relations between mechanisms, breakdowns, treatments, and prevention. In medicine, biological mechanisms explain what makes people healthy, but breakdowns in these mechanisms explain the occurrence of disease symptoms. Treatments overcome diseases by fixing the breakdowns. Similarly, the causal mechanisms I will describe can produce real information, but breakdowns in these mechanisms produce misinformation that must be fixed by reinformation tools. Repairing machines, bodies, and information requires knowing how they work. The prevention aspects of medicine and information are discussed later in this chapter, where I introduce the term "preinformation" to capture ways of avoiding the start and spread of misinformation. The mechanisms responsible for health and real information include social interactions as well as biological processes in bodies and brains.

This medical analogy for information reveals the causal relations among information and misinformation, which are ignored by other commonly used metaphors. People sometimes talk about information wars (infowars) and information epidemics (infodemics).[16] The information-war metaphor highlights the exercise of misinformation between combatants such as Russia and Ukraine, but it reveals little about how misinformation can be overcome.

Information spread is superficially like the spread of infectious diseases, but this comparison ignores the causal structure of misinformation. The representations that carry information differ from the germs that spread infectious diseases because they stand for situations in the world, are generated by goal-directed psychological mechanisms rather than random mutations, and are spread by social mechanisms rather than infection.[17] The war and epidemic metaphors are fine as rhetorical flourishes, but they provide much less insight into misinformation than the medical analogy with its deep causal structure. To make the medical analogy useful, we need a theory of the causal processes that produce information and misinformation.

THE AIMS THEORY OF INFORMATION AND MISINFORMATION

Information can be explained by mechanisms that fall under four general processes: acquisition, inference, memory, and spread (AIMS). By acquisition, I mean gathering information through interaction with the world, not by communication with other people, which falls under spread. Inference involves evaluating pieces of information and transforming them into new pieces that go beyond observations. Memory operates by storing information for future use and retrieving it when useful. Spread occurs when people transmit information to other people.

As an illustration, consider what happens when two people, Pat and Sam, meet for the first time. Pat immediately acquires perceptual information about Sam, such as approximate height, shape, skin color, and clothing. In addition, Pat may get information about Sam that is spread from a mutual friend Quinn, for example, being told that Sam is a baseball fan. Pat's information about Sam is subject to inferences about its quality, for example, through doubts about whether Pat's observation of Sam's height was accurate or through wondering whether Quinn actually knew Sam's sports interests. If Pat thinks that Sam is interesting and important, then the information that Pat has acquired about Sam will be remembered for recall on future occasions. Pat may then spread information about Sam to other people.

The four AIMS processes usually succeed in giving people real information, but breakdowns can lead to misinformation. Acquisition breaks if it uses unreliable perceptions, for example, if Pat's observations of Sam occur in bad lighting or while Pat is drunk. Inference breaks if Pat uses poor methods of generating new information, for example, through wishful thinking that Sam likes Pat. Memory breaks if Pat fails to remember an accurate perception of Sam or if Pat later has trouble remembering much about Sam. Spread breaks if a source is unreliable, for example, if Quinn is prone to lying. Chapter 2 looks at acquisition, inference, memory, and spread in much more detail, describing how their mechanisms contribute to real information, misinformation, and reinformation.

Real information is acquired from the world through reliable mechanisms of perception, systematic observation using instruments, well-designed experiments, and clinical trials. In contrast, misinformation fails to represent the world because it comes from biased observations, faulty instruments, sloppy experiments, and making stuff up, which originates from imagination rather than interacting with the world.

Inference allows real information to go beyond observations when it employs rigorous, evidence-based evaluation and causal reasoning that take into account the full range of data and alternative explanations. Misinformation results from inferences that are biased by personal motivations and deficient causal stories.

Memory supports real information when it uses proper methods of storage and retrieval that preserve accuracy and are accompanied by sound evaluations of the value and reliability of the data stored and retrieved. Misinformation results from faulty individual memories and distorted public repositories that fail to follow good standards of storage and retrieval.

Finally, real information is spread by trustworthy senders and skeptical receivers, who use repeated evaluations to determine what information is sufficiently real to be believed and transmitted to others. Misinformation spreads through a population as the result of irresponsible senders and gullible receivers who then turn into additional irresponsible senders.

Overall, real information springs from reliable acquisition of representations from the world, sound inferences that extend those representations, careful memory practices of storage and retrieval, and sensible spread of representations among a group of users. In contrast, misinformation results from shoddy acquisition, defective inferences, sloppy memory, and careless or malicious spread. Real information reflects reality, whereas misinformation ignores or distorts reality.

TABLE 1.2 Profiles of real information and misinformation

Process	Real information	Misinformation
Acquisition	Collecting by perception, instruments, systematic observations, and controlled experiments	Making stuff up, faulty observations, sloppy experiments
Inference	Evidence-based causal reasoning	Motivated reasoning, flawed causal reasoning
Memory	Evaluation-based storing and retrieving	Motivated storing and retrieving
Spread	Evaluation-based sending and receiving	Motivated sending and receiving

Chapter 2 provides details of the relevant mechanisms.

The mere definition of misinformation as information that is false or misleading is unhelpful for identifying it because it does not say anything about *when* information is false or misleading. Fortunately, the four processes identified by the AIMS theory of information provide a scrupulous guide to distinguishing real information from misinformation, as shown in the contrasting profiles in table 1.2. Motivated reasoning, which reaches conclusions based on personal goals and group identities rather than evidence, is one of the main drivers behind misinformation, as case studies will show. Memory and spread need to be constrained by evaluations concerning the accuracy and usefulness of information rather than by personal motivations.

OVERCOMING MISINFORMATION

These profiles provide powerful ways of recognizing misinformation, as we will see in applications to medical, scientific, political, social, and military cases. They also suggest ways of countering misinformation that amount to a reinformation tool kit. I propose eight techniques for converting misinformation into real information: spotting misinformation, locating sources, scrutinizing the motives of sources, factual correction, critical thinking, motivational interviewing, institutional modification, and political action. I now illustrate these techniques using fictional examples, with serious cases to come in chapters 4–8.

The first technique is to spot pieces of misinformation, initially by noticing claims that contradict what is already well known. For example, if Pat says that Sam is in New York City when you know from Instagram that Sam is in Toronto, then you have good reason to suspect Pat of spreading misinformation. Misinformation can also be detected according to the profiles in table 1.2, for example, when a claim is not backed by any well-collected evidence and therefore appears to arise by making stuff up. If you know that Pat has had no contact with Sam, then the claim about Sam's location is more likely to be imagined rather than based on observation.

The second technique is to locate the source of the misinformation, which in the current case is clearly Pat. Ideally, locating the source should come with knowledge of the credibility of the source, for example, Pat's history as a pathological liar.

The third technique is to examine the motives of the source with respect to the particular claim. Why is Pat saying that Sam is in New York City? Is Pat outright lying about Sam's location, or does Pat actually believe that Sam is in New York City because of motivated reasoning or being misinformed by someone else? Knowing what is behind the source of misinformation can influence how to counter it.

The fourth technique is the most straightforward way of overcoming misinformation: simply provide facts that correct it. For example, if you can show Pat the current Instagram postings of Sam from Toronto, then Pat may readily acknowledge being wrong.

More work is often required, however, to correct misinformation through the fifth technique, critical thinking. This method requires identifying thinking errors made by the sources or believers of misinformation and then replacing the errors with more logical reasoning. For example, suppose Pat claims that Sam is depressed and that this claim is based on Sam having a bad day. This claim commits the fallacy of hasty generalization and can be corrected by considering alternative explanations for Sam's behavior, such as a romantic disappointment.

The sixth technique for overcoming misinformation adopts a method more like psychotherapy than logic. Motivational interviewing uses empathy and open-ended questioning to get at why a person such as Pat is attached to misleading attitudes. Critical thinking sometimes helps to overcome misinformed beliefs, but it does not address underlying problems in emotional attitudes that may be the drivers of mistakes. For example, Pat may want to believe that Sam is in New York City because of some hopeless romantic desire.

The six techniques so far operate with individuals, but much misinformation may be unfixable by individual approaches to factual correction, critical thinking, and motivational interviewing. Social change may also be needed to slow the spread of misinformation by modifying institutions such as schools and businesses. If Pat gets much misinformation from an organization such as a company, then modifying the company to adopt truth-seeking policies might be desirable. Dealing with misinformation thoroughly may require modifying whole systems of educational, commercial, governmental, and military institutions.

Finally, misinformation can be limited by political action to encourage government authorities to stop the flow of falsehoods. Pat could vote for a political party that is committed to making important decisions based on evidence rather than making stuff up. We will see the inevitability of resorting to political action across all domains of misinformation. Chapter 8 provides detailed accounts of individual and social techniques for fighting misinformation.

PREVENTING MISINFORMATION

Medical experts know that it is better to prevent diseases than to wait until they need to be treated.[18] Healthy diets and exercise help to avert heart disease and diabetes, while avoidance of toxins like cigarette smoke and asbestos helps to prevent cancer. We need a term concerning information that is analogous to "preventive medicine," so I introduce the term "preinformation" to mean the general attempt to stop misinformation from arising and spreading. I propose four strategies of preinformation: gullibility reduction, teaching critical thinking, prebunking (preemptive debunking), and gatekeeping. Details about these strategies are provided in chapter 8.

People have a natural tendency to believe what they hear or read, which amounts to gullibility. Thus, my first preinformation strategy is to encourage people to be less gullible and more suspicious of information that is communicated to them. Always being suspicious would interfere with social life, but people can try to shift their default attitude from believing to questioning.

The second preinformation strategy teaches people to perform critical thinking and motivational interviewing. Courses in critical thinking are common in universities and should also be taught in high schools. Motivational interviewing based on empathic questioning would be more complicated to teach, but empathy can be taught even in schools. People might learn how to convince family members, friends, and acquaintances how to change their minds through gentle understanding.

Third, people could learn to adopt prebunking as a general preinformation strategy. Whereas debunking is a reinformation technique that operates by factual correction or critical thinking about claims already made, prebunking does not wait until misinformation is already in place but instead warns people that it is coming. Prebunking should prepare people to be questioning and skeptical about contentious matters, working as a kind of inoculation against infection by dangerous ideas.

Finally, a social preinformation strategy is effective employment of gatekeepers in relevant institutions. Professions including science, medicine, law, and journalism use gatekeepers such as editors and judges to constrain the spread of misinformation. Social media companies have failed to serve as effective gatekeepers against misinformation but could do much better if they tried. Ordinary individuals can also be encouraged to serve as gatekeepers in restraining themselves from passing on information just because it seems interesting and fun rather than accurate. But gatekeepers are most effective as social instruments that encourage institutions to be responsible in controlling misinformation. Artificial intelligence companies need to assume a better gatekeeping role to keep models such as ChatGPT from becoming sources of misinformation.

Evidence is sadly limited concerning the relative effectiveness of reducing gullibility, teaching critical thinking, encouraging prebunking, and employing gatekeepers. I hope that experiments can be conducted to determine which strategies, independently or combined, are most effective at thwarting misinformation.

FIGHT PLAN

How can we stop misinformation from killing or hurting people? Guidance should come from a strong theory of information and misinformation, which the AIMS theory provides. Information arises from four main processes: acquisition from the world, inference that extends perceptual information, memory that saves information for future use, and spread of information among individuals.

Each of these processes can be spelled out by classes of mechanisms specified in chapter 2. Misinformation results from identifiable breakdowns in these mechanisms, but it can be corrected through reinformation strategies that include critical thinking, motivational interviewing, and political action.

The AIMS theory also suggests how to characterize the concept of information using a technique more psychologically plausible than traditional definitions. Cognitive research on concepts suggests analyzing them with three factors: familiar examples, typical features, and explanations.[19] For example, the concept of movie star includes the *examples* of Emma Stone and Ryan Gosling, who possess *features* such as being good actors and being physically attractive, which help to *explain* why their films are successful.

The standard examples of information include everyday perceptions, more systematic observations of regularities in the world, and results of scientific experiments. The typical features of information include the processes of acquisition, inference, memory, and spread. The concept of information also provides explanations, for example, about how information can be useful in dealing with the world and how it sometimes fails through breakdowns in the four processes. These examples, features, and explanations provide a deep and useful characterization of the concept of information.

Some support for the AIMS theory comes from psychological evidence for its proposed mechanisms and from its ability to explain a range of empirical findings concerning misinformation. But the major test of the theory derives from its unified application to major domains that include medicine, science, politics, societies, and war, which will be discussed in chapters 4–8. Beforehand, chapter 3 spells out how motivated reasoning is a major driver of misinformation through its influence on cognition and emotion, including strong interactions with social identities.

Chapter 4 uses the AIMS theory and its mechanisms to explain the development of information and misinformation concerning COVID-19 and other medical problems. Chapter 5 examines the problems of misinformation that have hampered attempts to deal with climate change and other scientific issues. Political conspiracy theories that thrive on misinformation are the topic of chapter 6, which focuses specifically on QAnon and the great replacement theory of white supremacists. Chapter 7 considers how misinformation has been used to justify social inequality, for example, by false theories of poverty and racial differences. My last application in chapter 8 is the Russia-Ukraine war, where misinformation has been used to conceal motivations and plans. In all these domains, the AIMS theory provides guidance about strategies for overcoming misinformation.

Chapter 8 also serves as a manual for combating misinformation by using the Russia-Ukraine war to illustrate the key questions that need to be asked to save real information from its enemies. Threatening cases can be tackled by applying the AIMS theory of information versus misinformation, the reinformation tool kit for correcting misinformation, and the preinformation package for preventing misinformation in the first place.

The crucial distinction between real information and misinformation assumes the existence of reality independent of people's thoughts about it. This distinction has been challenged from many directions, so chapter 9 concludes the book by rescuing reality to allow the recognition and critique of misinformation. Truth has not been superseded.

My approach combines scientific explanations using psychological and social mechanisms with philosophical accounts of critical thinking and the ethics of social injustice. Philosophical discussions of truth and rationality blend with empirical findings about error-filled thinking and communication to suggest how to clear the miasma of misinformation that blocks the satisfaction of human needs. I hope that readers will come away with improved understanding of the difference between real information and misinformation, appreciation of the great harm that misinformation can cause, and a tool kit for correcting misinformation. The improvement should foster effective decisions based on real information that is both accurate and useful.

George Orwell's brilliant book *1984* parodied an authoritarian government by having it promote three absurd slogans: war is peace; freedom is slavery; ignorance is strength.[20] Current attempts to use antiscientific and antidemocratic misinformation align with three additional slogans: lies are truth; sickness is health; equality is bias. The diagnosis and treatment of the perils of misinformation can help to battle these noxious contradictions.[21]

CHAPTER 2

INFORMATION AND MISINFORMATION

How They Work

Scientific theories provide powerful ways of explaining, predicting, and controlling important aspects of the world. For example, Louis Pasteur's germ theory explained the origins of many infectious diseases and led to ways of predicting their development and treating them with antibiotics. We need a theory of information and misinformation that explains how they arise and can be managed. Otherwise, we would be as helpless in the face of misinformation and disinformation as the speaker in figure 2.1.

The acquisition, inference, memory, and spread (AIMS) theory explains information as resulting from four processes: acquisition from the world, inference that goes beyond observation to provide evaluations and causal explanations, memory that saves information for future use, and spread of information among people. Each of these processes operates by mechanisms with connected parts that interact to produce regular changes. Misinformation results from breakdowns in these mechanisms, and reinformation mends misinformation by fixing those breakdowns. This chapter describes generally how the AIMS theory explains information and misinformation, with detailed applications in chapters 4–8.

To ease understanding of the AIMS mechanisms, I illustrate them with fictional examples from personal medical information and romantic relationships. Chapter 4 provides deeper accounts of medical misinformation concerning COVID-19 and other diseases, while chapter 7 describes myths about relationships.

*"I do think twice before believing anything I read online.
I think, 'Really?' then 'o.k.'."*

2.1 Belief.

Source: www.CartoonStock.com. Reprinted by permission

ACQUISITION

We get most of our information from sources such as media and other people, as I will describe under the heading of spread. By acquisition, I mean the more basic process of obtaining information through interaction with the world, which people primarily accomplish by perceiving it through the senses of sight, hearing, touch, taste, and smell. You see and touch trees, hear thunder, and taste and smell bananas.

Technological developments over the past five thousand years have taken acquisition of information beyond perception through the use of powerful instruments such as weight scales, telescopes, particle accelerators, and DNA sequencers. Acquisition of information from the world has progressed from isolated, individual perceptions to systematic observations, controlled experiments, and statistical techniques that identify patterns in the changing environment.

Acquisition of information from the world requires two classes of mechanisms: collecting and representing. Collecting occurs when perception, instruments, and experiments generate data, but using data requires their representation in the mind and in external forms such as writing and computer databases. Collecting and representing are often effective at providing useful information, but breakdowns in their mechanisms can lead to misinformation.

Real Information

The simplest way of collecting medical information is perception, for example, using external senses such as vision and internal senses such as pain. I can look at my hand to see that the small cut I got recently has healed, and I can notice that the pain in my sore leg is nearly gone.

How perceptual collecting works is clear from mechanisms for the different senses that operate in humans and similar animals. Sight occurs when photons reflect off objects, enter the eye, and stimulate retinal cells to send signals to the back of the brain via the optic nerve.[1] These signals are interpreted by a succession of brain areas, which leads to visual experiences of objects and motion. Similarly, hearing occurs when sound waves stimulate a series of ear parts to send signals to the brain's auditory cortex, which processes them into coherent sounds such as words. Touch, taste, and smell also operate via the transmission of electrical signals from bodily sensors to neural processing in the brain. Perception sometimes makes mistakes, but not enough to undercut the conclusion that the senses provide real information about the world, as I argue in chapter 9.

The neural mechanisms for pain perception are also well understood. For example, skin receptors send neural signals to my brain, which processes them into the experience of pain. Perceptual collecting can also occur in robots such as those in driverless cars, which have cameras analogous to eyes but which also have different kinds of sensors including Global Positioning System (GPS) using satellites and laser imaging, detection, and ranging (LIDAR) using lasers.

TABLE 2.1 Some important instruments

Instrument	What is observed	Time of origin	Place
Ruler	Size, distance	ca. 2600 BCE	Sumeria, Indus Valley
Weight scale	Weight	ca. 2000 BCE	Indus Valley
Telescope	Distant objects	ca. 1600	Netherlands
Microscope	Small objects	ca. 1620	Netherlands
Thermometer	Temperature	ca. 1620	Italy
Ammeter	Electric current	1820	Denmark
Interferometer	Light waves	1877	United States
Particle accelerator	Subatomic particles	1932	England
Polymerase chain reaction (PCR)	DNA copies	1983	United States
DNA sequencer	DNA structure	1987	United States

I can also collect medical information about myself using instruments that serve as tools for measurement. I use an electronic scale to see whether I am maintaining a healthy weight. My thermometer can tell me if I have a fever. I track my heart rate using my Apple Watch to make sure that I am exercising strenuously. I also have a monitor that periodically reassures me that my blood pressure is in the normal range. I have not had to use the oximeter that I bought at the beginning of the COVID-19 pandemic that can tell me if my oxygen levels are dangerously low.

My family doctor employs additional instruments, including a stethoscope to check my heart and an otoscope to examine my ears. She orders periodic blood tests where instruments measure my levels of glucose and cholesterol. All these instruments interact with my body to generate measurements that surpass ordinary perception.

Collecting information improved enormously when people invented instruments like those in table 2.1.[2] Body parts such as hands and movements such as steps serve to measure size and distance, but rulers made of metal, wood, and ivory allow for more precise and intersubjective measurements. Instruments are machines that interact with perceptual mechanisms such as sight and touch to yield information that perception alone cannot produce. For example, telescopes provide information about objects such as stars that are far beyond the range of human vision.

Information collection became much more systematic after the Sumerians invented writing around 3000 BCE, which expedited recording of occurrences and changes in crops and stars. Systematic observation complements the use of instruments as a way of enhancing the collection of information. Perceptions can be haphazard depending on the situations and activities of the people doing the perceiving. Systematic observations provide order to perceptions, for example, when people conscientiously observe the motions of planets at regular times or the annual occurrences of seasons and plant growth.

Even for personal health, systematic observations have big advantages over sporadic perceptions and instrumental recordings. My medical records track changes in my health over three decades, serving to identify concerning trends.

Even better than systematic observations are experiments that manipulate the world to find out what changes result. Although experiments were sometimes carried out by ancient Greek and Arab thinkers, experimentation only became a common part of investigation in the seventeenth century through work by Galileo, Blaise Pascal, Robert Boyle, and others.[3] Experiments differ from systematic observations by manipulating factors that reveal differences among varying conditions. For example, nutritionists can systematically observe the health effects of eating vegetables by tracking people's behavior and diseases, but they can perform a controlled experiment by dividing a population into two groups and giving a specific vegetable such as broccoli to one of the groups. The group that does not eat broccoli becomes the control group. Instruments can be used to measure differences between the two groups; for example, scales can serve to compare the weights of members in the two groups. I occasionally do crude experiments on myself: for example, I sometimes reduce my consumption of carbohydrates to see whether I lose weight.

Experiments are not guaranteed to yield real information, but they have substantial advantages over ordinary perception and systematic observation. Observations admit an enormous number of factors that might be relevant to reaching general conclusions, for example, the day of the week or the locations of planets. But the manipulations in an experiment constrain the relevant changes in the world, and the measurements of them, to those relevant to specific hypotheses. Because experiments introduce a controlled change for a specific factor that varies across conditions, they can help to identify causal relations that go beyond general associations, for example, that eating broccoli causes less disease.

Four centuries of scientific research have shown how instruments and experiments produce evidence with the following five characteristics.[4]

1. Reliability: A source of evidence is reliable if it tends to yield truths rather than falsehoods, as in systematic observations using instruments such as telescopes and microscopes and in controlled experiments. A similar idea in psychology is the validity of tests, which depends on whether they measure what they are supposed to measure, for example, intelligence.
2. Intersubjectivity: Systematic observations and controlled experiments do not depend on what any one individual reports but are intersubjective in that different people can make the same observations and experiments. Pat, Sam, and Quinn should all be able to collect the same information.
3. Repeatability: A major source of the intersubjectivity of systematic observations and controlled experiments is that different investigators can get similar results at different times, replicating the original experiments or observations. Pat should be able to replicate Sam's experiments years later.
4. Robustness: Experimental results should be obtainable in different ways, such as using different instruments and methods. An example would be using different kinds of microscopes to provide similar insights into cell structure. Pat and Sam should be able to see the same things with different kinds of telescopes.
5. Causal relationship with the world: Evidence based on systematic observations or controlled experiments is causally connected with the factors in the world being studied. For example, telescopes and microscopes provide evidence because reflected light enters the eyes of observers, stimulates their retinas, and generates perceptions in accord with optical and neural processes that are causally regular in all observers.

Perception, systematic observation, instruments, and experiments are all generally reliable mechanisms of collecting information about the world. But they are not infallible, and each can fail in identifiable ways that generate misinformation. Nevertheless, they are immensely preferable to just making stuff up, which is a purely mental activity that requires no direct, causal interaction with perceived aspects of the world. Perception and observation depend on mechanisms where the interacting parts are brain areas and sense organs that receive signals from the world. Instruments and experiments expand these mechanisms by adding invented parts that measure the world and manipulate it.

Collecting from the world would be pointless if observations were immediately discarded, but use of information requires mechanisms of representing the results for scrutiny and future application. Representations can be external and publicly visible, like the sentences in this book, or they can be internal, like beliefs held in the minds of individuals. Representing is the process of producing external or internal structures that are intended to stand for aspects of the world. For example, I represent the result of weighing myself this morning by a written record in my notebook and by a digital record in the weight-tracking application on my iPhone.

Public representations are not just sentences because they can also be images such as pictures and graphs. For example, the information that people eat broccoli can be captured by the words just produced, by a picture of a bunch of people eating broccoli, or by a graph showing increase in broccoli consumption over time. Information about broccoli consumption can also be stored in a computer database with details about time intervals and broccoli sales. My iPhone uses data collected by my Apple Watch to generate graphs that portray my heart rate through the course of exercise, and it can even graph heart rhythms using its electrocardiogram (ECG) function.

The great advantage of public representations is that they can be accessed and evaluated by anyone. Their great disadvantage is that, in the absence of artificial intelligence, they are inert and do not, by themselves, carry out inference, problem solving, decision making, and learning, which are the functions that make people want information.

In contrast, mental representations naturally contribute to all these functions of information. Beliefs are verbal mental representations with the grammatical structure of sentences, as when my thought that broccoli is green corresponds to the greenness of broccoli in the world. Most people can also think with images such as envisioning a head of broccoli, the taste and smell of broccoli, the feeling of touching broccoli, or the sound of broccoli being spit out. Perception of a green head of broccoli naturally gets mentally represented as a picture of the green broccoli, which can then be translated into the sentence-like belief that broccoli is green. Also important are the emotional mental representations that combine with beliefs and images, for example, the loving or loathing of broccoli experienced by different people.

The neural mechanisms for all these kinds of mental representations are increasingly well understood.[5] For example, the belief that broccoli is green is a pattern of neural firing that combines the concepts of broccoli and green, which are also patterns of neural firing. In this mechanism, the external parts are the broccoli, the light that reflects off it, and the eyes that receive the light and begin to process it. The internal parts are the neurons that interact with each other by synaptic connections with neurotransmitters that control neural firing.

Focus on mental and neural mechanisms misleadingly suggests that collecting is just a matter of individual psychology. The importance of social mechanisms is most obvious in scientific experiments that often require a whole team of investigators. Almost all scientific publications have multiple authors, and some physics articles have hundreds of coauthors from many countries. Collaboration increases the reliability and productivity of scientific collection of evidence.[6] Many interacting institutions are required for the development of scientific knowledge, including research groups, academic departments, universities, granting agencies, publishers, and professional associations. Collaboration and institutional connections also make important social contributions to information collecting in other fields such as journalism, law, and government.

Representing can also be a social process when people collaborate to improve verbal and visual descriptions of the world. Biologists meet periodically to make collective decisions about classifications of animals, for example, that there are three species of elephants. Physicists similarly meet to change their terminology, for example, redefining "planet" to exclude Pluto. Geographers debate the merits of different kinds of map projections for representing world data. Institutions such as professional associations mediate changes in scientific representation that result from communications among individuals. Representational change also results from social processes in journalism, when organizations agree to move away from derogatory terms like "gypsy."

Acquisition of information works though people's interactions with the world, thanks to collection mechanisms that include perception, instruments, systematic observations, experiments, and social cooperation. Collecting generates representations that can contribute to problem solving, learning, and decision making. Unfortunately, acquisition can also produce misinformation that arises through faulty collecting and erroneous representing.

Misinformation

Perception is generally a good source of real information because it results from interactions between bodily senses and the world. But perception can go wrong in three different ways: mistakes, illusions, and hallucinations.

Perceptual mistakes occur because perceptions are not direct reflections of reality but rather arise through many layers of neural processing. For example, seeing a distant object as a fox requires groups of neurons to identify shapes and colors and to combine them into representations of an object that can be compared with representations of concepts stored in memory. If the fox is far away or the light is dim, the brain can easily misidentify the object as something more familiar, such as a large dog. Dreams include perceptual mistakes resulting from the brain's attempt to consolidate recent experience with stored memories, for example, when seeing a large dog during the day inspires a nighttime dream of a fox. I can make a perceptual mistake about my medical condition if I think I look sickly pale when the only problem is insufficient lighting.

Illusions are systematic mistakes that occur because of recognized flaws in perceptual judgments. More than one hundred optical illusions have been identified and explained as side effects of usually reliable mechanisms.[7] For example, the illusory triangle that people see in the middle of figure 2.2 results from the brain's normally effective strategy to create whole images from edges. Auditory illusions, such as hearing interrupted tones as continuous, result from overextension of normally effective neural mechanisms. The senses of taste, smell, and

2.2 Illusory triangle.

Source: Wikimedia Commons, Fibonacci, Creative Commons 3.0.

touch are also prone to recognized illusions. Such sensory illusions are sufficient to explode the naïve view that we always perceive the world as it really is, but they are consistent with the claim that perception gets things approximately right most of the time.

The third type of perceptual misinformation is hallucination, such as seeing pink giraffes or hearing voices of dead people. Hallucinations can result from various breakdowns in brain mechanism caused by dementia, schizophrenia, or drug consumption. For example, lysergic acid diethylamide (LSD) can induce intense images because of its effect on neural receptors for serotonin and dopamine. The possibility of mistakes, illusions, and hallucinations precludes an automatic pathway from perception to good information, but it allows for the general accuracy of perceptual experience, as chapter 9 argues.

Instruments such as telescopes greatly expand the range and accuracy of perception, but no instrument is perfect. Instruments can fail to observe and measure reality because they are broken or poorly calibrated. For example, a weight scale can get people's weight wrong because of problems such as defective sensors or high humidity. My blood pressure measurements become useless when the measurement tool's batteries are low. Powerful telescopes cannot prevent people from experiencing mistaken perceptions such as seeing men on the moon.

Besides making mistakes, instruments can fail because they are inherently incapable of detecting their intended aspects of reality. For example, Wilhelm Reich was an Austrian-American doctor who developed a device called an orgone accumulator that was supposed to detect and concentrate orgone, a kind of sexual energy. The machine was even claimed to treat cancer and schizophrenia. Such bogus instruments generate misinformation either by claiming to measure something that does not exist or by failing to measure accurately what does exist. Ouija boards are supposed to detect the presence of spirits but are moved only by the fraudulent or unconscious wishes of their users.

Systematic observations can also go wrong in various ways. Errors can occur in particular observations because of carelessness, and errors can also slip in as a result of recording the observations, for example, when a doctor enters mistakes in a medical record. Observers may inadvertently distort the data they collect because of personal biases, or they may intentionally create fraudulent records. Researchers on extrasensory perception have been accused of generating bogus data.[8]

Scientific experiments, done well in accord with the standards that have developed over centuries, provide excellent information about observable aspects of reality and make strong suggestions about underlying causal realities. But like all human practices, experimentation can be done poorly. The following are some ways to conduct bad experiments:

- Have a sample size so small that effects will likely fail to generalize to larger populations.
- Misuse statistics so that claimed effects are the result of mathematical manipulations rather than accurate detection.
- Use faulty instruments or imprecise procedures so that the data collected are not accurate.
- Fail to establish adequate controls so that the results may derive from factors that you have not taken into account.
- Allow researchers or assistants to tamper with observed values to get more interesting results.

Much concern has arisen about a replication crisis based on findings that influential studies in medicine and psychology are hard to repeat.[9] Drugs such as antidepressants that seemed initially effective in clinical trials turn out to help people less than expected over the long run. Some of the most interesting findings in social psychology, such as ones about priming behavior, have not stood up well to subsequent investigation. Researchers in medicine, psychology, and other fields have been working hard to change practices that have sometimes led to shoddy experiments. For example, investigators are encouraged to publish experimental designs and predictions in advance to prevent self-serving interpretations of data.

Science has witnessed wonderful experiments that have contributed enormously to physical, chemical, biological, and social knowledge. The difficulties of conducting well-controlled experiments and the occasional occurrence of experiments that are bad because of incompetence or fraud should not discourage the use of experimental evidence in making important theoretical and practical decisions. Experts in each field are familiar with the appropriate standards for distinguishing good experiments from bad and for selecting studies that are likely to provide real evidence. In medicine, for example, researchers usually rank randomized, blind, controlled clinical trials over studies that merely summarize naturally occurring cases, although the latter can also provide worthwhile evidence. Like perception and instrumental observation, experiments are not guaranteed to yield truths, but when they are done properly, they are far superior to guessing or making stuff up.

Another way that collecting can break down is through failure to recognize the presence of *noise*, which is unwanted variability.[10] Judgments in fields such as law, business, medicine, and politics abound with noise, for example, when physicians give widely discrepant diagnoses in patients with the same symptoms. Noise is more random than bias, which skews judgments in particular directions. Noise generates misinformation when people see patterns when they should notice variability.

Failures of collecting inevitably lead to failures of representing that vary with different formats. Written sentences and mental beliefs can end up false because they make claims that do not correspond to reality. Pictorial representations such as images, graphs, and diagrams can also distort reality. Figure 2.3 shows a famous picture of a rhinoceros produced by the German artist Albrecht Dürer in

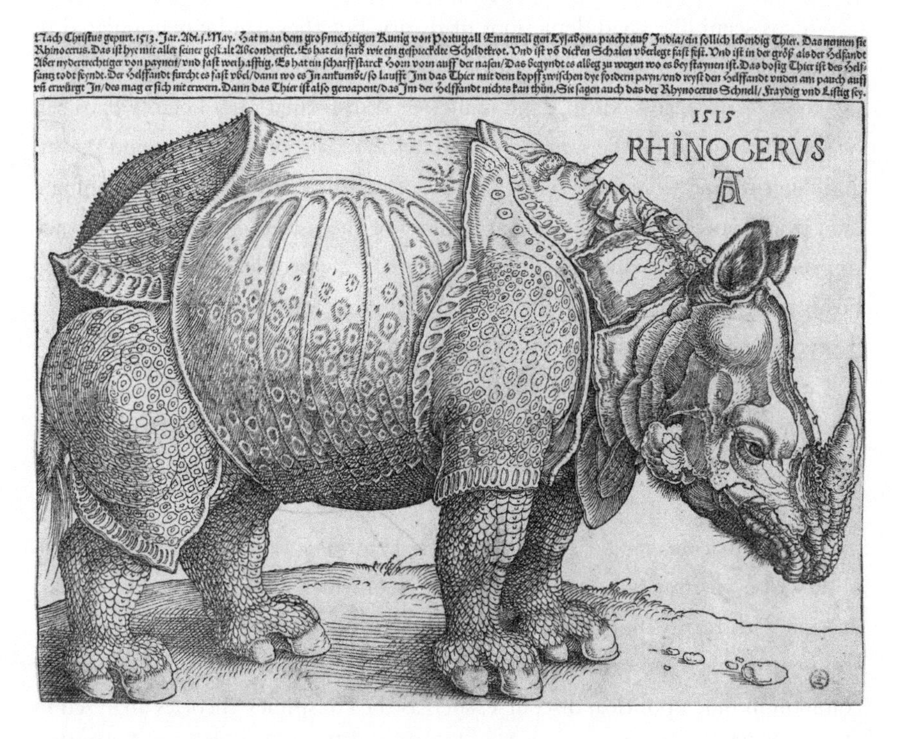

2.3 Albrecht Dürer's misrepresentation of a rhinoceros.

Source: Wikimedia Commons, public domain.

1515, based on reports of others' rather than on his own observations. This representation was taken as accurate for centuries, but it contains numerous mistakes; for example, it suggests that the animal has armored plates.

Another case of visual misrepresentation is the distortion introduced by the Mercator projection in world maps, which makes Greenland look larger than South America. Pictures and corresponding mental images can be powerful ways of representing the world accurately, but they can also sink to misinformation. Graphs are often effective ways of presenting data, but they can misinform when misused, for example, by distortion of scales that exaggerate differences.[11] Photographs seem real, but they can introduce cognitive and emotional distortions, for example, when a picture of Hillary Clinton slipping was hyped as evidence of ill health.[12] Technology is bringing new ways of generating visual misinformation through programs such as DALL-E and Stable Diffusion that can turn verbal instructions into plausible pictures, for example, of a monkey riding a rhinoceros. Similarly, ChatGPT and other generative language models can turn instructions into short essays of varying degrees of accuracy.

Videos seem like vivid depictions of reality, but they can be manipulated to misinform. The *Washington* Post warns people to watch out for videos with missing context, deceptive editing, and malicious transformation.[13] Bogus videos can be spotted by finding the source, reviewing the record of the source and poster, and determining where and when the video was filmed. Videos are becoming a dangerous source of disinformation because of the use of artificial intelligence to produce deepfakes that can convincingly portray people as saying and doing bogus things. Google and Meta have produced programs that can generate videos from text instructions such as "teddy bear washing dishes."

Computer databases are convenient ways to store vast amounts of information, but they may contain errors because of incompetence or fraud. Hence misinformation can also result from breakdowns in computer mechanisms for representing the world.

Misinformation can arise from defective forms of perception, systematic observation, instruments, and experiments, all of which require some shoddy connection with reality. But the easiest way to generate misinformation is to ignore reality altogether and just make stuff up. Such fabrication of claims has become common in much political discourse, such as the conspiracy theories analyzed in chapter 6. These theories are not based on any reality-based evidence but rather on bare assertions that gain credibility from repetition on social media.

Misinformation by defective acquisition often depends on institutions designed to generate data for ideological purposes. In Nazi Germany, the Kaiser Wilhelm Society (predecessor of today's prestigious Max Planck Institutes) generated data to support racist claims about heredity.[14] Tobacco and fossil fuel companies have supported research groups that collect data congenial to their goals of avoiding government control.[15]

Harry Frankfurt adapted the term "bullshit" to mean utterances by people who do not care at all about the truth but just make things up to suit their own purposes.[16] Such utterances do not result from failures of acquisition strategies, such as perceiving the world, but rather by purposely ignoring such strategies. Convincing people to appreciate the value of accurately representing the real world is the first step toward accomplishing reinformation with respect to acquisition.

Reinformation

Overcoming misinformation requires different strategies depending on the origins of the errors and on the mental state of the people afflicted by it. Methods of accomplishing reinformation align with the four central processes of the AIMS theory: acquisition, inference, memory, and spread.

With respect to acquisition, the primary goal is to establish the difference between misinformation and real information about the world. This difference requires showing that a world exists independent of people thinking about it and that claims are true if they correspond to the world. Such independence is obvious in the serious domains I am considering, such as pandemics and climate change. You cannot make people well just by wishing them to be well, any more than you can make the pandemic go away just by trying to think it away. Disease and death are parts of the world that need to be recognized as real so that we can cope with them. Thus, we must reject philosophical claims that reality is mind-dependent, truth is not truth, people can have their own alternative facts, and we are living in a post-truth world. Chapter 9 provides a vigorous defense of the reality of reality.

How can we deal with bullshitters who will say whatever they want to accomplish their ends? Philosophical arguments about reality have little effect on psychopathic narcissists, so the best available strategy is simply to reduce their influence on normal people. Such reduction can occur by stopping channels of information, as Facebook and Twitter did in 2021 by removing the accounts of Donald Trump. Just as effective is the political strategy of ensuring that such people are denied the power to control others by spreading misinformation.

Fortunately, psychopaths are only about 1 percent of the population, and most people can be open to education concerning good and bad ways of acquiring information through interaction with the world. People generally understand the difference between perceptions that are illusory and ones that are accurate because they take place under normal conditions with a well-functioning brain and a world situation suited to being perceived. Some perceptions are not to be trusted when the brain is suffering from disordered functioning induced by drugs or illness, or when the situation is not suitable for good observation because of darkness. Good information comes from dependable perception, which sometimes breaks down to produce misinformation.

Other points about the acquisition of information about the world are less familiar but should be teachable. People know in general that some machines are more reliable than others; for example, they want a reliable car that does not frequently break down. They should be able to understand that some instruments are more effective than others at detecting reality, including thermometers, stethoscopes, and blood pressure monitors that are standard parts of medical practice. Systematic observations that track changes over time have clear advantages over anecdotal ones because they support correct generalizations instead of isolated stories.

The most difficult challenge in helping people distinguish misinformation from real information concerns the difference between good and bad experiments. Bad experiments produced by biased or incompetent researchers fail by having insufficient observations, defective instruments, improper controls to reduce the likelihood of confounding factors, and faulty statistical and causal interpretations. Experimental results published in reputable journals are not guaranteed to be true, but they stand a greater chance of being informative if the papers describing them have gone through careful review by experts in the field. On the other hand, a slapdash experiment whose results are just thrown onto a website is more likely to be a source of misinformation.

Fixing misinformation often requires multiple techniques, just as curing disease often requires multiple treatments. For example, treating cancer frequently requires a combination of surgery, radiation, and chemotherapy, each of which would be ineffective on its own. Similarly, treating misinformation that results from acquisition breakdowns can require a combination of (1) philosophical defense of the robustness of reality and (2) education concerning the differences in effectiveness of ways of acquiring information from the world, such as the advantages of instruments and experiments over making stuff up. Some acquisition errors can be corrected by pointing out claims that are just made up rather than based on controlled experiments. More complicated errors require techniques of critical thinking and motivational interviewing to be explained in connection with the process of inference. In extreme cases of psychopathic bullshitters, only political action to constrain their power is likely to slow the spread of misinformation.

Institutional modification can work to fight misinformation by operating socially at a level between individual changes in thinking and governmental action. All institutions maintain themselves by sharing values, norms, rules, and practices among their members.[17] We can work to develop truth-seeking institutions that combat ones operating with lies.

The primary social institution in most cultures is the family, which can be a locus of reinformation if parents and children interact in the interests of truth rather than dogma and authority. By instruction and example, parents can teach their children to appreciate the advantages of gathering evidence over making stuff up. By the age of ten, my sons could ask questions like, "What's the control group for that experiment?" It helped, of course, that their mother was an experimental psychologist, but all parents can encourage discussion and the resolution of arguments by facts rather than by unsupported opinions.

Schools can also serve as engines of reinformation. From kindergartens to universities, students can be encouraged to recognize the differences between (1) information acquired by sound processes such as observation and (2) misinformation generated by biased imagination. Critical thinking should be recognized not just as a course but as a process that permeates education. Like good parents, good teachers should be authoritative rather than authoritarian. Like effective families, schools can encourage unabashed truth seeking rather than blind obedience.

Schools can also be powerful locations for teaching people how to avoid errors of misrepresenting. Students can be taught to detect and avoid the use of misleading graphs, inaccurate maps, photoshopped pictures, and videos created by deepfake techniques. Verbal misrepresenting can also be highlighted through examples of loaded language, such as Russia's description of its invasion of Ukraine as a mere "military operation."

Other institutions can also be modified to improve their capacity for truth seeking and overcoming misinformation. An institution is not just a group of individuals because it can have emergent properties that belong to the whole but not to any of the parts. These properties arise by interactions of its members with each other and with other institutions.[18] For example, a political party can have explicit policies and practices produced by the interactions of its members in conferences and elections. Societies depend on an astonishing array of institutions: legal, medical, economic, religious, charitable, political, military, and so on. All are capable of adopting policies that encourage their members toward acquiring and acting on real information and toward correcting misinformation.

INFERENCE

Acquisition is interacting with the world to gain knowledge of it. For nonhuman animals, perception is the source of almost all their information, but humans have the capability of going beyond the information provided by perception. They have the capacity to make inferences, reaching conclusions that are not simply derived from their environments. Such inferences are crucial for human activities ranging from social interactions to science, but they provide additional paths for misinformation to arise. Perception requires some inference to interpret scenes and to recognize objects, and systematic observation uses simple inferences to form generalizations. But I am concerned here with more complex inferences that go well beyond perceptions of the world.

The two most important classes of inference mechanisms are evaluating and transforming. Evaluating takes an existing representation and determines whether it is accurate and useful. For example, I periodically question whether my current weight is healthy, evaluating my belief with respect to expanding medical evidence on longevity. Transforming is more ambitious in generating and assessing novel representations. Once when I had a mild cough, I thought that I might have COVID-19, which then failed evaluation on the basis of lack of evidence. Let us now look in more detail at how evaluating and transforming can lead to real information, and how failures of these mechanisms can lead to disinformation.

Real Information

Evaluating operates when minds scrutinize representations already formed to determine whether they are likely to be true and useful. Bad evaluating leads people to reach conclusions that are counter to the truth and to their personal goals. Evaluating takes an existing representation and adds an assessment to it. For example, suppose that Pat has a romantic interest in Sam but thinks that Sam might already be in a relationship. Pat has to evaluate the hypothesis that Sam is romantically engaged in two ways: is there good evidence that the hypothesis is true, and if it is true, how should it affect Pat's actions?

People evaluate their representations to determine if they are accurate and important. I use the term "accurate" rather than "true" to cover nonverbal representations such as pictures and emotions and to allow for degrees of accuracy. Representations need to be fairly accurate to be useful in dealing with the world. A representation is important if it is relevant to people's goals. Collecting information that has no contribution to understanding and flourishing is a waste of time, energy, storage capacity, and communication. For example, I could spend thousands of dollars getting a whole-body medical scan that would provide copious data about my organs without much relevance to my health.

In the mechanism of evaluating, the parts are the human evaluator and the inspected representation, such as a contentious belief. Interaction of these parts produces an assessment, such as the judgment that the belief is accurate and important. For example, if my doctor suggests that I might be diabetic, I have to examine the evidence for this conclusion and weigh the medical significance.

The most important kind of reasoning that contributes to the evaluating of causal claims is what philosophers call *inference to the best explanation*.[19] A doctor can infer that a patient is diabetic because that hypothesis explains blood sugar levels plus other evidence such as symptoms that include frequent urination and fatigue. But the doctor should also consider alternative hypotheses that might explain the symptoms and ensure that diabetes is the best explanation of all the evidence.

The results of evaluating have an influence on the operation of other information mechanisms. Uncertainty about evaluation may spur additional collecting to resolve open issues, for example, getting additional blood sugar tests. A claim that is evaluated as accurate becomes worthy of storing in stable forms such as human memory, print, and electronic databases. A positively evaluated claim can be used in transforming information by further inferences.

Accuracy is not the only consideration that affects whether a representation is worthy of being collected, stored, transformed, and sent. My home office has two desks in it, but that information is unimportant to most people. In contrast, information about disease treatments is important and spreads rapidly because millions of lives are at risk. The degree of accuracy required of information depends on its practical importance, with greater exactness required for some purposes such as determining optimal doses of disease treatments.[20]

In principle, assessing the importance of a piece of information could be performed by the process favored by economists, maximizing expected utility, but the relevant probabilities and utilities are rarely known. Instead, people assess the importance of information by emotional reactions in accord with Norbert Schwarz's theory of feelings as information.[21] Pieces of information can grab our minds by stimulating emotions such as interest, excitement, enthusiasm, amazement, astonishment, fear, surprise, outrage, disdain, anger, or contempt. These emotions reflect different ways that a piece of information can be relevant to people's goals. For example, Pat may judge both the importance and plausibility of Sam's romantic interest through emotional reactions such as excitement and nervousness. Anger is a powerful signal of relevance, which is exploited by Facebook's algorithms for recommending news feeds.[22]

Evaluating is performed primarily by individuals, but it can also be a social process where people and institutions work together. Medical consensus about the causes and best treatments of diseases can be achieved by people interacting at conferences or through institutions: Cochrane is an international nonprofit organization dedicated to improving health decisions using evidence acquired by scientific research.[23] Social evaluating is more effective than individual evaluating when interactive communication counters individual biases such as motivated reasoning.

Transforming is a more ambitious inference than evaluating because it requires generating new representations, not just evaluating old ones. Transforming can be carried out by many kinds of deductive and inductive inference, including logical reasoning, generalizations from experience, and forming causal hypotheses. Inductive inference is inherently risky and threatens to introduce calamitous misinformation, but it is legitimate when used properly. For example, suppose that one day Sam is friendly with Pat by providing coffee. Pat might then generate a number of hypotheses about why Sam did this: Sam is just in a good mood, Sam likes Pat, or Sam is romantically interested in Pat. Comparative evaluation of conjectures can lead to a transformation of Pat's beliefs, for example, by adding the new belief that Sam is a romantic possibility. Medical diagnoses transform representations to produce new ones: the hypothesis that Quinn is diabetic combines representations of Quinn and diabetes.

In the human mechanism of transforming information, the parts are the person, the initial representations, and the generated representation that results from interactions among the person and the representations. Specific mechanisms of transformation include deductive inference and many kinds of inductive learning, such as generalization from examples, causal inference, and analogy.[24] Here are some inductive inferences that occur in romantic relationships:

- Generalization: many people today meet partners via internet dating, so dating online is worthwhile.
- Causal inference: Sam is treating me well, so Sam must have a kind and generous personality.
- Analogy: Sam is similar to my parents, so this relationship will work out.

Later chapters in this book present more complicated examples of inductive inference in the domains of medicine, science, politics, society, and war.

Such inferences operate in individual brains, but they can benefit from communication where different people provide different pieces of information that go into a transformative breakthrough. Francis Crick and James Watson transformed knowledge by proposing the double helix structure of DNA, which combined pieces of knowledge from physics and biology that they each contributed.[25] Hence transformation can be a social process as well as an individual one.

Misinformation

Inference can generate misinformation when it proceeds on the basis of flawed evaluating or erroneous transforming. Scientists, statisticians, and philosophers have established high standards for evaluating scientific claims by comparing alternative hypotheses with respect to all the available evidence. But these standards are not always followed by scientists, let alone by people unfamiliar with scientific norms. Mechanisms of evaluating can break down because of neglecting evidence, ignoring alternative hypotheses, and preferring hypotheses because of personal motivations.

Purveyors of misinformation who make stuff up rather than collecting from the world are loath to use evidence to evaluate their claims. For example, proponents of miracle medical cures prefer to ignore debunking studies. In some cases, neglect of evidence goes with full-scale rejection of science as a source of knowledge based on preference for other sources, such as religious doctrines and political ideology. People generally are subject to a confirmation bias in which they prefer to collect evidence for views that they already hold.[26] For example, if Pat believes that Sam is a lawyer, Pat will be inclined to find more evidence for this belief and ignore evidence to the contrary.

Evaluating claims is supposed to be impartial, but psychologists have documented people's tendency to fit conclusions to their personal goals.[27] This tendency of motivated reasoning is not just wishful thinking but occurs when people collect or select evidence that helps them reach the conclusions that they want. For example, coffee drinkers are less likely to believe claims that caffeine causes cancer. Because Pat is romantically interested in Sam, Pat will want to interpret even mild behaviors of Sam as signs of a reciprocal interest. Motivated reasoning is such a major source of inferential misinformation that it deserves its own chapter (chapter 3 in this book) that connects it with emotions and group identities.

In contrast to motivated reasoning, fear-driven inference occurs when people believe misinformation because it scares them and thereby attracts their attention.[28] For example, if Sam ignores Pat, Pat may interpret the behavior as suggesting that Sam does not like Pat, even though Pat really wants to be liked. Pat ends up with the opposite of the desired belief because being rejected by Sam is too upsetting to be forgotten.

Motivated reasoning is easy to recognize because people so frequently bias their thoughts toward their desires in many areas, such as relationships, health, and finances. More puzzling is why people are inclined to believe what scares them, but I will describe this occurring in political conspiracies and other domains. It seems ridiculous that people should be doubly irrational in believing falsities that make them miserable. But believing the worst is common in hypochondriacs who cannot help thinking that a freckle is cancer or in anxious parents who conclude that something horrible has happened to their children who are only moderately late. The psychological mechanism behind fear-driven inference is that a serious threat draws so much mental attention that it seems to be true.

The evaluation of the importance of information by emotional reactions sometimes encourages misinformation. Schwarz developed the influential theory that psychological information consists not only of beliefs but also of moods, emotions, bodily sensations, and experiences about cognition. These feelings make valuable contributions to judgments, decisions, and assessments of current situations. Schwarz also acknowledges, however, that feelings can also lead us astray, as the following examples of misguided emotions show:

- Love can indicate a valuable attachment to a good person, but infatuation based on excitement can lead people to overlook serious flaws in a budding relationship.
- Anxiety and fear are useful when they alert people to dangerous situations such as pandemics, but they can be debilitating when false beliefs and physical overreactions generate phobias that limit people's lives.
- Anger sometimes works as a signal that someone has blocked your important goals, but it can also be an impediment to resolving critical life issues.
- Sadness often concerns real losses, but it can descend into severe depression that fails to recognize that life still has value and hope.
- Happiness usually signals that one's vital needs are being satisfied, but it can be bogus and perilous when derived from drugs such as cocaine, amphetamines, or alcohol.
- Disgust valuably steers people away from putrid food, but it can be socially destructive when directed for purely cultural reasons at social groups or nonstandard behaviors.

To understand how emotions can be misinformation, we need to connect a theory of emotion with a theory of information. Psychological theories are not just statistical associations between variables but can go much deeper by specifying mental and neural mechanisms.

My theory of emotions explains them by neural mechanisms that combine representations of situations, appraisals of those situations concerning their relevance to personal goals, and physiological changes such as heart and breathing rates.[29] Emotions are informative when the appraisals accurately assess the impact of the situations on personal goals, for example, whether an anxiety-producing situation really is dangerous. The physiological changes should also be proportional to the situation rather than the result of defective genetics or body-altering drugs.

Misinformation in emotions results from defective appraisals, distorted physiology, or both. Emotions as information fail primarily through breakdowns in the mechanisms for evaluation, including both appraisal and physiology. For example, grief resulting from the loss of an important relationship is an informative kind of sadness because it is based on natural physiological changes and realistic appraisal of the loss. But chronic clinical depression is usually misinformative because it results from mistaken estimations of the hopelessness of life and from faulty physiology such as neurotransmitter problems.

This account of emotions as misinformation easily adapts to other kinds of feelings such as bodily sensations. Pain is normally informative about damage to body parts, but breakdowns occur when pain is referred from one body part to another; for example, when a heart attack produces jaw pain. In the extreme case of phantom limbs, pain seems to occur in body parts that no longer exist. Then pain is misinformative because of breakdowns in the neural mechanisms for collection and transformation of information. Similarly, taste often provides good information about the value of a food, but enjoying processed foods such as a Big Mac hamburger is misinformative because their appeal depends on unhealthy amounts of fat, salt, and low-fiber carbohydrates. As with sugary drinks, *yummy* can be feeling-as-misinformation.

Groups and institutions should use interactions among people to make inference effective by pooling information, but evaluating and transforming can both be socially distorted. Groupthink occurs when individuals share a desire for harmony and conformity that can produce a group decision that is inferior to what the group members would do on their own.[30] Political, military, and other organizations are prone to inferences where group identity takes precedence over real information, for example, when members of a political party put loyalty to their leader above truths about their country.

Reinformation

The usual strategy for overcoming misinformation derived from inferential breakdown is to correct the inference using a twofold strategy: (1) identify the flawed evaluating or transforming and (2) educate people to make better inferences using logically appropriate reasoning strategies. Together, these steps offer an approach to critical thinking that has been advocated by philosophers and psychologists.[31] Step 1 is assisted by the chronicling of more than a hundred kinds of inferential error tendencies that philosophers call fallacies and psychologists call cognitive biases.[32] An identified inferential error can be corrected by using a normatively correct reasoning strategy such as generalization or reasoning with probabilities.

Suppose that Pat has leapt to the conclusion that Sam is romantically interested based on tiny bits of evidence such as providing coffee. Sam is actually not interested in Pat, so Pat has acquired misinformation about their romantic prospects. The critical thinking approach to reinformation could proceed by pointing out to Pat the lack of strong evidence for Sam's interest and the strong possibility that Pat reached the romantic conclusion based on motivated rather than evidence-based inference. Pat wanted to believe in a romantic connection and was inclined to interpret a few observations as evidence for this conclusion. Instead, Pat can be encouraged to consider alternative explanations for Sam's behavior and reach the more objective conclusion that Sam is not romantically interested. We will see that this approach, based on identifying inferential flaws and correcting them with better patterns of reasoning, has many useful applications to problems such as medical errors. Critical thinking has sometimes proved successful in correcting people's inferences.[33]

Adam Grant suggests an alternative technique for changing minds that is based on psychotherapy-inspired methods for altering beliefs and behaviors.[34] The technique, called motivational interviewing, was developed in the 1980s to help people with alcohol problems and has since been applied to addictions that include smoking and drugs.[35] Motivational interviewing is partly based on psychotherapy in the style of Carl Rogers, with use of empathy and support, but differs in being short (one or two meetings) and directed at a specific goal, such as controlling alcohol consumption.

Here is how motivational interviewing could be used by an interviewer to deal with people gripped by misinformation:

1. Understand people's concerns by asking them open-ended questions and empathizing with their concerns.
2. Be affirmative, reflective, and nonjudgmental about their concerns.
3. Identify discrepancies between people's current and desired behaviors and goals.
4. Summarize the issues and inform people while respecting their autonomy.

This method is not guaranteed to change people's minds, but its success with many problematic behaviors suggests that it is worth trying as an antidote to misinformation.

For example, suppose Quinn is worried that Pat is involved in a destructive relationship with Sam. Quinn could take a critical thinking approach and point out to Pat that Sam treats Pat badly, logically applying the general principle that romantic relationships should make people feel better rather than worse. This approach might work, but it is just as likely that it will make Pat annoyed with Quinn and drive Pat closer to Sam.

As an alternative, Quinn could attempt the motivational interviewing approach and ask Pat general questions about the relationship and respond empathically to Pat's expressions of dissatisfaction with Sam. Empathy shows that Quinn has had similar emotions in similar situations. Without making harsh judgments, Quinn could point out discrepancies between how Sam treats Pat and how Pat wants to be treated, while acknowledging Pat's freedom to make relationship decisions.

Whether such motivational interviewing would be more effective at improving Pat's life than logical argument is an empirical question. But various reasons suggest using a technique that is more akin to therapy than logic. Throwing an argument at people is an adversarial process designed to show that they are wrong. In contrast, motivational interviewing poses behavior change as a collaborative process. One of the major determinants of the success of psychotherapy is the establishment of an alliance between a client and a therapist.[36] Arguing with people is likely to make them oppositional, whereas empathic interviewing encourages rapport and increased appreciation of opposing views rather than sharp rejection.

Use of empathy rather than cold logic gets at the emotions and motivations that are behind people's beliefs and practices. Brains lack firewalls between cognition and emotion, and much psychological and neurological evidence supports the view that human thinking intermixes thoughts and feelings. Motivational interviewing respects such blending, while pure logic dismisses it as irrational. Changing minds is as much about emotional change as it is about belief revision.[37] Logic has no way of disarming motivated reasoning, whereas motivational interviewing can identify people's goals and help people see how they are distorting their inferences. Motivational interviewing can also help people to appreciate how their goals might be served by beliefs and practices that are in line with evidence. By intervening with people's emotions, motivational interviewing might also be able to help with feeling-based misinformation.

Motivational interviewing might not be as successful in correcting misinformation, however, as it is in overcoming addictions. Motivational interviewing assumes that people with problems such as alcohol overconsumption have some motivation to change, which makes them at least somewhat ambivalent about their behavior. Empathic conversation works with their motivation and ambivalence to shift their beliefs and attitudes. But people who are dogmatically misinformed may be totally lacking in motivation to change their beliefs, and their absence of ambivalence leaves no room for the interviewer to work with them.

Another problem with motivational interviewing is its labor intensiveness due to dependence on one-to-one interactions. However, chatbots are increasingly being used in psychotherapy, and applications to automated motivational interviewing are underway.[38] Critical thinking may also benefit from automation.[39]

The best hope for changing beliefs in people who are avid antivaxxers, climate change deniers, or political conspiracists would be to find in them some belief, attitude, or action that is incompatible with their firm convictions. This incompatibility would provide a wedge of ambivalence that could generate some internal motivation to change. Then logic might provide some of the impetus to change, making logic and empathy collaborative rather than competitive.

Still, alliance, emotion, and motivation might mean that motivational interviewing can do a better job of correcting misinformation than logical argument. The soft glove of empathic interviewing is more appealing than the bludgeon of logic. I hope to see experiments that examine which approaches are most effective in changing people's minds about COVID-19 vaccines and other social issues such as climate change and political conspiracies. Chapters to come outline, for each of these domains, how critical thinking and motivational interviewing might combine to bring about reinformation.

Another way to convert feeling-based misinformation into real information is to reflect consciously on the underlying sources. Was the information collected by reliable processes rather than distorted perceptions? Was the information carefully evaluated or merely assumed? Were there any physiological flaws behind the acquisition and evaluation? Was the information received through misleading social processes such as emotional contagion in a riot? Given how our brains work, we have no prospect of abandoning feelings as sources of information, but we can watch for mechanism breakdowns that sometimes make emotions and other feelings misinformative.

Reinformation is a social as well as an individual process. I described how social institutions, from families to political parties, can be encouraged to acquire information by objective collection rather than by making stuff up. Similarly, institutions can be modified to support sound inferences by their members based on good strategies such as statistics and inference to the best explanatory hypotheses. Inferences should be made on the basis of evidence rather than on deference to unqualified authorities who might include parents, teachers, and chief executive officers. Appeals to authority are not always fallacious because some people do possess real expertise that warrants listening to them. The norms of institutional culture should include identifying genuine authorities to avoid having loyalty trump reality. As chapter 8 describes, democratic legal institutions have rules of evidence that serve to limit biased inferences. Schools at all levels can encourage students to make objective inferences based on thorough evaluations and to detect inferences based on fallacies and biases. News organizations can have policies such as fact checking that encourage evidence-based inferences.

Institutions can make people less inclined to use motivated reasoning than isolated individuals. Each of us has personal goals that tend to distort our conclusions, but working as part of a team can lessen these distortions. Suppose that Pat, Sam, and Quinn are a business team that makes inferences about what to do. Pat may naturally be biased by personal goals, but Sam and Quinn have different goals so they will not succumb to Pat's motivated reasoning. This benefit will not accrue if Pat is unduly charismatic or powerful, but a more egalitarian institution can foster inferences that are less distorted by personal motivations.

Institutions under autocratic control may not be amenable to modification in the direction of good inference strategies. Political action to constrain their inferences may be required, as in decisions by social media companies concerning control of misinformation. Chapter 8 considers possible political constraints on corporate decisions to support real information over misinformation. Improving inferences in the direction of real information can require reforms in a whole system of individuals and institutions.

MEMORY

Information in humans, animals, and computers is most valuable when it can be saved for future purposes. Memory is the general process of saving and reusing acquired information, and it is performed by two classes of mechanism: storing and retrieving. When these mechanisms work well, information keeps on providing benefits for improving lives. But both storing and retrieving are subject to breakdowns that can systematically contribute to misinformation. Memory-based reinformation that fixes poor storing and receiving requires critical thinking and other techniques.

Real Information

Storing is a mechanism that requires the interaction of three parts: a representation that is stored, a person who stores the representation, and a location where the representation is stored. The key interaction is that the person places the representation in the location, just as a car owner places a car in a garage.

For humans and other animals, the primary form of storing is memory accomplished in brains through formation and modification of synaptic connections between neurons. Human memory differs from that in other animals because it includes linguistic representations in addition to sensations, perceptions, images, emotions, and procedures. Human memory differs markedly from computer storage in being much more selective and reconstructive.[40] A computer memory takes any information it receives and returns it in basically the same form. But humans store only a small selection of their daily experiences based on their importance, with memory consolidation during sleep providing a means of transferring salient items into permanent storage.

That storage is not exact because new memories are assimilated with old ones through learning that adjusts synapses. These selective and reconstructive characteristics make human memory seem defective and inefficient, but they are effective means of ensuring that memories will be goal-relevant and integrated with previous experience. I do not remember all my medical data, but I can usually recall the numbers most relevant to my health. Similarly, Pat need not store and retrieve every detail of interactions with Sam; she just remembers an approximation of their most important interactions.

Individual memories are stored in individuals' brains and are not accessible by other people. Memory became social when cultures began to use storytelling, songs, and shared poems to produce group memories that could be stored and retrieved by small groups of individuals. The invention of writing in ancient Sumer and other civilizations less than six thousand years ago rendered social memory much more powerful because individuals could store what they knew in a form that could be retrieved by others. Today, social memory operates in many forms, including paper (books, journals, magazines, and newspapers); websites; videos; other electronic sources such as Twitter, Facebook, and YouTube; and computer databases. These venues allow people to store and retrieve information. The mechanisms of human information storage have expanded to include interactions with manufactured parts that are often more exact and long-lived than brain-based memories. Storing of information is often a social process supported by many valuable institutions, including libraries, governments, and corporations.

Retrieving is a mechanism where a person who has stored a representation in a location can get the representation back. As with storing, the parts of the mechanism are the representation, the person, and the location, which all interact when the person retrieves the representation from the location. Specific retrieval mechanisms operate in human brains as memory recall and in computers through search procedures in databases.

The primary goal of retrieving is to pull from storage representations that are relevant to the holder's goals. Human memory is not as capacious or fast as electronic databases, but it has some useful capabilities. Unlike computers, brains are not limited to syntactic matches but can use semantic capabilities that rely on neural encoding of meanings derived from the sensory origins of some concepts and from the ability of brains to combine meaningful representations into more complex ones.[41] For instance, the word "blue" acquires meaning from human visual experience and can contribute to the meaning of the expanded combination "blue sky." In brains, syntactic and semantic information are both encoded in the same way: by synaptic connections between neurons that are modified by association learning, reinforcement, and inference. Memory retrieval in brains is partly driven pragmatically by the role of emotions, which serve as cues.[42] For example, when Pat acquires interesting information about Sam's life, Pat will not toss the information away but rather store it by forming synaptic connections between neurons. These neurons operate in groups that encode observations and evaluations such as that Sam is attractive. On later occasions, Pat can remember this information by reactivating the relevant neurons. Institutional retrieval can operate through physical objects such as books and files, or digitally by computer databases.

Misinformation

Storing might seem to be an innocuous process, but it can go wrong in both human memory and computer databases. Storing is an important information mechanism because holders cannot keep everything in active consideration and need to place representations in permanent locations for future use. Misinformation can result from distortions in representations stored and from choosing items for storage that are lacking in accuracy or importance.

Human memory is not like keeping a photograph or computer file; rather, it is subject to distortions that Daniel Schacter calls the "seven sins."[43] Memory is transient because it deteriorates over time, for example, when Pat begins to forget facts about Sam. Memory is subject to absent-mindedness when insufficient attention is paid to what needs to be remembered, for example, when Pat cannot recall Sam's phone number. Blocking occurs when another memory interferes with one that is desired, for example, when Pat confuses Sam's email address with that of another friend. Misattribution is failure to recall the source of a piece of information, and the similar sin of suggestibility occurs when people are manipulated to remember what did not happen. These two sins are responsible for well-documented errors in eyewitness testimony in legal trials when people err in identifying accused criminals.[44] The sin of bias occurs when memories are distorted by feelings and beliefs, for example, when Pat is inclined to remember only good things about Sam, a process that can contribute to motivated reasoning. Finally, the sin of persistence occurs when disturbing information such as trauma is recalled when it is not desired.

Memory is selective rather than comprehensive because short-term memory is limited in the amount that can be held during attention and because only some of the daily experiences are transferred from short-term memory to long-term memory. Ideally, humans would select their long-term memories based on the criteria of accuracy and importance, but emotional salience often functions as a stand-in for both. Thus, people are inclined to remember what arouses their positive emotions such as excitement and negative emotions such as fear. This emotional bias makes people inclined to remember what they want to hear such, as when Pat remembers only Sam's good behavior. I find it easier to remember my good and problematic medical test results rather than the ordinary ones. Human memory storage is also subject to decay and distortion because of confusion of old memories with new ones and loss of cognitive function through aging and dementia.

Misremembering is caused not only by poor storing but also by defective retrieving fostered by defects in human memory, which in turn is driven by availability, superficial similarity, and emotion. People's memories are often spurred by available superficial cues rather than by relevance to goals. For example, people find it easier to generate words that start with "k" than words that have "k" as the third letter.[45] In analogical problem solving, people are more likely to remember cases that are superficially similar to a problem to be solved than cases that have really useful analogies.[46] Emotion can serve as a misleading retrieval cue when it pulls out of memory events that match a current situation more by strong emotions than by useful information. Storing, retrieving, and forgetting can all be motivated by personal goals.

Bad retrieving encourages misinformation because it fails to recover evidence that might challenge false beliefs or other misrepresentations. The biases and selectivity of memory retrieval discourage useful revision of beliefs. For example, if Pat is inclined to store and retrieve only bad things about Sam, Pat may make bad decisions about the quality of their relationship. Motivated remembering and forgetting distort the information needed for future decisions.

Institutional storing and retrieving are also subject to breakdowns based on the prejudices of their organizers. The German Nazis burned books to prevent saving of information associated with Jewish writers or conflicting ideologies. Contemporary democratic governments allow freedom of information requests but do not always cooperate in retrieving the desired information. Hence the social organization of memory can sometimes tilt it toward misinformation.

Reinformation

Work on critical thinking attends to bad and good ways of making inferences, but it largely ignores bad and good ways of storing and retrieving memories. This neglect must be overcome to provide a full account of information and misinformation, which have a large memory component, as we shall see in discussing COVID-19, climate change, political conspiracy theories, inequality, and war.

I described the two-stage critical thinking approach to mending misinformation resulting from bad inferences: recognizing the thinking errors that lead to bad inferences and correcting them using good inference patterns. My discussion of reinformation based on acquisition identified bad practices for collecting information from the world that could be corrected by following good practices of perception, use of instruments, systematic observation, and experimentation.

A similar approach works for fixing misinformation that results from memory mechanisms of storage and retrieving. To take a simple example, Pat and Sam may realize that some of their problems result from having different recollections of their previous interactions. Pat may recall them having a pleasant conversation that Sam remembers as tense. They can understand these differences as resulting from Schacter's seven sins such as insufficient attention paid to events that are remembered differently. Pat and Sam may recognize that they are each inclined to motivated remembering and forgetting.

Lessons about how to use memory more effectively derive from decades of psychological studies of the errors of eyewitness testimony in criminal trials. Jurors are inclined to accept witnesses' recall of what they observed, but many false convictions have been recognized when DNA evidence exonerated people who had been convicted through eyewitness testimony. Research by Elizabeth Loftus and others identified a *misinformation effect* where misleading questions implant or contaminate memories of what actually happened.[47] For example, if Quinn asks Pat, "What color was Sam's coat?" Pat may misremember that Sam was wearing a coat.

Psychologists have developed recommendations for overcoming memory mistakes in eyewitness testimony.[48] Some of these are peculiar to the legal system, for example, ways of changing prisoner lineups to make it less likely that witnesses will make false identifications. But some of the recommendations work across all the domains in which misinformation arises. Here are some suggestions for how improved memory practices can aid reinformation:

- Beware of false memories that can be introduced by manipulations.
- After an event, obtain information quickly before memory can decay.
- Rely heavily on external memories such as contemporaneous notes.
- Appreciate the importance of corroborating memories with other kinds of physical evidence.

Policies such as these can help to make the storage and retrieval of memories the basis of real information rather than misinformation. Similar policies can be applied to make computer databases serve as reliable memories, for example, by evaluating what gets stored.

Memory is controlled by institutions as well as individuals. Storage and retrieval are controlled by organizations such as libraries, media companies, and cloud-computing facilities. These organizations should have policies that encourage the storing and retrieving of real information rather than misinformation. No library can store all available material, and the criteria for selecting books, journals, and documents should include the quality of their evidence and inferences.

Science indexes are good examples of memory tools that can encourage reinformation, in contrast to general search engines such as Google and Bing. When I get interested in a new topic, I use Google Scholar to find the most frequently cited publications on that topic. Frequent citation of an article or book is no guarantee of truth, but it does indicate that the publication has been taken seriously by people whose work has been screened by journals or publishers that employ some kind of review. In contrast, a high ranking on Google may indicate only popularity among people with no interest in accuracy. Google is constantly adjusting its search algorithm and a valuable institutional modification would be tilting it more toward facts.

SPREAD

For most people, the major source of information and misinformation is not acquisition from the world, inference, or memory but rather spread from other people. Spread requires two main classes of mechanisms: sending, where an agent transmits a message to one or more recipients, and receiving, where the recipients transfer the message into their own processing systems. When sending and receiving work well, accurate and useful information spreads among agents. But breakdowns in these mechanisms foster misinformation, which can be corrected by recognizing the problems and working to fix them. I get large amounts of valuable medical information from good sources such as journals and reliable newspapers, but I am constantly wary of misinformation spread by irresponsible politicians and journalists.

Real Information

The mechanisms of acquiring, inferring, and memory operate mostly in individuals, but the usual point of information is to communicate it to others using social mechanisms of sending and receiving. In social mechanisms, the parts are multiple agents, connected into groups by social bonds and communication channels that enable them to interact with each other.[49] Communication involves a sender and one or more recipient. The agents, channels, and representations interact to produce the sending and receiving of the representation. For example, Pat may spread information to Sam and Quinn by talking to them, writing to them, or sending electronic messages. More broadly, Pat may send information through the press, radio, television, government bulletins, or academic journals. Schematically, the communication mechanisms amount to:

Sender + representation → channel → receiver → representation received

The representations sent, often called messages, are usually verbal but can also include pictures, graphs, diagrams, videos, and large data sets. Receiving is a mechanism where messages cause changes in the brain processes of a receiver.

Because no sender wants to send everything to everyone, the mechanism of evaluating contributes crucially to the mechanism of sending. The holder of a piece of information has first to decide whether the representation is important and accurate enough to send to others. Second, the sender has to decide which recipients should receive the representation. Careless senders often proceed on the basis of emotional stimulation rather than careful assessments of accuracy and importance.

Receiving is similarly selective because people are unlike electronic equipment in not receiving all the messages that are sent to them. As with sending, the parts of the receiving mechanism are the representation, the agent that sends it, the agents who receive the message, the channel by which the message is sent, and the representation that is established in the receivers. Unless a receiver is gullible or inefficient, the receiver needs to evaluate the proposed representation by asking questions such as the following. Is the sender trustworthy? Is the message credibly based on good sources? Is the message consistent with other information held by the recipient? Is the message sufficiently important, that is, relevant to appropriate goals, to warrant storing in the recipient's limited brain or computer? Is the message sufficiently accurate and important to warrant sending to additional recipients? When I receive an email announcing a medical breakthrough, I sometimes discard it immediately because of a bogus source or incendiary message. But if the information looks accurate and important, I save it and consider passing it along to friends who I think will be interested.

Receiving information is not just accepting testimony as fact but requires evaluating the quality, importance, and sources of what is sent. For example, when Sam gets an email, message, text, tweet, or other communication from Pat, Sam should not simply accept what Pat says as true. Instead, Sam should evaluate the message by considering Pat's accuracy track record, the original source of the information, and how well the message fits with the rest of Sam's knowledge.

Misinformation

The normally successful mechanism of sending from one agent to others can break down in two main ways: through what messages are sent and through who receives them. Spurious sending and broken receiving can encourage the spread of misinformation, for example, when a careless friend sends me untested ideas about diet that I do not scrutinize skeptically.

Ideally, people would send only those messages deemed accurate and important, but evaluations can easily be distorted through carelessness and motivated reasoning. Faulty evaluations result in sent messages that suffer from errors or triviality. Studies of social media have found that misinformation is often passed without the senders paying much attention to whether it is true.[50] Platforms such as Facebook, Twitter, and Instagram are huge sources of misinformation about COVID-19, climate change, and politics, as later chapters will document.

The word "incontinent" today usually relates to urinary problems, but it has another, original meaning: lack of self-restraint. The way information spreads through society today suffers acutely from lack of restraint, leading to the proliferation of misinformation and disinformation on important topics that include COVID-19, climate change, conspiracy theories, and the Russian invasion of Ukraine.

Information incontinence was only a minor problem before the development of social media platforms in the 2000s. Previous to that time, people sometimes showed a lack of restraint in spreading misinformation to others through gossip and rumors, but the spread was limited by the slowness of communicating by talking or writing. Technology has allowed instant communication with millions of people through social media such as Facebook, Twitter, YouTube, TikTok, Instagram, WhatsApp, WeChat, Reddit, 4chan, and 8kun. Chapter 8 describes how to mend the internet and social media to reduce information incontinence.

The second way sending can break is by messages going to the wrong people. A misdirected phone call is a trivial example of getting the wrong recipient, but more serious are cases where technologies like email, Twitter, Facebook, and Instagram make it easy to send messages to thousands or even millions of people. Resending is also easy so that people get the same messages back, with repetition encouraging illegitimate confidence in their quality.

Communication systems that transmit meaningful information about important matters are subject to reception problems including distortion, misunderstanding, and lack of critical assessment. For example, Quinn may be too tired or lazy to wonder about the quality of Pat's message and simply believe what it says or sloppily pass it on to other people.

When an agent receives a message from other people, the message may be distorted through the communication process, for example, when some of the dots and dashes in a Morse code message are switched or when a video feed breaks up. Moreover, the recipient of an intact message may fail to understand it. People may receive via email, texting, or social media a communication about a scientific study that used a double-blind controlled clinical trial. But many people do not know why such trials are more credible evidence than celebrity testimonials, so the content of the message is not valued. Health and government officials bear a heavy responsibility to present scientific evidence in ways that the general population can understand. Otherwise, people will lack real information that they can use to counterbalance misinformation.

Critical thinking requires that people subject incoming messages to evaluation of their sources and plausibility, but limitations of time and cognitive capacity incline people simply to believe what they are told.[51] Such naïve reception is unproblematic when people have reliable sources, but much medical, scientific, and social information is received from unreliable sources such as self-serving merchants and politicians. If people forego evaluation of incoming messages or if evaluating is distorted by motivated reasoning or other biases, then automatic receiving can contribute to misinformation.

Institutions support misinformation when they encourage sloppy or malevolent sending and careless receiving. Culpable institutions include foundations that promulgate pseudoscience and social media that care more about engagement-driven advertising than controlling the spread of falsehoods.

Reinformation

The main techniques for overcoming the spread of misinformation are critical thinking, motivational interviewing, institutional modification, and political action. The study of critical thinking is typically confined to individuals, but the huge impact of spread on misinformation shows the need for generalization to "critical communicating." Just as critical thinking combines recognition of error tendencies with replacement by better inference strategies, critical communicating should combine recognition of communication errors with advocacy of improved ways of ensuring that spread avoids misinformation. My description of spurious sending and bad receiving shows what to watch for in avoiding communication mistakes.

But how can we communicate better? The key is to recognize the centrality of the mechanism of evaluating that I discussed as part of the process of inference. Using techniques such as inferring the best explanation of the evidence, senders and receivers can assess both the cognitive plausibility and the practical importance of messages.

Senders should apply such evaluations before transmitting messages to others. For example, before Pat sends a message to Sam and Quinn, Pat should ask whether the message is sufficiently interesting and backed by evidence to warrant sending to them. Similarly, when Sam and Quinn get the message from Pat, they should evaluate the sender and the content of the message before believing or resending it. Communication then becomes evaluation-based rather than automatic, thus hindering the transmission of misinformation.

A more indirect way of improving people's communication practices would be to employ motivational interviewing. Suppose Pat is an indiscriminating user of social media and routinely sends out messages spreading falsehoods and dangerous values. Quinn could confront Pat with arguments that point out practices of careless receiving and spurious sending, but it might be more effective if Quinn took a more empathic approach. Like a nondirective psychotherapist, Quinn could ask Pat open-ended questions about the use of social media. Quinn would use empathy to understand Pat's communication behavior by finding emotional analogies between what Pat does and Quinn's own practices.[52] This kind of empathy works by noticing similarities between the situation of another person and yourself and transferring your own emotional experience to the other. Perhaps Quinn could recognize in both a tendency to get excited about a social media message on Facebook or Twitter and quickly send it out to contacts without scrutiny of its origins and credibility. If Pat and Quinn meet in person, Quinn could also operate with more visceral kinds of empathy that involve mirroring Pat's body language and facial expression. If Pat feels understood and cared for by Quinn, as often happens in the psychotherapeutic alliance, then Pat may acquire Quinn's gently communicated concerns about spreading misinformation.

Trust is a crucial contributor to reliable transmission of information. Trust in persons or organizations is a feeling of confidence and security that is more than an estimate of the probability that they will behave as expected because trust is also an emotional reaction that can be based in part on bodily reactions and evaluation.[53] Motivational interviewing can help to uncover why trust is awarded to some senders rather than others, and critical thinking can help to evaluate whether the sender is genuinely trustworthy.

Critical thinking and motivational interviewing deal with individuals, but controlling spread also requires social interventions. People often receive information through the organizations they belong to, such as families, schools, companies, clubs, and political parties. These organizations should take responsibility for reducing the spread of information between their members. I have two email accounts, one from the University of Waterloo and one from Google Mail, and both organizations have filters that reduce the amount of spam sent to my accounts. Institutions naturally want their members to communicate with each other, but they should strive to ensure that the content of these communications shuns misinformation. Schools, companies, government agencies, news media, and social media should consider it part of their missions to limit the spread of falsehoods.

Political action is often needed to regulate individuals and organizations that contribute to the spread of misinformation. Social media platforms such as Facebook and Twitter have contributed hugely to the spread of misinformation about COVID-19, climate change, and political conspiracies, as chapters 4–8 document. Because these platforms are different from previous media, governments have been slow to regulate them. Chapter 8 outlines various measures that governments can take to regulate social media, such as breaking up monopolies and making social media companies accountable for the irresponsible transmissions of their users.

Effective reinformation requires understanding the targeted audience to figure out what strategies are most likely to be effective. Factual correction may work for people who are simply mistaken, but critical thinking will be more effective when the mistakes arise from identifiable thinking errors such as fallacies. Motivational interviewing may be a better strategy to help well-meaning people with confused values. For dealing with flat-out, unconscientious liars, political action to control their influence may be the only recourse.

THEORY, EXPLANATION, AND CONTROL

Information and misinformation are pervasive phenomena that call for description using evidence collected by observations and controlled experiments. For both intellectual and practical reasons, we need to go beyond mere descriptions of phenomena to provide explanations of them and ways of controlling them. Explanation and control require theories that show the underlying causes of phenomena and suggest useful changes.

In physics and other mathematical sciences, theories often consist of mathematical equations that can be used deductively for prediction. But in many sciences, including biology and psychology, theories specify mechanisms, which are combinations of connected parts whose interactions produce regular changes. This chapter has described eight mechanisms organized under four general processes: acquisition, inference, memory, and spread. I have named the mechanisms with gerunds ending in "ing": collecting, representing, evaluating, transforming, storing, retrieving, sending, and receiving. The corresponding nouns (collection, representation, evaluation, transformation, storage, retrieval, transmission, and reception) are misleading because these nouns can also refer to the results of processes, such as a collection of items that result from the process of collecting.[54] Mechanisms do the explaining, not the resulting entities like collections and representations.

The fictional illustrations of Pat, Sam, and Quinn I used as examples provide no evidence for the mechanisms of the AIMS theory, whose cogency comes primarily from its ability to illuminate the development and breakdown of information in a wide range of domains. Before tackling these domains, chapter 3 looks in more detail at motivated reasoning as one of the major springs of misinformation. Additional evidence for the AIMS theory comes from its explanation of psychological effects concerning misinformation, which are reviewed in the online supplemental material. That supplemental material also reviews alternative theories of information and analyzes interactions among information mechanisms, breakdowns, reinformation techniques, and domain areas.

CHAPTER 3

BELIEVING WHAT YOU WANT

Motivated Reasoning, Emotion, and Identity

Ziva Kunda's concept of motivated reasoning is central to explanations of the increasing prevalence of misinformation. For example, the chapter "What's Wrong with People" in Steven Pinker's book *Rationality* attributes irrationality primarily to motivated reasoning.[1] Similarly, Jason Stanley's book on propaganda ascribes belief in political ideologies to motivated reasoning.[2] Motivated reasoning is essential to my explanations of how misinformation works in medicine, science, politics, inequality, and war, so a deeper examination is required. The phenomenon has even been noticed in cartoons, as in figure 3.1.

Kunda's 1990 paper "The Case for Motivated Reasoning" has been cited more than 10,000 times, according to Google Scholar, with more than 1,000 citations just since 2021! Its influence has steadily expanded beyond social psychology to fields that include political science, economics, communication, and philosophy.[3] I have a personal connection because Ziva and I were married from 1985 to her death from cancer in 2004. Our three collaborative publications include a 1987 conference paper on mechanisms of motivated inference.

All of us are prone to motivated reasoning in important domains:[4]

- Romantic relationships: my lover treats me poorly but will change.
- Parenting: my child hates school but will settle down and straighten out eventually.
- Medicine: this pain in my chest is indigestion not a heart attack.
- Politics: the new leader will be the country's savior.
- Sports: our team has been losing, but we're going to play great today.
- Law: the evidence against my hero is serious, but my hero couldn't have done it.

3.1 Cartoon spoof of the motivated reasoning Olympics.

Source: Randall Munroe, xkcd.com: a webcomic of romance, sarcasm, math, and language, https://xkcd.com/2167/. Reprinted by permission, Creative Commons Attribution-NonCommercial 2.5 License.

- Religion: life is hard, but my caring God will lead me to eternal bliss.
- Economics: current inflation is just a temporary deviation that can be controlled.
- Research: the article I'm writing is my best ever and will get into a top journal.

In all these domains, our motivations incline us to believe falsehoods.

After some historical background, this chapter analyzes the cognitive, emotional, and social mechanisms that contribute to motivated reasoning. Understanding these mechanisms suggests ways of countering the negative influences of motivation on the search for real information.

A BRIEF HISTORY OF MOTIVATED REASONING

I remember Ziva telling me around 1983 about her plan to do a PhD thesis on the effects of motivation on inference; my reaction was—what a cool topic! At the time, the dominant view in psychology was that thinking errors arose because of the cognitive biases investigated by Daniel Kahneman and Amos Tversky.[5] A different, older view that explained irrational behavior as the result of dissonance between beliefs and attitudes had fallen out of fashion. I was teaching informal logic and telling my students about an array of logical fallacies, but I never thought to warn them about the more powerful effects of motivation and emotion.

I do not recall what inspired Ziva to look at motivation, but her talks on the results of her dissertation research always started with, "Let me tell you about my mother." The same story appears in her 1999 textbook, *Social Cognition*, where she describes how her mother, a heavy smoker, dismissed an article on the effects of smoking on pregnancy by pointing to her tall sons.[6] Ziva was well aware that the idea that motivation and emotion influence judgments was not new, but she originated the terms "motivated inference" and "motivated reasoning". Her experiments provided the first solid evidence of motivational biases in everyday reasoning, for example, when people evaluate medical studies. She also developed the first articulated theory of motivated reasoning as resulting from cognitive processes of biases in memory search and evidence collection.

Precursors of Motivated Reasoning

I have since encountered numerous ancestors of the idea of motivated reasoning. The earliest is from the fifth century BC when the historian Thucydides wrote about the enemies of Athens: "Their judgment was based more upon blind wishing than upon any sound prediction; for it is a habit of mankind to entrust to careless hope what they long for, and to use sovereign reason to thrust aside what they do not desire."[7] Aristotle in the fourth century BC discussed akrasia, translated as incontinence or weakness of will, which occurs when the domination of reason by emotion leads to bad actions.[8] Writing about Alexander the Great in the first century AD, Arrian wrote: "Accordingly, as is usual in such cases, not knowing the facts, each man conjectured what was most pleasing to himself."[9] Thus the basic idea behind motivated reasoning has been known for two thousand years.

A particularly lucid description of motivated reasoning was presented in 1620 by Francis Bacon in his pioneering treatise on scientific thought, *Novum Organum*: "The human understanding is no dry light, but receives an infusion from the will and affections; whence proceed sciences which may be called 'sciences as one would.' For what a man had rather were true he more readily believes . . . Numberless, in short, are the ways, and sometimes imperceptible, in which the affections color and infect the understanding."[10] Many other commentators such as John Stuart Mill and Jon Elster have noticed the human susceptibility to wishful thinking.[11] David Pears pursued Aristotelian themes of weakness of will and self-deception under the heading of *motivated irrationality*.[12] Ari Kruglanski introduced the term "motivated cognition" to cover other ways in which goals affect thinking.[13] Related ideas include motivated thinking, positive illusions, optimism bias, desirability bias, and wishful thinking bias.

Terminology: Inference or Reasoning?

In her 1987 experimental paper, Ziva introduced the phrase "motivated inference," but her 1990 theoretical paper changed it to "motivated reasoning".[14] I do not know why she made the change, and until 2011 I mis-cited the 1990 paper as "The Case for Motivated Inference." From the perspectives of formal logic and dictionary definitions, the words "inference" and "reasoning" are synonyms because they result from patterns of verbal argument.

However, the last century of psychological research suggests a different perspective because of the importance of unconscious inference that operates in perception, emotion, and nonverbal images such as pictures. Reasoning is usually deliberate, slow, conscious, verbal, unemotional, and serial—one step at a time. In contrast, the brain carries out inferences that are often automatic, fast, unconscious, nonverbal, emotional, and parallel, with billions of neurons working simultaneously.[15] So inference and reasoning are not the same, and we cannot equate motivated inference and motivated reasoning.

Which is the better way of describing motivated thinking? Consider the examples at the beginning of this chapter in domains such as romance and medicine. Sometimes we may be consciously aware and verbally explicit in thinking along these lines, but our motivations often affect our judgments without our realizing it. For example, people denying that their medical symptoms are not serious usually have no idea that their motive to be healthy is swamping their assessment of evidence. Aside from the New Age manifesters discussed below, people rarely explicitly argue: I want something; therefore, I will get it.

Psychologists have found evidence for the phenomena of *wishful seeing* and *wishful hearing*, where motivations influence perceptions such as the size of desired objects.[16] Such nonverbal, unconscious, motivated perception is inference rather than reasoning. Thus, I think that Ziva's original term "motivated inference" is usually the most appropriate, with "motivated reasoning" better reserved for more public, social occurrences of motivated distortions of thinking and communication. However, the term "motivated reasoning" has become standard, as shown by Google Scholar citations and Google Ngram. Thus, I will stick with the usual terminology but caution that the psychological mechanisms for motivated thinking should recognize how it is often different from public, verbal, argumentative reasoning in being unconscious, nonverbal, and emotional.

We cannot assume just one mechanism for motivated reasoning. I will consider several: cognitive, emotional, neural, and social. Ziva emphasized a cognitive mechanism based on memory, and I have written about an emotional mechanism based on coherence. This chapter adds a different emotional mechanism that considers the role of specific emotions in provoking motivated inference. I also consider social causes of motivated reasoning connected with institutions and identity. Recognizing the full range is important for correcting misinformation, which requires different strategies depending on the information breakdowns that led to it. Similarly, curing diseases requires different treatments based on the biological breakdowns that cause them: an infection requires different drugs depending on whether it is bacterial, fungal, or viral. Diseases also have social causes, including poverty and family disintegration, and psychological causes such as stress.

Constraints on Motivated Reasoning

Ziva often emphasized that people do not usually believe whatever they want to believe because they feel some need to constrain their inferences by evidence: "People motivated to arrive at a particular conclusion attempt to be rational and to construct a justification of their desired conclusion that would persuade a dispassionate observer. They draw the desired conclusion only if they can muster up the evidence necessary to support it."[17] More than a hundred years earlier, John Stuart Mill made a similar point: "We cannot believe a proposition only by wishing, or only by dreading, to believe it. The most violent inclinations to find a set of propositions true will not enable the weakest of mankind to believe them without a vestige of intellectual grounds—without any, even apparent evidence."[18]

I think these generalizations apply to most people, but contemporary counterexamples have emerged. Some of the cases of misinformation described in later chapters seem to lack even a "vestige of intellectual grounds." Examples include the bizarre claims of QAnon described in chapter 6 concerning Democrat pedophile rings, and Donald Trump's claims that the 2020 presidential election was rigged. I conjecture that the rise of misinformation in recent years has resulted from a storm of unconstrained motivated reasoning that combines three interacting forces.

The first is the dominance of strong leaders with personalities that show signs of narcissism and psychopathy. For narcissists, the only goals that matter are their own personal benefits, so all reasoning is motivated reasoning. Psychopaths lack empathy and moral concern for other people, which also turns reasoning into personally motivated reasoning. Strongmen like Mussolini, Hitler, Stalin, Trump, and Putin are so self-focused that they rarely feel constrained by evidence.[19]

The second influence that encourages the spread of misinformation by unconstrained motivated reasoning is the rise of media organizations that operate without the usual constraints of evidence and journalistic ethics. This influence comes in two flavors, depending on whether they are controlled by the state or operate in countries where freedom of the press still operates. In countries like Russia and Hungary, the major media, including television stations and newspapers, are controlled by the central government so their only constraints come from the motives to obey their rulers. In democratic countries, some media outlets are so ideologically aligned with political movements that their utterances are constrained by the ideology rather than by evidence and ethics. American examples include Fox News, Breitbart News, and Infowars.

The third class of misinformation influence relying on unconstrained motivated reasoning consists of social media such as Twitter, Facebook, and TikTok. The spread of misinformation is often unconstrained by anyone's accuracy goals because people tend to pass on to others whatever they find engaging. Social media algorithms encourage such irresponsibility because rampant spread increases their ad revenues.

The motivated promotion of misinformation by irresponsible news organizations and uncontrolled social media can provide the illusion of evidence. Normally, hearing a piece of news from multiple sources should increase confidence in its truth, but if the pieces are just the same lies recirculated on Twitter and other platforms, then such confidence is misplaced. People confuse repetition with evidence.

Another area of unconstrained motivated inference is the New Age idea of *manifesting*, the practice of thinking aspirational thoughts to make them real.[20] In 2006, Rhonda Byrne published a book called *The Secret* that claimed that thinking positive thoughts was sufficient to make positive things happen to you, which she glorified as the "Law of Attraction."[21] The 2023 version is Lucky Girl Syndrome on TikTok, with young women professing that they can get anything they want, and with over 200 million views.[22]

Being motivated and setting goals are great aids to helping people accomplish what they want, but manifesting and the law of attraction imply, contrary to evidence, that just wanting something should be enough to get it.[23] Perhaps New Age followers think that videos about manifesting are evidence in the form of testimonials that it actually works. I bet that Ziva would now recognize such New Age ideas, irresponsible media, and strongmen as exceptions to her hypothesis about how people constrain their motivated reasoning.

MECHANISMS OF MOTIVATED REASONING

Is motivated reasoning the result of cold, dispassionate thinking or hot, emotional imagining? In various individuals, motivated reasoning can run cold or hot through different cognitive and emotional mechanisms. Motivated reasoning is also influenced by social mechanisms operating in institutions and other groups.

Cognitive Mechanisms

In her articles and textbook, Ziva explained motivated reasoning as primarily the result of biased memory search. "To construct justifications for desired conclusions, we search through our memory for beliefs and rules that support these conclusions directly and use existing knowledge to construct new general beliefs and theories from which our desired conclusions can be derived."[24] I programmed a computational model of this process that incorporated motivated memory search into a computational model called processes of induction (PI).[25] In the expanded model, called Motiv-PI, inference depends in part on the relevance of a potential conclusion to the person's goals. If the relevance is positive in that the conclusion promotes the goals, then a search is conducted for supporting evidence and a lower threshold for inference is allowed: if you want the conclusion, then you require less evidence. But if the relevance is negative in that the conclusion goes against the goals, then a search is conducted for contrary evidence and a higher threshold for inference is required: if you do not want the conclusion, then you require more evidence. For example, coffee drinkers require less evidence to infer that caffeine is healthy, but they require more evidence to infer that caffeine is unhealthy.

Ziva and I called our conference paper "Hot Cognition," and she used the same title for the chapter on motivation in her textbook, but this model is remarkably lacking in any mention of emotions and moods. Goals are purely cognitive entities, which ignores their major contribution to emotions: we desire the accomplishment of our goals, we feel happy when we accomplish them, and we feel sad when we fail. The goals that are most commonly assumed to generate motivated reasoning, such as maintaining self-esteem and social relationships, are interconnected with positive emotions that include joy, pride, and gratitude, and with negative emotions that include fear, anger, and shame. The narrowly cognitive model of motivated reasoning based on goal-directed selection of evidence only captures one way in which motives can influence thinking.

Emotional Mechanisms

In the 1990s, I got increasingly interested in emotion when I was spurred by a student's interest in empathy as emotional analogy. By the late 1990s, I had developed a new theory of emotional coherence that extended neural network models of constraint satisfaction to explain how people's emotional attitudes could be explained by positive and negative valences that reflect desirability and dislike.[26] The early version only allowed beliefs to influence desires rather than allowing desires to influence beliefs, so they did not cover motivated inference, but I soon extended the model to allow valences to influence beliefs. Much later, I came up with a neural theory of specific emotions, such as happiness and sadness, which I will here apply to motivated reasoning.

Emotional Coherence

My first application of emotional coherence to motivated reasoning was explaining why the jury acquitted O. J. Simpson of murdering his ex-wife despite substantial evidence of his guilt.[27] I described how the jury's decision was not just

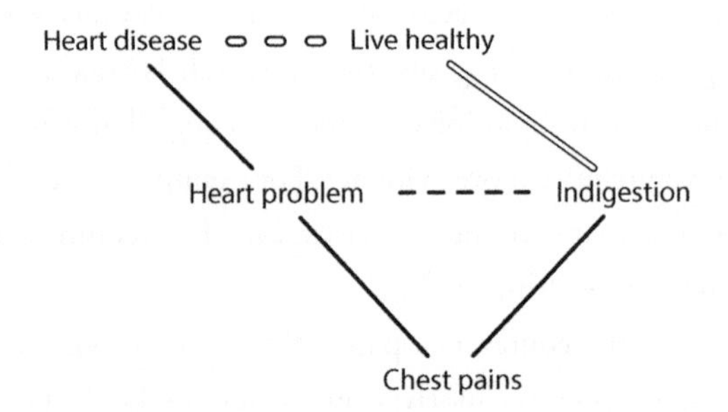

3.2 Emotional coherence of explanation of chest pains. The solid lines indicate explanation; the dotted lines indicate incompatibility. The solid double line indicates goal accomplishment, and the dotted double line at the top indicates goal incompatibility.

wishful thinking but rather combined assessment of the evidence with emotional reactions that included liking O. J. Simpson and disliking the Los Angeles Police Department (LAPD), who were well known to be prejudiced against Blacks. I also modeled an experimental result of Ziva's concerning motivated application of stereotypes: whether people apply a Black stereotype to a boss depends on whether the boss praised or criticized them.[28] I will show in chapters 4–8 that emotional coherence can similarly explain motivated reasoning in five major cases of misinformation. Psychological experiments have found experimental evidence for emotional coherence as operating in human thinking.[29] The theory of emotional coherence inspired a diagramming technique called cognitive-affective mapping that will also be useful in these cases.[30]

Simple diagrams about a fictional medical example illustrate the operation of emotional coherence. Suppose that Quinn has chest pains that occur during exercise. Possible explanations of the chest pains are that Quinn has exercise-induced indigestion or that Quinn has heart disease that could presage a heart attack. Quinn naturally prefers the hypothesis that the chest pains are the result of indigestion because heart disease is much more threatening, as shown in figure 3.2. Another way of portraying Quinn's values is by using the cognitive-affective map in figure 3.3, which shows interconnections among likes and dislikes. Both kinds of diagram illustrate how values can distort inference through emotional coherence.

I do not know what Ziva thought of my use of emotional coherence to explain motivated reasoning because in the early 2000s, we were busy raising our sons

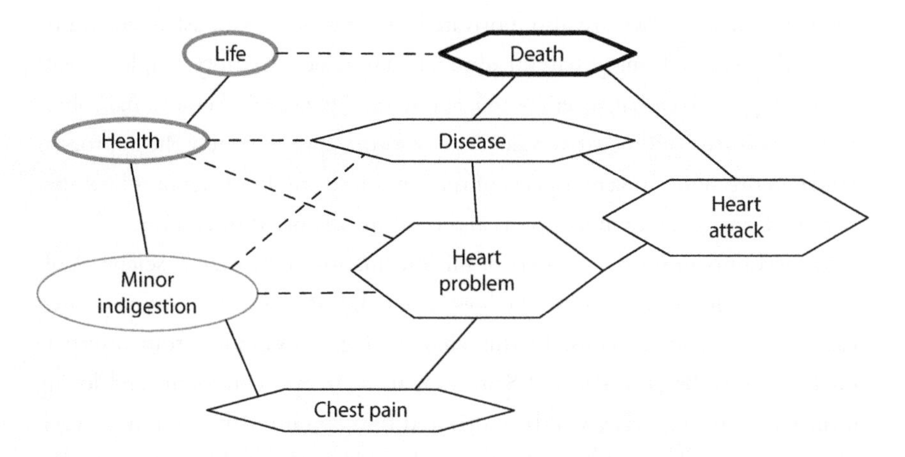

3.3 Cognitive-affective map of values about health. Ovals indicate desired values, hexagons indicate negative values, solid lines indicate emotional associations, and dotted lines indicate emotional incompatibility.

and managing her increasingly serious illness. Perhaps she would have incorporated it into the later editions of her textbook that she never got to write. Unlike her narrowly cognitive model based on biased search, my coherence account uses valences for liking and disliking to introduce emotional influences into motivated thinking. But emotional coherence neglects how specific emotions can affect inference and reasoning.

Specific Emotions

My recent realization that specific emotions can generate motivated reasoning was anticipated 400 years ago by Francis Bacon. I earlier quoted his anticipation of motivated reasoning, but I omitted this description of a biased person: "Therefore he rejects difficult things from impatience of research; sober things because they narrow hope; the deeper things of nature, from superstition; the light of experience, from arrogance and pride, lest his mind seem to be occupied with things mean and transitory; things not commonly believed, out of deference to the opinion of the vulgar."[31] In this list, biases come from the emotional reactions of impatience, hope, superstition (fear), arrogance, pride, and deference (humility).

The involvement of emotions in motivation is clear because people want to be happy, proud, thankful, serene, confident, interested, amused, hopeful, loving, inspired, and so on. We are also motivated to avoid being sad, ashamed, guilty, resentful, frustrated, angry, hateful, afraid, lonely, rejected, empty, helpless, inadequate, disgusted, embarrassed, bored, and so on. My model of emotional coherence uses positive and negative valences to capture approximately the approach/avoid and promote/prevent aspects of these motivations, but it ignores how specific emotions generate motivations and resulting distorted inferences.

My accounts in future chapters of misinformation in medicine, science, politics, society, and war will describe how such emotions contribute to motivated reasoning in these domains. In this chapter, I give examples from interpersonal relations. Because Pat and Sam want to be happy, confident, and loving in their relationship, they will be motivated to maintain high opinions of each other, which might actually make their relationship better.[32] For example, Pat may interpret an odd habit of Sam's as a harmless quirk rather than as a heinous affliction. Benjamin Franklin advised couples to keep their eyes wide open before marriage but half-shut afterward.

On the other hand, if Pat and Sam's relationship sours and they are threatened with feeling angry, afraid, and rejected, then the motivation to avoid these emotions may lead to inferences about their own strengths and the weaknesses of their partners. For example, fear of the relationship ending can encourage the cognitive search for evidence that the partner is flawed and the coherence-based accentuation of miserable moments.

Some patterns of motivated reasoning have the following structure:

Emotion → motivation → biased inference

Fear involves consideration of loss of life or other important goals, which provides motivation to either avoid the loss or to diminish its importance, biasing conclusions about the likelihood or costs of loss. Anger is directed at an agent viewed as responsible for bad occurrences, which provides motivation to get back at or denigrate the agent, biasing conclusions about the agent's character and prospects. Similarly, hate is based on an appraisal that someone has caused serious harm, which can motivate beliefs about likely retribution.

On the other hand, gratitude is directed at an agent viewed as responsible for good occurrences, which provides motivation to support the agent, biasing conclusions about the agent's character and behavior. Hence specific emotions can contribute to the mechanisms of motivated reasoning, as we will see in major cases of misinformation.

Can values constitute misinformation? If values were just subjective preferences corresponding to individual whims and inclinations, then values could not be false, inaccurate, or misleading, so they could not be misinformative. I view values as emotional mental representations that can be objective if they correspond to human biological and psychological needs.[33] For example, health care is a human need because people can die without it, so the concept of health care deserves the positive emotion that is usually attached to it. This value constitutes information that should help to guide decisions about governmental obligations to improve human lives. Chapter 2 argued that feelings could be information and misinformation, and values come with feelings, so they can also be both information and misinformation. Chapter 9 provides further defense of how values can be objective even though they are emotional. It follows that objective values can provide real information, and distorted values can lead to disinformation.

Neural Mechanisms

When Ziva developed her ideas about motivated reasoning in the 1980s, social psychologists largely ignored the operations of the brain. But that decade saw the beginning of brain scan studies of cognitive processes and applications of computational neural networks. Ziva and I published a much-cited neural-network model of stereotype application in 1996, but we never discussed neural mechanisms for motivated reasoning.[34] Today, enough is known about relevant brain areas and more biologically realistic neural networks to sketch how motivated reasoning operates in human brains.

In 2006, Drew Westen and his colleagues published a pioneering functional magnetic resonance imaging (fMRI) study that found that motivated reasoning is associated with emotional processing, including the ventromedial prefrontal cortex, cingulate cortex, and insula.[35] These correlations support the view that motivated reasoning has a substantial emotional component.

Brent Hughes and Jamil Zaki summarize the neural correlates of motivated cognition, including wishful seeing as well as self-serving inferences.[36] They describe how motivation may automatically and effortlessly influence cognition by reducing neural activation in neural structures associated with deeper information processing, such as the orbitofrontal prefrontal cortex and lateral prefrontal cortex. These areas contribute to decision making and evaluation of the self and people close to us.

Hughes and Zaki do not say how the brain undergoes this shift, but other studies suggest possibilities. Motivation for potential rewards is associated with neural activity in the nucleus accumbens, also known for correlations with wanting and liking.[37] Motivation associated with potential threats is associated with neural activity in the amygdala, also known for correlations with fear and anxiety.[38] Perhaps intense neural processing in the nucleus accumbens and amygdala undermines deeper cortical processing and thereby makes people susceptible to motivated inferences, a breakdown from normal functioning that stimulates useful actions to achieve desirable goals and avoid threatening outcomes.

Brain scans tell us only about the neural areas that correlate with thought processes and do not provide the mechanisms by which neurons generate thoughts. Fortunately, theoretical neuroscience has advanced sufficiently to describe such mechanisms, even for cognitions such as beliefs and emotions.[39] We can sketch how motivated inference works for a simple example where Pat believes that Sam is honest because Pat wants to maintain a romantic relationship with Sam.

Mental representations of individuals like Pat and Sam and concepts like honesty and romance can all be explained as patterns of firing in groups of neurons.[40] Beliefs are also neural patterns that the brain constructs by binding other patterns together, for example, when Pat's belief that Sam is honest binds the representations for Sam and honest. Goals are neural patterns that bind representations of states such as Sam loving Pat with neural patterns for the emotion of wanting. Emotions are also neural patterns that bind together representations for (1) a situation such as Sam loving Pat, (2) an appraisal such as that this situation suits Pat's goals, and (3) physiological states such as Pat's heart rate.[41] The appraisal and physiology indicate potential rewards for Pat of Sam loving Pat, activating neural groups in the nucleus accumbens.

Motivated reasoning occurs in this case because of further binding in Pat's brain of the belief that Sam is honest with the goal of Sam loving Pat. The connection between the goal and the belief generates neural activity in brain areas that correlate with acceptance of beliefs such as the medial prefrontal cortex.[42] Hence, by a process that amounts to inference rather than reasoning, personal goals and emotions in the brain bring about a motivated conclusion.

To be plausible, this sketch could be fleshed out with sufficient detail to allow a computer model of how motivation can lead to inference, similar to an existing model of how intentions and emotions lead to action.[43] But the sketch provides a start at explaining motivated reasoning by neural mechanisms that go beyond correlations with brain areas.

Social Mechanisms

Motivated reasoning results from psychological processes that occur in individual brains, but it is also influenced by group interactions. In social mechanisms, the parts are people, the interactions consist of communications between individuals, and the results are the mental states and actions of individuals and groups.[44] How can communication lead to the spread of motivated reasoning in groups of individuals?

One social contributor to group motivated reasoning is that people tend to gather information from their own groups.[45] For example, people get information from their friends and Facebook affiliates, who tend to be people with whom they have common goals. This communication spreads goals and emotions as well as beliefs. Hence social bias leads to information bias that can fuel motivated reasoning. Biased selection of information leads to skewed samples that interact with motivated interpretation of the biases.

Another social mechanism of group motivated reasoning is the transmission of emotions within members of a group. Emotions are important for motivated reasoning because they mark goals as situations to approach or avoid. Positive emotions such as happiness and pride indicate situations to be approached, pursued, and promoted, for example, finishing a job. Negative emotions such as sadness and fear indicate situations to be avoided, fled, and discouraged, for example, being fired. Hence, the social transmission of emotions also transmits goals and generates motives, potentially contributing to biased inference.

Emotional communication occurs in ways that include contagion, group rituals, and empathy.[46] Emotional contagion occurs when one person mimics the bodily expression of another and thereby generates physiological inputs that contribute to emotions. For example, crying babies tend to make other babies cry. If Pat is excitedly smiling and gesturing and bubbling about a political candidate, then Sam may unconsciously mimic these bodily occurrences and acquire some of Pat's enthusiasm for the candidate. Mirror neurons that fire in one brain with patterns similar to those found in the brain of an observed person may contribute to emotional contagion.[47]

Emotional contagion helps to explain the social power of religious and political rituals that can be strongly motivating.[48] For example, people at a political rally who are cheering and shouting together may pick up on one another's energy and attitudes. Religious rituals also require people to duplicate each other's behaviors such as praying, kneeling, and singing, which can coordinate their emotions and hence their motivations.

A more complex kind of emotional communication is empathy, in which people imagine themselves in the situations of others and thereby generate similar experiences. For example, if Pat is puzzled by Sam's political enthusiasm, Pat could imagine being in Sam's precarious economic situation and experiencing something like the fears that motivate Pat to support an extreme candidate. Hence, group interactions such as exchange of emotions and biased information sampling can encourage motivated reasoning.

Emotional communications that influence motivation and inference are heavily influenced by institutions including families, social clubs, schools, government agencies, and political parties. Institutions have policies, rules, values, and norms that affect the thinking and behavior of their members. If Pat and Sam are members of a political party, they will be motivated to conform to or at least consider its policies. I belong to the Canadian New Democratic Party so I pay close attention to the proclamations of its leaders and conventions. Institutional motivations include wanting to support an organization and to believe that it is performing as it should. Such motivations can distort reasoning, for example, in failures to recognize that the organization has become corrupt. The political epithet "my country right or wrong" reeks of motivated reasoning. Institutions also affect people's personal identities.

IDENTITY

Ziva observed that people with high self-esteem tend to put a positive, self-enhancing spin on social information, recalling events that make them look superior.[49] Social and political identities more generally influence people's motivations and hence their reasoning about themselves and events in the world. For example, Daphna Oyserman and Andrew Dawson describe how the British referendum on whether to leave the European Union was influenced by group-based identities.[50] People who identified themselves as patriotically British were more likely to vote for Brexit, while people who identified themselves as European were more likely to vote against it. Understanding the impact of identity on motivated reasoning requires figuring out how identities influence motivations, emotions, and inferences. Then we can consider how to counter the distortions of identity-based motivated reasoning.

What Is Identity?

People's identities are how they think of themselves with respect to social groups such as families and personal characteristics such as gender. How identities work can be explained through noticing how the self results from the interaction of neural, mental, molecular, and social mechanisms.[51] Psychologists have discerned more than eighty phenomena concerning the self. People represent themselves to themselves using concepts and images, for example, when I think of myself as a Canadian philosopher, cognitive scientist, and father. Self-representing also requires evaluations concerning what attributes and behaviors are admirable or blameworthy, for example, viewing being hard working as a strength. Self-understanding concerns what a person can and cannot do, for example, attempting self-help, and how a person develops. Hence, social and personal identities are based on complexes of representations, evaluations, and expectations about the self.

How Does Identity Influence Motivations and Emotions?

Some self-representations are too insignificant to have a motivational impact, for example, when I think of myself as a resident of Ontario. But characteristics that come with strong positive evaluations are highly motivating, for example, when patriotic Britishers are proud to be British rather than European and emphatically do not want to change. The evaluations and attitudes about change packed into self-representations generate powerful motivations to maintain characteristics viewed as positive and to resist characteristics viewed as negative, such as continuing to be proudly British rather than embarrassingly European, respectively.

These evaluation-based motivations bring emotions that encourage actions. For example, positive emotions associated with being patriotically British include strongly liking one's country, being proud about the national heritage, and being happy and grateful to have been born there. The same identity can also connect with negative emotions such as fear or hatred of foreigners, anger at foreign control, and even disgust at foreign food. Motivations to foster positive emotions and avoid negative emotions spur actions such as voting for anti-immigrant politicians and being mean to foreigners.

How Does Identity Influence Reasoning?

Identity can influence reasoning through all the mechanisms I described. When people's identity includes a social or individual characteristic that they evaluate positively, then they will be motivated to retain thinking of themselves as having that characteristic. Then the cognitive mechanism of selective search will make them inclined to remember and find information that supports that characteristic. For example, wanting to be British will make people think of what they like about being British, for example, their native food and music.

Identity can also contribute to motivated reasoning via emotional coherence. During the Brexit referendum, the Leave side spread misinformation in support of the claim that the United Kingdom would be better outside the European Union, for example, that leaving would free up large funds for the National Health Service. People's motives to be British and to have a well-functioning health system combined to help them believe that Brexit would be beneficial, even though there was no evidence that Brexit would benefit health. People who strongly identified themselves as British and non-European found Brexit compelling without thinking through the cost-benefit calculations that revealed its pitfalls.

I mentioned Steven Pinker's citation of motivated reasoning as the major source of irrationality. His second major source is myside bias, which Pinker interprets as the tendency of people to reason to conclusions that enhance the correctness of their political, religious, ethnic, or cultural tribe. For example, conservatives want to support the beliefs of other conservatives, and liberals want to support the beliefs of other liberals. Pinker's version of myside bias is akin to identity-based motivated reasoning.

Pinker's version is narrower than the original interpretation by Keith Stanovich of myside bias that "occurs when we evaluate evidence, generate evidence, and test hypotheses in a manner favorable toward our prior opinions and attitudes."[52] Stanovich's myside bias is equivalent to personal motivated reasoning, whereas Pinker applies it to cases where people reason to support their tribe's interests even when they go against personal motivations. However, people's vital needs to belong to and identify with social groups usually motivate them to support beliefs and values of the group. Institutions such as families, political parties, and nations contribute to personal identities, emotions, and motivations in ways that can distort inferences.

Goals based on group identities are a major source of motivated reasoning about health, climate, politics, inequality, and war, as chapters 4–8 show. Correcting identity-based distortions uses the same strategies as dealing with motivated reasoning in general.

MENDING MOTIVATED REASONING

Motivated reasoning is based on legitimate personal goals such as success, health, being liked by others, and maintaining self-esteem, as well an on social goals that come with group identities. Thus, we can ask generally, How can the thinking errors instigated by motivated reasoning be reduced or corrected?

Psychologists have noticed that some cases of motivated reasoning seem to be good for people, for example, when optimism stabilizes a relationship, helps people look after themselves in old age, or gives an athlete confidence to produce a stellar performance.[53] Ziva recognized such cases but had strong counterexamples, such as the disastrous overoptimism of Jews who thought that the antisemitic measures of Nazi Germany were temporary. My case studies of motivated reasoning about COVID-19, climate change, conspiracy theories, inequality, and the Russia-Ukraine war provide further examples of deaths and suffering boosted by motivated reasoning.

A naïve evolutionary argument might claim that motivated reasoning is so prevalent that it must have some adaptive value, presumably because people who do it are better at surviving and reproducing. However, not all prevalent features of human behavior should be explained as adaptive because alternative explanations include random drift, cultural learning, and side effects. For example, all human cultures use fire, but fire use is not a neural adaptation but instead became universal because of its usefulness and cultural communication. Diseases such as infections, heart problems, and cancer are common but not adaptive: they are side effects of breakdowns in biological mechanisms such as cell division that are adaptive. Similarly, motivated reasoning is best viewed as a side effect of breakdowns in normal brain processing, not as inherently adaptive.

In an article on how rationality is bounded by the brain, I explain the tendency of people to succumb to motivated reasoning as the result of the tight neural integration of cognition and emotion.[54] This integration is often beneficial because it keeps people's thinking focused on what is important to their well-being, as indicated by emotional significance. Motives and goals are appropriately relevant to deciding what to do because we should choose actions that are most consistent with our goals.

Integration of cognition and emotion causes problems when people's motivations swamp their ability to draw conclusions based on good evidence. People need to consider their goals when they decide what to do, but problems arise when such inferences leak into inferences about what to believe, which should be based on evidence. Other limitations of the brain that hinder rationality are its slowness and restricted size, along with imperfections in attention and consciousness. We should look for ways to diminish the negative effects of motivated reasoning rather than assume for psychological or biological reasons that it must be desirable. Motivated reasoning is more commonly a bug rather than a feature.

The best ways of countering the negative effects of motivated reasoning are based on the reinformation techniques outlined in chapters 1 and 2. First, we should encourage people to recognize their susceptibility to motivated reasoning. Even the finest logicians and statisticians can succumb to motivated reasoning on topics closest to their hearts, such as family, health, and success. People find it easier to recognize motivated reasoning in others than in themselves, but comparisons of our own thinking to that of others may help us to realize when we are influenced by personal goals rather than evidence. Advice from other people is useful because they do not share our personal goals and should therefore be more likely to recognize our errors based on motivated reasoning. For example, when I see other people overestimating the benefits of dubious dietary supplements, I should become more aware of my own tendencies in that direction.

Second, people can be encouraged to enhance their accuracy goals and realize that motivated reasoning is often only directed at short-term goals such as immediate pleasure rather than long-term goals such as health. Desiring truth and long-term benefits can help tame the appeal of motivated reasoning. People can be encouraged to spend more time and effort on deep processing that focuses more on evidence and less on motivation. For medical problems, answers can be sought from evidence-based websites such as the Mayo Clinic rather than from hype-based enterprises on social media.

Third, critical thinking can help people to recognize alternatives to motivated reasoning, provided by evidence-based reasoning patterns such as statistical inference and inference to the best explanation. People who realize that their inferences are unduly goal-based rather than evidence-based can shift to forms of reasoning that are much better at tracking the truth.

Fourth, we can work with other people using the therapy-like techniques of motivational interviewing. Asking open-ended questions and empathizing with people can help them understand why they are prone to making conclusions based on personal goals rather than evidence. One of the major benefits of therapy is to help people regulate their emotions through techniques that include modifying goals, beliefs, and physiology.[55] Changing goals and emotions can remove or reduce the motivations that distort reasoning. Motivational interviewing is enhanced by paying attention to the vital needs that furnish goals that can distort reasoning.[56] Emotion and humor can serve to boost people's inclinations to recognize and challenge misinformation.[57]

Fifth, institutional modification can be directed at mitigating the social causes of motivated reasoning. Bad institutions have policies and values that encourage people to maintain their identities by accepting misinformation, for example, when a political party takes a stance against mandatory vaccinations. But good institutions encourage people to consider social goals about the needs of the general population, which reduces the distortion of inferences by individual goals. If Pat and Sam only think about themselves, they will be inclined to motivated reasoning based on their personal goals. But if they altruistically consider the goals of others, then they will have to consider a broader range of evidence than what suits themselves.

Benevolent institutions such as nonauthoritarian families and political organizations consider many voices that compensate for individual motivated reasoning. Rich people are naturally motivated toward economic doctrines that support low taxation and minimal governmental intervention, but parties with broader membership will naturally consider a fuller range of interests. Changing institutions to be responsive to general needs is therefore important for reducing the social distortions of motivated reasoning.

Finally, dealing with other people's motivated reasoning must sometimes resort to political action to prevent their bad inferences from propagating to others. Monitoring and restricting incontinent platforms such as Twitter and YouTube can reduce the spread of beliefs that reflect only the personal goals of their initiators rather than evidence. Chapter 8 describes how gatekeepers such as editors can work to inhibit misinformation. Motivated political misinformation can be reduced by preventing narcissistic psychopaths from gaining power through propagandized elections or coups.

Similar techniques should help people to resist identity-based motivated reasoning. People's personal and group identities are so central to their lives that eliminating motivations based on them would be impossible. We cannot tell individuals to stop caring about themselves and their important social groups such as families, communities, and nations. Identity-based motivated reasoning cannot be eliminated, but it can nevertheless be mitigated by warning people about the dangers of such reasoning and encouraging them to place a higher value on accuracy goals and to engage in the more rigorous reasoning processes advocated by critical thinking and responsible institutions.

For example, British Brexit voters needed reminding that their decision should be based on evidence of the costs and benefits of leaving the European Union, not just on a strong national identification. They also needed reminding that the politicians extolling the advantages of exiting were motivated by personal gains and libertarian ideologies, which explained their exaggerated claims about the National Health Service better than any available evidence for them. The two-stage critical thinking approach of (1) revealing thinking errors such as motivated reasoning and (2) replacing them with good reasoning patterns might be effective. Plain factual correction is unlikely to be of much use in dislodging people from views that their identities place close to their hearts

Given the centrality of identities to people's feelings, a more effective technique might be the therapy-like approach of motivational interviewing. Nonthreatening questioning and empathy could be used to understand how people are thinking about the aspects of themselves and their groups that matter to them. For example, Brexit voters could have been asked about what is valuable about being British and then sympathetically encouraged to think more carefully about the consequences of Brexit for their lives and the lives of others in the country.

Finally, institutional modification and political action can help to control the negative effects of identity-based motivated reasoning. The Brexit referendum could have been more closely regulated to disallow propagandistic disinformation such as lies about the National Health Service. Regulation could also have inhibited the use of social media such as Facebook to spread lies that tapped into group identities to support Brexit. Identity-based motivated reasoning will always be hard to stop, but a combination of critical thinking, motivational interviewing, and social action might slow it down.

MOTIVATED IGNORANCE

Motivated reasoning leads to false or misleading beliefs, including the denial of important truths. Another kind of motivated thinking leads to no beliefs at all when personal goals and emotions incline people to remain ignorant. Knowledge is power, but ignorance is bliss.

Here are examples of motivated ignorance. Mary thinks her husband might be cheating but is reluctant to check his emails. John has a lump in his armpit but does not go to a doctor. The tobacco companies work to suppress research on the effects of smoking on cancer. Governments eliminate environmental research that might show the negative effects of oil sands. Health departments stop doing tests that track the spread of diseases. A major role of educational and governmental institutions should be to discourage ignorance and the motivations that maintain it.

People easily succumb to motivated ignorance when their goals lead them to avoid learning potentially valuable information.[58] Such avoidance of knowledge naturally happens with respect to important personal topics such as relationships and health. Motivated ignorance also operates socially and politically, when organizations such as governments and companies work to ensure that some kinds of knowledge remain unavailable, as has happened with respect to climate change and the dangers of tobacco.

Motivated ignorance is generally harmful to individuals and societies, but sometimes it can be rational, that is, when knowledge that might be acquired would be predominantly damaging to human wellbeing; for example, we may be better off not knowing how to convert viruses into ones that can cause pandemics. Some medical tests like prostate specific antigen (PSA) tests cause more harm than good because they lead to false diagnoses and unnecessary treatments.

Motivated ignorance is different from motivated reasoning, which occurs when people's goals distort the conclusions they reach. The result of motivated ignorance is the avoidance of any conclusions altogether, which happens in all the domains that incline people toward motivated reasoning:

- Romantic relationships: I'd rather not know about the affair.
- Parenting: I don't want to know about my teenagers' social life.
- Medicine: Going to the doctor is just a waste of time.
- Politics: The leader's personal life is no business of mine.
- Sports: The star's drug use is his own problem.
- Research: It's better not to check my citation count.

- Law: Don't ask, don't tell.
- Religion: God is a mystery.
- Economics: Better not to know about corruption.

In addition to these individual impacts, motivated ignorance also operates systematically in society. The book *Merchants of Doubt* describes how some businesses, governments, and even scientists have worked to maintain ignorance about strategic defense, acid rain, the ozone hole, global warming, pesticides, and the risks of tobacco and secondhand smoke.[59] Motivated ignorance sometimes takes the shape of motivated skepticism, where the desirable skepticism that demands appropriate evidence is replaced by a global skepticism that rejects all knowledge claims. Combatting motivated ignorance and skepticism requires pointing out the value of real information for dealing with important personal and social issues, along with the craven motivations of those who attempt to deny people the relevant knowledge.

PATTERNS OF MOTIVATED REASONING

In preparation for applications in chapters 4–8, we can identify common patterns of motivated reasoning based on personal goals, group identities, and specific emotions. Some of these patterns may combine and overlap, for example, when national identities are strong enough to shape personal goals such as military service and emotions such as pride. The personal goals and group identity patterns result from mechanisms of cognitive search and emotional coherence, whereas the emotions pattern is tied to specific emotions.

The motivated ignorance pattern does not produce a conclusion but instead blocks the acquisition of any real information at all. Motivated reasoning usually concerns finding support for existing claims; motivated invention explains the generation of dubious views that then become fodder for motivated reasoning. Motivated invention is the major explanation of the bogus acquisition method of making stuff up. Firms can be hired to generate disinformation for nefarious commercial and political purposes.[60] Science and engineering also engage in invention, but their legitimate motivations are explanation and human benefits, which depend on subsequent validation by evidence.

The following are six patterns of motivated reasoning. A seventh pattern is added in chapter 7 concerning goals to justify political systems.

PERSONAL GOALS PATTERN

- People have personal goals such as self-esteem, success, health, and romance.
- Some beliefs seem to enhance those goals without evidence.
- Other beliefs seem to run counter to those goals.
- Therefore, people will tend to accept the goal-enhancing beliefs and reject the goal-limiting beliefs.

GROUP IDENTITY PATTERN

- People identify with a social group and support its goals.
- Some beliefs enhance those goals and foster the desired identification.
- Other beliefs conflict with those goals and identification.
- Therefore, people tend to accept the identity-supporting beliefs and reject the identity-threatening beliefs.

EMOTIONS PATTERN

- People are susceptible to pleasant emotions such as happiness and pride as well as to unpleasant emotions such as fear and anger.
- Therefore, people will be inclined to accept beliefs that encourage pleasant emotions and discourage unpleasant emotions.
- Conversely, people will be inclined to reject beliefs that encourage unpleasant emotions and discourage pleasant emotions.

PERCEPTIONS PATTERN

- People's personal goals are fostered more by some visual and auditory perceptions than others.
- Therefore, people are more likely to have the encouraging perceptual experiences than the discouraging ones.

IGNORANCE PATTERN

- Acquiring evidence and real information may have unwanted effects on personal goals, group identities, or emotions.
- Therefore, people prefer to remain ignorant and collect no information.

INVENTION PATTERN

- An individual has a strong goal such as financial success or political domination.
- The individual puts existing concepts together to invent a new idea that would support the accomplishment of the goal.[61]
- Therefore, the individual passes the idea along to other people independent of evidence.

Awareness of these patterns is important for fighting the contributions of motivated reasoning to misinformation. Recognition that people are thinking in line with the personal goals pattern or others paves the way for remediation by critical thinking, motivational interviewing, institutional modification, and political action.

Motivated reasoning is not the only cause of the flight of falsehoods, which can also result from mistakes such as faulty experiments, sloppy causal inferences, and laziness.[62] But these six patterns of motivated reasoning lead people astray in many domains, including the medical problem of COVID-19.

CHAPTER 4

PLAGUES

COVID-19 and Medical Misinformation

The COVID-19 crisis (the term "COVID-19" is short for "coronavirus disease of 2019") has generated profound information about the causes, treatments, and preventions of the disease. But the pandemic has also spawned a torrent of misinformation, such as COVID-19 is a hoax, masks are an unnecessary imposition, and hydroxychloroquine and ivermectin are effective treatments. The development of vaccines has saved many lives, but many more could have been saved if misinformation had not discouraged people from getting vaccinated. On one estimate, the United States in June through September 2021 had ninety thousand deaths that could have been prevented with vaccination.[1]

The AIMS theory of information explains the medical successes of COVID-19 research and public health measures. The four processes of *a*cquisition of information from the world, *i*nference that extends these observations, *m*emory, and *s*pread have enabled governments and medical officials to forestall even worse outcomes. But these efforts have been hindered by cascades of misinformation that can be understood as breakdowns in the AIMS mechanisms. Reinformation about COVID-19 and other medical problems can be accomplished by a combination of factual correction, critical thinking, motivational interviewing, institutional modification, and political action.

REAL INFORMATION ABOUT COVID-19

Fighting disease requires real information about symptoms, causes, and treatments. We are fortunate that an enormous amount has been learned about COVID-19 since it was first recognized in December 2019.

Acquisition

Good decisions about COVID-19 depend on acquisition of real information about the disease, through interactions with the world using observations, instruments, and experiments. Collecting and representing this information has generally been effective, but breakdowns in these classes of mechanism have spurred misinformation, such as claims that the risks of vaccines outweigh their benefits. Acquisition of information is just the start of identifying the causes and treatments of disease, which also requires inferences.

The most basic information about COVID-19 is that particular patients have the disease, which starts with perceptions that include hearing them cough and reporting symptoms such as tiredness, loss of taste or smell, sore throat, aches and pains, and chest pain. Visual perception reveals other symptoms such as rash, diarrhea, red eyes, shortness of breath, and loss of speech. Touching a patient's forehead gives a rough indication of fever.

Using instruments expands perceptual abilities, for example, when a thermometer gives a much more exact measurement of a patient's fever. Pulse oximeters attached to fingers measure the saturation of oxygen carried in red blood cells, thereby providing a numerical correlate of shortness of breath observed by sight or sound. Patients with difficulty breathing can be scanned using X-rays or computed tomography (CT) images to determine the extent of lung infections caused by COVID-19 or pneumonia.

Symptoms recognized by perception or instrumental measurement provide some evidence that a patient has COVID-19, but some symptoms are also compatible with less lethal viral diseases such as influenza and varieties of the common cold. Fortunately, tests for COVID-19 were developed quickly soon after the recognition of the disease. By February 2020, the virus that causes the disease, eventually named severe acute respiratory syndrome coronavirus 2 (SARS-CoV-2), was identified by electron microscopy, and its genetic composition was determined by sequencing technology.[2] This knowledge allowed the rapid development of a polymerase chain reaction (PCR) test to detect genetic material from the virus responsible, indicating that a patient whose nose swab yields a positive PCR test has COVID-19 rather than another viral disease. Knowledge about the source of COVID-19 developed through powerful instruments that included microscopes, genetic sequences, and PCR, all of which provide reliable interactions with the world.

No test is completely reliable, but the PCR procedure is an instrument that rarely yields false positives (saying incorrectly that a patient has the disease), and only about 20 percent of the time yields false negatives (saying incorrectly that a patient does not have the disease).[3] Other tests look for protein fragments (antigens) from the virus or antibodies against the virus, but these are not as accurate as the PCR test. The perils of testing and the vagaries of instruments are illustrated by the initial failure in February 2020 of the U.S. Centers for Disease Control and Prevention (CDC) to produce a reliable test, making it hard to recognize that COVID-19 was already spreading rapidly through the country.

Viruses are too small to be viewed by ordinary microscopes, but electron microscopes can produce images like the one shown in figure 4.1. Electron microscopes have been used for decades to examine viruses, and they reliably transcends the limitations of human vision by magnifying sixty thousand times.

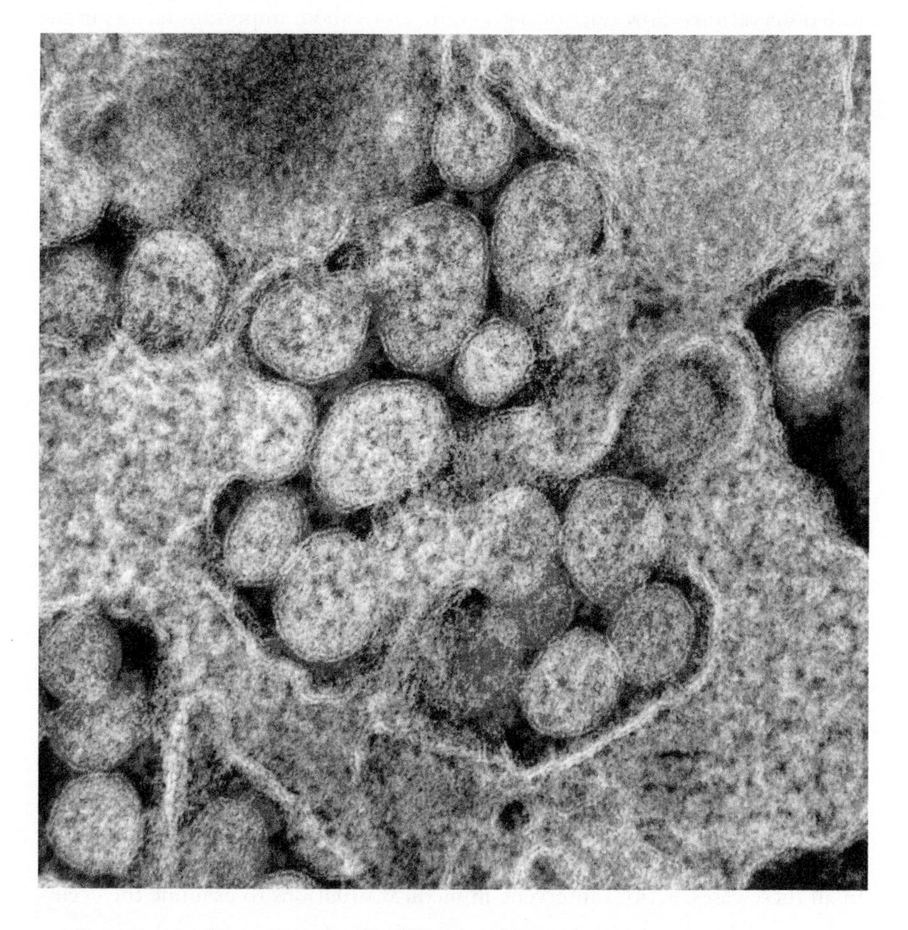

4.1 Transmission electron micrograph of SARS-CoV-2 virus particles.
Source: Courtesy of U.S. National Institute of Allergy and Infectious Diseases, public domain.

Single observations concerning particular patients tell us little about COVID-19. Researchers often note that the plural of the word "anecdote" is not "data."[4] This remark highlights how scientific research is not a collection of stories but rather depends on compilations of observations that are systematic in several ways. First, the observations take place in a well-defined population, such as people in the United States that are diagnosed with COVID-19. Second, the observations concern specified outcomes, such as people diagnosed with COVID-19 without symptoms, people with mild symptoms, people with severe symptoms that require hospitalization, hospitalized patients requiring treatment in intensive care units, patients placed on ventilators, and patients who die. Third, systematic observations yield numerical quantities, such as the percentage of people diagnosed with COVID-19 who die of the disease. Fourth, such observations allow statistical generalizations about important factors in the course and treatment of the disease. For example, systematic observations can identify the proportion of people who die of the disease and who have contributing conditions such as obesity and diabetes.

Thus, systematic observations are a valuable source of real information about COVID-19 and other medical problems. Such studies can identify informative correlations between conditions and disease development and between treatments such as ventilators and disease outcome. But they are limited in what they can tell us about the causes of disease and the real effectiveness of treatments. For such inferences, we need information derived from experiments.

Experiments are systematic observations that employ manipulations to divide populations into two or more groups.[5] Here are some examples of COVID-19 manipulations that provide evidence relevant to causal inferences:

- Give one group of patients a treatment such as a drug and give another group another drug or no treatment at all.
- Have one group of a general population wear masks and another group not wear masks.
- Have one group vaccinated with a specific vaccine and another group get no vaccine.

In all these cases, actions intervene in medical situations to examine the occurrence of expected events such as fewer cases of the disease or fewer deaths.

Experimentation alone is not enough to provide the best information relevant to making causal inferences. In the 1940s, beginning with trials for treatments for the common cold and tuberculosis, medical experimentation became more effective through the adoption of clinical trials that are randomized, controlled, and blind. A randomized trial ensures that the different groups of people in an experiment are selected randomly rather than by researchers who may be biased, for example, in assigning more healthy patients to get a drug. Whether a participant in a trial gets a drug or vaccine can be determined by a coin flip rather than by a biased expectation of whether the participant will benefit from it.

A controlled trial is designed to ensure that various causal factors can be ruled out as interfering with the results, for example, if the patients in a drug condition tended to be younger than the patients in a nondrug condition. A blind experiment helps to eliminate other sources of bias in the expectations of the patients or the researchers: in a double-blind design, researchers do not learn which condition patients are in until the experiment is finished. We have no absolute guarantee that randomized controlled, blind studies will yield correct results, in keeping with the general fallibility of science. But randomizing, controlling, and creating blind experiment designs help to reduce the occurrence of confounding factors that interfere with the good causal inferences discussed below.

Clinical trials of drugs are divided into four phases described by the U.S. Food and Drug Administration (FDA):[6]

- A Phase I trial tests an experimental treatment on a small group of often healthy people (twenty to eighty) to judge its safety and side effects and to find the correct drug dosage.
- A Phase II trial uses more people (one hundred to three hundred). While the emphasis in Phase I is on safety, the emphasis in Phase II is on effectiveness. This phase aims to obtain preliminary data on whether the drug works in people who have a certain disease or condition. These trials also continue to study safety, including short-term side effects. This phase can last several years.
- A Phase III trial gathers more information about safety and effectiveness, studying different populations and different dosages, using the drug in combination with other drugs. The number of subjects usually ranges from several hundred to about three thousand people. If the FDA agrees that the trial results are positive, it will approve the experimental drug or device.

- A Phase IV trial for drugs or devices takes place after the FDA approves their use. A device or drug's effectiveness and safety are monitored in large, diverse populations. The side effects of a drug may not become clear sometimes until more people have taken it over a longer period of time.

Such trials have found that vaccines, produced by Pfizer, Moderna, AstraZeneca, and other companies, are effective at preventing the spread and reducing the severity of COVID-19. Carefully conducted trials have similarly shown that some drugs, including monoclonal antibodies, antivirals such as molnupiravir and paxlovid, and the antidepressant fluvoxamine, work to reduce the severity of the disease.[7]

Perceptions, instruments, systematic observations, experiments, and clinical trials have yielded vast amounts of real information about the course, causes, and treatments of COVID-19, and this information has been published in millions of articles just in 2020–2022, according to Google Scholar. The methods conform with the five characteristics of evidence noted in chapter 2:

1. Reliability: Well-conducted COVID-19 experiments produce truths that stand up well to future examination. COVID-19 research is too new for much replication to have been performed, but natural experiments with widespread vaccine use have largely reproduced the results of the original clinical trials.
2. Intersubjectivity: COVID-19 systematic observations and controlled experiments can be performed by different people in different countries, not just individuals reporting their own opinions.
3. Repeatability: Original experiments with vaccines have been frequently duplicated with different age groups.
4. Robustness: Investigation of the novel coronavirus has received convergent results with diverse instruments that include electron microscopes, genetic sequences, and PCR duplication.
5. Causal correlation with the world: Thanks to perception and the use of instruments such as oximeters, experiments concerning COVID-19 have sufficient interaction with the world to make it likely that the information they yield is real.

These five standards of evidence yield a rough ordering of methods for gaining medical information, from stronger to weaker: clinical trials, experiments, systematic observations, solitary observations based on instruments, and solitary observations based on unassisted perception.[8] COVID-19 has combined different sorts of research, ranging from epidemiological studies of disease transmission to microbiological studies of vaccine genetics, to provide a coherent understanding of the disease.

Mechanisms of collecting would be useless without ways of representing the collected information. Ongoing use requires representations by structures that can survive in minds, physical devices, or computational resources. Findings such as the effectiveness of vaccines and drugs are usually represented verbally in sentences that occur on paper, in computer programs such as Microsoft Word and web browsers, and beliefs in human minds.

Visual representations are also useful for conveying information with spatial structure, such as the picture of the novel coronavirus in figure 4.1. Other pictures vividly portray characteristics of the disease such as the international diffusion. For example, the World Health Organization (WHO) provides a map of the world where clicking on a country reveals its total cases and deaths, and the *New York Times* has a map where clicking reveals a country's current daily rate of cases per hundred thousand.[9] The WHO and other websites provide detailed graphs that show the occurrence of the disease over time. The Worldometer website provides a more comprehensive table that reports total cases, deaths, and other figures for more than two hundred countries.[10] Hence information is usefully represented in words, pictures, and databases.

Attention to the mechanisms of collecting and representing allows characterization of data as the products of reliable collecting mechanisms such as perception and experimentation, formed into representations of the world. For example, perceptual representations in the brain are data and so are statistical results of controlled experiments. Important COVID-19 data include the national statistics about cases and deaths, and the results of clinical trials concerning the effectiveness of drug treatments and vaccines.

The acquisition of real information is more than an individual, psychological process. Collecting and representing knowledge about COVID-19 depends on many groups and institutions, including laboratories, universities, companies, and government agencies.

Inference

For COVID-19 and other diseases, collecting information from the world provides essential data about the occurrence, development, and treatment of illness.

Unlike other animals, humans are not limited to what is perceived and can use inference to transcend perception with defensible conclusions about unobservable causes and mechanisms. Viruses, for example, are not perceivable by bodily senses but require microscopes that make heavily theoretical assumptions about the existence and behavior of electrons. Even observational judgments about diseases require ongoing evaluation concerning their coherence with evolving bodies of evidence. Hence, maintaining real information under threat of misinformation requires frequently evaluating even the most observational findings and close scrutiny of the results of transforming representations into ones that go beyond observations with causal leaps. Developing and maintaining knowledge about COVID-19 and other diseases requires continuous evaluating of representations and frequent transforming.

Representations need to be approximately accurate to be useful in dealing with the world, and accuracy of evidence and explanations is crucial for the effectiveness of claims about the treatment and prevention of COVID-19. A representation is important if it is relevant to people's goals. Collecting information that has no contribution to understanding and flourishing is a waste of time, energy, storage capacity, and communication. The purpose of collecting information about COVID-19 is protecting and improving human health, and these goals require continuous inferences based on rapidly accumulating data. Thus, evaluating and transforming deserve to be counted among the mechanisms of information.

Specific evaluation mechanisms apply to particular forms of representation and collection. The parts are (1) the human evaluator, such as a scientist or journalist, and (2) the evaluated representation, such as a contentious belief. Interaction of these parts produces an assessment, such as the judgment that a belief is true and important.

Consider the claim that dexamethasone (a steroid that decreases inflammation in the lungs) increases patient survival. This claim was evaluated by clinical trials in which patients were randomly assigned to receive the drug or no treatment, and their twenty-eight-day mortality was assessed. Efficacy of the drug was shown by the difference between the 22.9 death rate for patients given dexamethasone and the 25.7 percent death rate for the patients given usual care.[11]

Death is observed by visual and tactile observation of lack of vital signs such as a pulse. Evaluating perceptual observations operates by determining whether the observations are made under normal conditions, such as good light for visual observations and lack of background noise for auditory observations, using well-functioning sense organs.

Claims that go beyond the senses require more complex evaluating, for example, causal claims such as that SARS-CoV-2 infection causes COVID-19. Ordinary people may evaluate the claim that A causes B simply by noticing that B follows A, but epidemiologists have identified multiple factors for assessing causal claims, including temporal precedence of A before B, strength of association of A with B, and coherence of the claim that A causes B with general biological theory.[12] Treatment and prevention claims about COVID-19, such as that wearing masks is helpful, can similarly be evaluated based on their coherence with relevant evidence, as shown below.

The results of evaluating have an influence on the operation of other information mechanisms. Uncertainty about evaluating may spur additional collecting to resolve open issues. A claim that is evaluated as accurate becomes worthy of storing in stable forms such as human memory, print, and electronic databases. A positively evaluated claim can be used in transforming information by further inferences, for example, when the conclusion that SARS-CoV-2 causes COVID-19 influences decisions about preventing the disease through masks, vaccines, and disinfecting surfaces. Most important, evaluating a claim as both true and accurate provides good reasons for transmitting it to others and thereby spreading real information.

Information about COVID-19 treatments is important enough to spread rapidly because millions of lives are at risk. Excitement about the success of COVID-19 vaccines occurs because they promise to save lives, whereas fear about the spread of SARS-CoV-2 variants is prompted by concern about greater infection and lethality.

Evaluating is a form of inference that works with representations that already exist, but transforming generates new representations and evaluates whether they are worth keeping. Kinds of transformative inference about COVID-19 include generalization, forming hypotheses about causes, and analogy.

In COVID-19, generalization occurs when practices that work with individual patients are judged to work with others, for example, putting people in a prone position to help breathing.[13] Decisions to approve vaccinations are also based on generalizations that the vaccines that worked well in clinical trials will also work in larger populations.

COVID-19 advances have required causal inferences to conclusions such as the disease is caused by SARS-CoV-2, mask wearing helps to prevent disease spread, and immune responses produced by vaccines cause dramatic reduction in disease incidence and mortality. Analogical inference has frequently been used in conclusions about COVID-19 through comparisons with previous pandemics such as SARS and the 1918 influenza disaster.

As a specific application of evaluating and transforming, consider the causal inference that wearing masks is an effective way of reducing the spread of COVID-19. Causal reasoning can be understood as a process of inference to the best explanation performed by simultaneously satisfying multiple constraints. The goal is to see whether a hypothesis that A causes B is part of the best overall explanation of all the evidence, taking into account alternative hypotheses such as that A and B are only accidentally correlated. Causal reasoning is then a matter of maximizing explanatory coherence rather than calculating with probabilities, as shown in many examples of scientific and legal reasoning.[14]

The main positive constraints for explanatory coherence are that causal hypotheses explain the evidence for them and that these hypotheses can in turn be explained by deeper hypotheses concerning the relevant mechanisms. For example, the hypothesis that wearing masks prevents disease explains various pieces of evidence such as that less spread of disease occurs in mask-wearing countries such as China. Scientists can explain the effectiveness of wearing masks by identifying the underlying mechanism: the virus spreads on droplets through the air and masks block the droplets. Hence the claim that wearing masks prevents disease spread gets coherence both from what it explains and from what explains it.

The main negative constraints operating in explanatory coherence concern the alternative causal hypotheses that compete with each other. For example, the hypothesis that wearing masks prevents disease has to compete with the hypothesis that masks do not prevent disease, which purports to explain other pieces of evidence such as that medical personnel get sick despite wearing masks. A hypothesis is accepted or rejected based on its overall coherence with all the evidence and all other hypotheses, where coherence is a matter of satisfying as many constraints as possible.

Figure 4.2 shows the structure of the explanatory coherence assessment of the mask hypothesis. The hypothesis that wearing masks causes a reduction of spread of COVID-19 gets its explanatory coherence from four directions: what

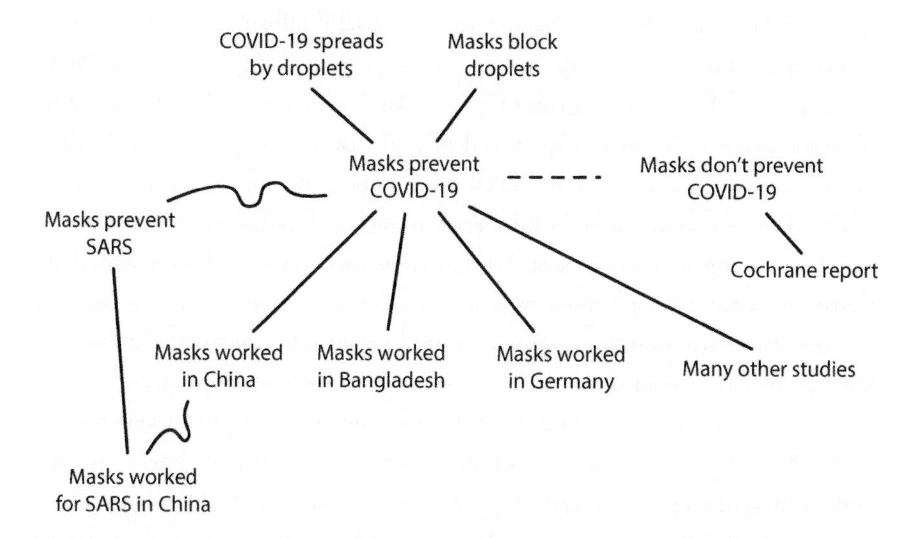

4.2 Explanatory coherence of the hypothesis that masks prevent COVID-19. Straight lines indicate positive constraints based on explanation. The dotted line indicates a negative constraint based on contradiction. The wavy lines indicate analogous explanations.

it explains, what explains it, analogy with SARS and other infectious diseases, and competition with the claim that masks do not prevent COVID-19. Causal explanations are identified by linguistic cues such as "prevent" and "reduce." The most important pieces of evidence explained by the hypothesis that masks reduce COVID-19 are that masks have been used effectively in Bangladesh, Germany, the United States, and other countries.[15] The major challenge to the effectiveness of masks comes from a 2023 Cochrane review of seventy-eight randomized controlled studies of the use of masks against respiratory infections, but almost all these studies did not concern COVID-19.

In addition to what mask wearing explains, we have a plausible explanation of why mask wearing works because masks block the droplets that spread the virus. Especially in the early days of the pandemic, health officials drew heavily on analogies with other infections such as SARS and Middle East respiratory syndrome (MERS): just as masks explained reduction in cases of SARS, so masks explain reduction in cases of COVID-19. Finally, the hypothesis that masks prevent COVID-19 outcompetes the alternative claim that masks do not cause prevention, which is only supported by the lack of a randomized clinical trial in the general population that would show that masks prevent the disease.

For planning and decision making, the most useful information concerns the future, both about what to expect if no actions are taken and what are the likely consequences of different courses of actions. For pandemics, policymakers need plausible predictions about the spread of a disease in a population including cases, hospitalization, and death. Even more important, they need predictions of the likely effects of public health measures such as lockdowns, mask wearing, social distancing, and vaccines. Such predictions are complicated inferences that transform current beliefs about factors such as current incidence of the disease and infection rates into new representations of the future course of the disease. Such predictions require too many factors to be done by simple hand calculations so computer models are employed to generate inferences about future events. Similar models are important for predicting the course of climate change under different scenarios of carbon production, as discussed in chapter 5.

For COVID-19, computer models produced by epidemiologists have been important for policy decisions from early days. A group led by Neil Ferguson at Imperial College London began in January 2020 to use a variety of mathematical techniques to build computer models to predict the course of the pandemic and the effects of different strategies for dealing with it.[16] These models unavoidably make assumptions such as the reproduction rate of viral infections and the proportion of infected people who will die, and risk the danger of the existence of unknown factors that would invalidate the model's predictions. Nevertheless, the models are frequently revised to account for the most recent data about the pandemic, and they provide the best means available for gaining approximately real information about future developments. A computer model was used to estimate that the first year of COVID-19 vaccinations prevented 14.4 million deaths.[17]

Computer models contribute to epidemiology, climate science, cognitive science, and other fields in several ways. To produce a computer program, hypotheses must be made explicit and mathematically exact, and the running of the program shows a minimal feasibility of the assumptions. Simulations reveal the consequences of hypotheses that can be compared with what happens in the world, for example, concerning the spread of coronavirus variants. Computational experiments are performed by varying hypotheses and parameters, for example, concerning virus mutation rates. Simulations can reveal failures of predictive hypotheses as well as unexpected interactions and consequences.

Sometimes good inference techniques are not sufficient to answer important questions, such as the origin of the novel coronavirus that causes COVID-19.

We can dismiss wild theories such as that the virus came from outer space on a meteor, but two serious explanations of where the virus came from are available. Initially, most scientists believed that COVID-19, like SARS in 2003, originated in bats who passed it on to live animals sold in a wet market in Wuhan, from which it spread to humans. Partly because of lack of access to China, however, scientists have been unable to fill in the causal chain in the explanation by identifying which bats and which animals propagated the virus.

The incompleteness of the animal spread hypothesis left the door open for the alternative explanation that the virus originated in a Chinese laboratory such as the Wuhan Institute of Virology. This institute does conduct research on coronaviruses and in 2017 linked the virus that causes SARS to horseshoe bats in Yunnan. The lab-escape theory proposed that this institute was conducting research on a collected or manipulated virus that infected lab researchers and then spread to the general population in Wuhan and the rest of the world.

The problem with this hypothesis is that the intermediate links in the causal chain that it proposes have not been substantiated. No evidence shows that the Wuhan institute was manipulating or studying the novel coronavirus or that lab researchers were thereby infected and spread the virus to others. Because both the animal-spread theory and the lab-escape theory are seriously incomplete, the appropriate conclusion might be that neither is the best explanation of the pandemic origin, and we should refrain from making any inference until the gaps in at least one of the explanations are filled in.

In July, 2022, however, two studies were published by the journal *Science* that appear to tip the balance toward the animal-spread explanation.[18] The earliest known cases of COVID-19 in December 2019 were geographically centered on the Wuhan market that sold several species of animals susceptible to coronaviruses. This centering is easily explained by spread of the disease through the animals, but the alternative lab-escape theory has to make additional assumptions about the virus spreading from the lab to the market. Genetic studies show that initial cases comprised two distinct viral lineages that can be explained by spread from multiple animals. Nevertheless, in 2023, some U.S. officials continued to support the lab-escape theory, which requires additional evidence about how the leak occurred and resulted in the newly recognized characteristics of viral transmission. Some political support for the lab-escape hypothesis results from motivated reasoning to blame China for the disease.

Memory

Memory is important for retaining information in individuals and organizations and is accomplished using mechanisms of storing and retrieving. Dealing with COVID-19 has put huge loads on the memories of individuals and social organizations to deal with the massive amount of information generated concerning causes, treatments, and preventions. The COVID-19 pandemic has required people to make important decisions about where to work, how to shop, and how to manage interactions with other people. When we make such decisions, we rely on information that we have stored in our individual memories. For example, when I decided whether to eat inside a crowded restaurant during the pandemic, I needed to remember what I had learned about the incidence of COVID-19 in my community, its vaccination rate, certificates of vaccination required for entering restaurants, and how the novel coronavirus can spread through the air from people who are not wearing masks because they are eating. I should have stored all this information when I first encountered it because it was emotionally important to my goal of staying healthy.

I need to make sure that I do not pick up information from unreliable sources such as gossip, social media, biased websites, and ideologically motivated news sources. Instead, my memories about COVID-19 usually result from reliable sources such as articles in reputable scientific journals, TV channels I trust such as CBC and CNN, and fact-oriented newspapers such as the *Guardian, New York Times*, and the *Economist*. Thus, mechanisms of evaluating are crucial for enabling an individual to store real information for future use.

How do I know that these sources are reliable? Have I succumbed to the propaganda of organizations that critics such as antivaxxers dismiss as the "mainstream media"? My sources have two characteristics that distinguish them from purveyors of misinformation. First, they employ reliable methods, ranging from the experiments that are reported in scientific journals to the consultation of multiple reputable witnesses that legitimate news media employ. Second, they have a decades-long track record of usually (but not perfectly) making statements that hold up to subsequent scrutiny. Unlike agents who make stuff up for ideological reasons, the sources I trust have a long-standing and mostly fulfilled commitment to getting the facts right.

When information needs to be retrieved for purposes of decision making and other inferences, evaluating is again important. Human memories usually do not come tagged with the sources from which they originate, so when I recall what I think is a fact about the pandemic, such as the percentage of people who get COVID-19 who eventually die of it, I need to ask myself whether the memory is correct. Maybe I confused an earlier report with a more recent update, or maybe I happened to hear the report from a less reliable source. Given the limitations of human retrieval, I may have forgotten other information that is relevant to my decision. Thus, retrieval should be attended by careful evaluating on top of the evaluating that should have accompanied storage.

Organizational memory also needs to be subject to high standards of evaluating to contribute to the retention and propagation of real information. Organizations such as WHO, CDC, and Worldometer have produced curated websites that contain reliable and useful documents and data. Institutions are important both for generating information and for storing it.

WHO maintains a rich website with information on aspects of COVID-19, including disease outbreaks, clinical trials, vaccines, and global research results.[19] For example, it provides a searchable database of hundreds of thousands of peer-reviewed articles published in scientific journals. This website also has a myth-busters page that debunks more than thirty kinds of misinformation on topics ranging from 5G mobile networks spreading the coronavirus to vitamins as a cure for COVID-19.[20]

The CDC also has a well-organized COVID-19 website that stores a wide range of valuable information about current case numbers, vaccinations, and a wide range of scientific research particularly as relevant to the United States.[21] Helpful advice is provided for health-care workers, schools, and other groups. Many other countries such as Canada and the United Kingdom provide useful websites to inform their residents.

The full store of the hundreds of thousands of scientific articles on COVID-19 can be thought of as an organizational memory of researched information generated with remarkable speed. Finding relevant information is not always easy, but search engines such as Medline and Google Scholar provide helpful tools. The quality of published research articles varies from stellar to shoddy, as discussed further below concerning the spread of information. But aside from rare causes of outright fraud, this research results from interaction with the world and is subject to appropriate standards of evaluation. In contrast to the misinformation literature, the scientific literature rarely results from making stuff up. So storing and retrieving of individual and social memories can work well for the maintenance of real information.

Spread

The fourth process central to the AIMS theory is spread, which operates by mechanisms of sending and receiving. COVID-19 communication involves senders such as scientific researchers, company administrators, and public officials. All of these are sometimes receivers, who also include the general public. The channels for COVID-19 information include: conversation; electronic messages; the press, including television; government bulletins such as the weekly *Morbidity and Mortality Review of the CDC*; and journals such as *Nature*.

The key to ensuring spread of real information is to incorporate evaluating into sending and receiving, just as memory needs to incorporate evaluating into storing and retrieving. Evaluating for spread is similar to the process used in deciding what information to store, but the purpose is inherently social, concerning the other people that would benefit from the information. Benefit requires ensuring that the information sent is in formats that can be understood and used by the recipient. For example, the CDC has a newsroom where it releases information intended for a general audience, but it also offers science briefs that summarize research results.[22]

Peer review in science journals provides a model of how evaluating can help to control sending. Journals are not just mechanisms for storing information but are also important channels for getting information from one or more researchers to others and the general public. Before an article is published in a reputable journal, it is reviewed by editors and two or more referees to determine its accuracy and suitability. Peer review is not perfect because flawed or fraudulent articles may still find their way to acceptance.[23] But like medical treatments and vaccines, peer review only needs to work most of the time to provide a valuable constraint on what information is communicated to other people. Scientific peer review contrasts starkly with sending methods that operate with absolutely no scrutiny of whether a message is worth sending, such as the social media discussed later as a major source of misinformation.

Not all journals are peer reviewed and people need to be cautious about the increasing numbers of predatory journals that have serious-sounding titles but will publish any article whose authors pay a fee. Because I published an article on the cognitive science of COVID-19, I get invited to publish quickly on COVID-19 in bogus journals, and thousands of authors have succumbed to such opportunities.[24]

Because of the desperate need for timely information about the pandemic, authors frequently have their papers appear quickly as preprints on servers such as medRxiv.[25] This practice dodges the scientific peer review process and increases the risk of misinformation. Press releases about unreviewed research have the same deficiency, putting an additional burden on readers to be skeptical about medical claims that are made in preprints or press releases rather than peer-reviewed journals.

Complementary to sending, careful receiving is also a crucial mechanism for the transmission of real information. COVID-19 researchers and medical practitioners constantly face questions about the quality of the information sent to them. Even apparently good channels such as publication in a major journal can prove defective, as when an early study in *Lancet* of the effectiveness of hydroxychloroquine for treating COVID-19 was retracted because it was based on faulty evidence.[26] Some senders can be quickly identified as offering bogus information for illegitimate political purposes, as when Donald Trump and his medical adviser Scott Atlas touted unreliable treatments and strategies.

Receiving information should not be just accepting testimony as fact but requires evaluating the quality, importance, and sources of what is sent. Such evaluating can help ensure that real information is absorbed while misinformation is blocked. Journalists and the public can strive to follow scientific practice in giving more credence to peer-reviewed publications than to random posts. Effective spread of real information depends on conscientious individuals and also on responsible institutions ranging from schools to social media.

MISINFORMATION ABOUT COVID-19

Misinformation occurs when information processes break down. COVID-19 misinformation results from breakdowns in the mechanisms that produce acquisition, inference, memory, and spread.

Acquisition

Real information can be collected using reliable methods that include perception, systematic observations, effective instruments, controlled experiments, and clinical trials. Misinformation avoids these painstaking methods through use of the much more efficient technique of just making stuff up. Humans have long been proficient in making stuff up, as seen from the cultural ubiquity of folklore stories, fairy tales, urban legends, and religious myths. Whereas the basic methods for information collection require interaction with the world in ways that can be repeated by other people, the generation of misinformation by making stuff up requires only the imagination of a single individual who is driven by personal or social goals rather than with concern for accuracy and truth. Motivated invention about COVID-19 occurs when someone concocts claims about its occurrence, causes, and treatments that have no connection with evidence.

Many claims about COVID-19 are purely imaginative results of making stuff up, including the following:

- COVID-19 is a hoax.
- COVID-19 will fade quickly.
- COVID-19 is a Chinese conspiracy.
- 5G technology is a cause of COVID-19 symptoms.
- SARS-CoV-2 originated in U.S. army labs.
- COVID-19 can be treated with bleach.
- COVID-19 vaccines make people magnetic.
- COVID-19 vaccines alter people's DNA.
- COVID-19 vaccines make people infertile.

These claims originated without any contact with the world, raising the question of why anyone would believe them. Answers come below from considerations of inference and spread.

Making stuff up is a total breakdown in reliable acquisition mechanisms, but partial breakdowns can also generate misinformation. I do not know of cases where misperceptions or faulty instruments contributed to COVID-19 misinformation, but it can easily arise from observations that are not properly systematic. Disorderly observations are ones where data are collected from only a few individuals or only part of a population. For example, someone who comes down with COVID-19 and follows internet advice to treat it by inhaling hydrogen peroxide fumes may report feeling better, but this anecdote is useless compared to a clinical trial that randomizes assignment of subjects into control groups. Nevertheless, disorderly observations can easily be used to support faulty causal inferences analyzed below.

Other misinformation is based on more subtly defective collecting of data that occurs in sloppy experiments. Early claims that hydroxychloroquine is an effective treatment of COVID-19 were based on poorly conducted studies with small samples and weak controls.[27] Later high-quality studies found no benefit, replacing misinformation by real information. Similarly, the antiparasitic drug ivermectin was considered as a possible treatment because of its antiviral activity, and some tiny, poorly controlled trials suggested it might have some effect on COVID-19. But careful reviews of the available evidence concluded that the studies done were not strong enough to support use of ivermectin.[28] Even the manufacturer of ivermectin, Merck, cautioned against its use as a treatment for COVID-19.[29] Nevertheless, the mechanisms of social spread discussed below encourage people to try an evidentially useless and potentially harmful treatment.

The other class of mechanism for acquisition is representing, which fails when public or private representations do not accurately correspond to the world. The most obvious examples of misinformation concerning COVID-19 are false sentences such as the examples of making stuff up listed above, which form false beliefs in people's minds.

Besides sentences, information can consist of other mental and physical representations such as images and emotions. Figure 4.3 is a much-reproduced image of SARS-CoV-2 prepared by the CDC that gives a good idea of the structure of the virus. But its shapes and color make it less accurate than the image of viruses shown in figure 4.1, which is a photograph produced using an electron microscope. The graphical reconstruction in figure 4.3 exaggerates the spike structures.

Nevertheless, figure 4.3 does not count as misinformation because the exaggeration of spikes highlights how the virus gains entrance to human cells and how vaccines derived from spikes are effective in getting the immune system to attack the virus.[30] Accuracy of representations is not absolute but rather depends in part on the intended purpose of the representation: the number π is not exactly 3.14 but that approximation works for many purposes. On the other hand, some media reports about the virus used distorted depictions of the structure of the virus and therefore qualify as misinformation.[31]

Sensations and perceptions are usually accurate but can generate misinformation when mechanisms are disrupted by internal distortions such as lysergic acid diethylamide (LSD) or environmental factors such as strange lighting. Some COVID-19 patients report that foods lost their taste and smell because the virus

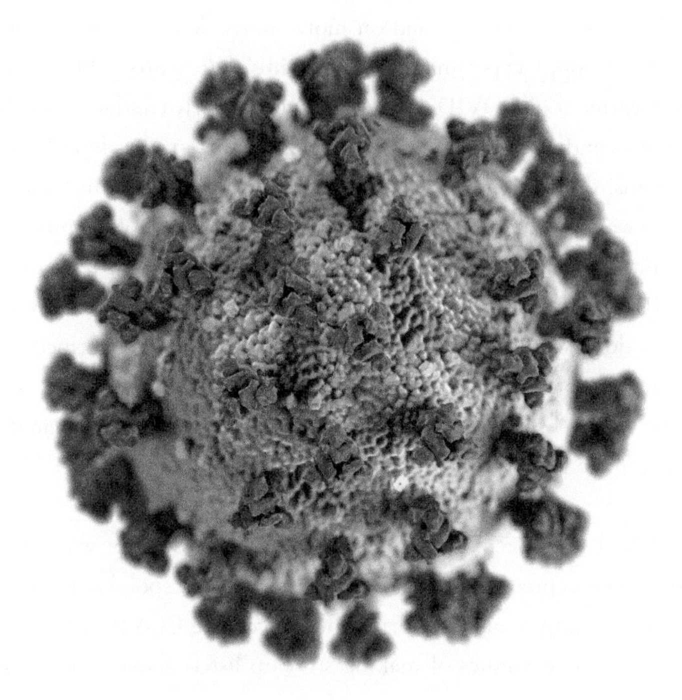

Figure 4.3 Graphical representation of SARS-CoV-2.
Source: Centers for Disease Control and Prevention, public domain.

had infected their olfactory cells. So these patients suffered from misinformation about food.

Emotions are informative when they incorporate accurate assessments of the goal relevance of a situation. For example, fear of the spread of variants of SARS-CoV-2 is appropriate because some are more infectious and more lethal than the original version. Other appropriate emotions include surprise that COVID-19 could be spread by people with no symptoms, sadness that millions of people have died, pride in the rapid development of effective vaccines, happiness in the effective rollout of vaccinations in some countries, disappointment in slow vaccinations in most of the world, and frustration with people who refuse to follow public health advice about slowing spread of the virus.

But reactions to COVID-19 have also included emotions that constitute misinformation. Some people are afraid that governments are lying to people to implement social control. Others are angry that the virus has been intentionally spread by the Chinese. Some leaders have been unduly confident and happy that minimal measures can control COVID-19. Extreme antivaxxers hate leaders who mandate vaccinations. These are all examples of feelings as misinformation.

Emotions turn into misinformation because of disruptions in the mechanisms by which the brain generates emotions: representation of a situation, bodily responses to the situation, and cognitive appraisals of the goal relevance of the situation. Bodily responses can be disrupted by illness, drugs, and involvement in extreme social events such as riots. Appraisals of a situation can be disrupted when accuracy goals are swamped by personal goals such as making money through selling bogus products to treat the virus, a source of motivated invention.

Values are emotional attitudes toward important concepts such as freedom and well-being, and they can be objective when they reflect fundamental human needs.[32] Objective values are sources of information about what matters to people and how their needs can be met. Misplaced values generate misinformation when they are wrong about what makes human lives worth living. For example, some opponents of COVID-19 mandates take freedom to be the only value worth considering so they insist that people cannot be required to take vaccines, disclose their vaccine status, quarantine when sick, or wear masks. Such people are misinformed about freedom when consideration of human needs shows that other important values include human health and social responsibility for the well-being of others. So misinformation applies to values and not just facts, in keeping with the U.S. surgeon general's description of it as misleading. Chapter 9 defends the objectivity of values that qualifies them as real information.

Blame for acquisition of misinformation about COVID-19 is attributable to irresponsible individuals and also to institutions that foster them. Organizational sources of shoddy collecting and representing include sloppy laboratories, ideological think tanks, and biased government agencies.

Inference

Purveyors of misinformation who make stuff up rather than collecting real information are loath to use evidence to evaluate their claims. For example, proponents of the clinical efficacy of hydroxychloroquine tended to ignore the flood of debunking studies. In some cases, neglect of evidence goes with full-scale rejection of science as a source of knowledge based on preference for other sources such as religious doctrines and political ideology. Chapter 5 discusses the differences between science and pseudoscience.

Outlandish causal claims about the origin of COVID-19 in 5G transmission or U.S. army research ignored much more plausible claims about the spread of the SARS-CoV-2 virus from animal transmission or from laboratory escape. Consideration of alternative hypotheses is also important for evaluating the results of low-quality experiments that can readily be explained as resulting from chance or bias. Misinformation about negative effects of vaccines and mask wearing requires ignoring alternative explanations of why some people have problems with vaccines and masks.

Motivated reasoning operates in COVID-19 when politicians downplay the risks of the disease to improve their own chances of reelection, for example, when the governors of North and South Dakota failed to discourage a large motorcycle convention in July 2020. One source of misinformation about COVID-19 treatments and vaccines is people who are motivated to sell their own bogus treatments as more natural than vaccines. People in general are motivated to think that the pandemic will not horribly affect their own lives so they latch on to whatever misinformation encourages them to socialize, travel, and work as much as they want.

Disagreement about science does not always indicate motivated reasoning because scientists can legitimately dispute the quality of experimental findings and their interpretation. The theory of explanatory coherence shows how scientists and ordinary people can reach rational conclusions about COVID-19 by evaluating causal hypotheses with respect to evidence, underlying mechanisms, and alternative hypotheses. Emotional coherence distorts this process by allowing personal goals to enter into the evaluation of hypotheses and evidence.[33] Goals have an emotional valence that involves a positive or negative attitude concerning their satisfaction. For example, the goals of having fun and maintaining freedom have a positive (desirable) valence because they are associated with emotions such as happiness and joy. In contrast, the prospects of staying home and getting sick have a negative (undesirable) valence because they are associated with emotions such as sadness and fear. These goals are relevant to making decisions but they are not supposed to influence judgments about what is true. In terms of standard decision theory, your utilities should not affect your probabilities.

But human brains do not make decisions based on maximizing expected utility, an idea first sketched in the eighteenth century and only developed in the twentieth. Current understanding of brain mechanisms reveals numerous interconnections among areas responsible for cognition such as the prefrontal cortex

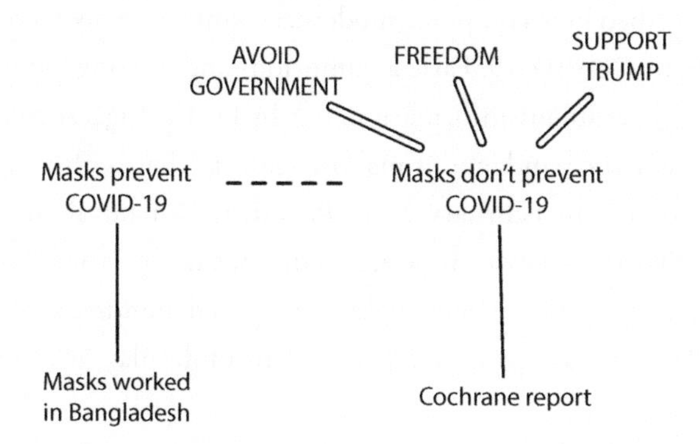

4.4 The emotional coherence of rejecting mask wearing. Despite limited evidence, the hypothesis that masks are ineffective is preferred because it fits with goals shown in capital letters. The double lines indicate emotional associations, the solid lines indicate explanatory links, and the dotted line indicates contradiction.

and areas responsible for emotions such as the amygdala and nucleus accumbens. In line with these findings, the theory of emotional coherence allows constraints concerning goal satisfaction to distort judgments about the acceptability of causal hypotheses. In the motorcycle convention case, personal and political goals lead people to discount evidence about the seriousness of COVID-19, which conflicts with strong goals such as allowing freedom, avoiding government interference, stimulating the economy, and supporting President Trump.

Figure 4.4 diagrams the emotional coherence of rejecting the claim that wearing masks reduces COVID-19 based on the values of freedom, avoiding government, and supporting Trump. These goals ought to be irrelevant to assessing the evidence that masks are effective, but motivated reasoning provides illegitimate support for a causal hypothesis that should be evaluated solely on explanatory coherence. Emotional coherence expands the constraints used in explanatory coherence by adding (1) positive constraints that come from satisfying goals, and (2) negative constraints that come from incompatible actions.

As with explanatory coherence, emotional constraint satisfaction can be computed by a program that implements hypotheses, evidence, and goals by artificial neurons; implements positive constraints by excitatory links; and implements negative constraints by inhibitory links. The result is a psychological, neural, and computational explanation of why some politicians and ordinary people have been inclined to make irrational causal judgments about COVID-19.

Earlier I described how computer models account for many factors to make predictions about COVID-19. False assumptions and missing factors can lead such models to generate misinformation, but by far the biggest source of misinformation about the pandemic is making stuff up that is also supported by motivated reasoning. In February 2020, President Donald Trump predicted out of the air that the disease "is going to disappear" by April.[34] Strong U.S. government action at that time might have saved hundreds of thousands of lives, but Trump was strongly motivated to underplay any threat to his reelection.

Besides emotional coherence, motivated reasoning can result from selective memory search and from specific emotions. Selective memory distorted inference about COVID-19 when people used anecdotes about patients who took ivermectin and felt better to conclude rashly that the drug is effective. Anger and fear have probably also contributed to motivated reasoning about the disease, for example, when people who felt that their freedom was threatened got so angry about having to wear masks or get vaccinated that they became all the more confident that masks and vaccinations are useless. Fear can promote motivated reasoning about cures when people who are worried about dying from COVID-19 gain hope from bogus treatments such as ivermectin. A Korean study found that people expressing high anger were more likely to disseminate misinformation about COVID-19, especially if they were conservative.[35]

Motivated reasoning can be based on group identities as well as personal goals. As the pandemic developed, people's behavior became tied to their group identities in ways that sometimes violated medical sense. Americans who thought of themselves as Republicans or Trump supporters or freedom lovers connected these identities with dangerous behaviors such as avoiding masks and vaccinations. The effects of identity are not always negative; for example, the penchant of Canadians to think of themselves as responsible citizens rather than rugged individualists contributed to a higher vaccination rate and a lower death rate compared to the United States.

Another pattern of motivated reasoning that operated in the COVID-19 pandemic was motivated ignorance, when some U.S. states and Canadian provinces restricted access to tests so that bad news about infection rates became unavailable. For example, when the omicron infection wave hit Ontario in December 2021, the provincial government drastically reduced PCR testing, making it hard to track the magnitude of that and subsequent waves.

Motivated reasoning is probably the most common source of misinformation dependent on failed inference, but other thinking errors can also contribute. In fear-driven inference, people believe misinformation because it scares them and thereby attracts their attention. This thinking error is different from motivated reasoning that responds to fear by generating positive illusions; people who use fear-driven inference remain scared. Conspiracy theories are often based on fear as much as on motivation, as when people succumb to anxiety about COVID-19 as the imagined result of evil deeds by China, the U.S. army, Bill Gates, or pharmaceutical companies. When people believe without evidence that COVID-19 vaccines can render them infertile, magnetic, or impregnated with microchips, their belief derives less from motivating personal goals than from intense attention captured by scary scenarios.

In addition to inferential errors encouraged by motivated and fear-driven inference, various errors in transforming representations contribute to the acceptance of misinformation about COVID-19:

Fallacy of false cause (post hoc ergo propter hoc): When event B happens after event A, people tend to think that A causes B. For example, when a vaccination is followed by a severe reaction such as a blood clot, people assume that the vaccination caused the reaction without considering other factors such as genetic predisposition or chance.

Fallacy of false authority: People tend to believe what leaders say even if the leaders are not experts in the relevant domain. For example, the views of many Americans about COVID-19 are heavily influenced by the pronouncements of Donald Trump and other Republican leaders who are not medical authorities.

Overconfidence bias: People tend to be surer of conclusions than the evidence warrants. For example, despite the terrible unpredictability of the spread of COVID-19, politicians have tended to think that they have it under control.

False consensus effect: People overestimate the degree to which other people agree with them. For example, opponents of COVID-19 and other vaccines may think that most people are like them.

Analogies are often useful for solving problems, but they can also be a source of misinformation. Early comparisons of COVID-19 with SARS and influenza provided some good ideas about infection and prevention, but such comparisons also encouraged the dangerously false belief that disease contagion from asymptomatic patients was unlikely. Even more dangerous is when critics of public health restrictions used the bogus analogy that the new disease is no worse than the common cold, although COVID-19 has killed millions of people.

Cognitive error tendencies like these incline people to acquire and hold on to erroneous beliefs about COVID-19. A useful exercise would be to survey all known fallacies and cognitive biases to identify more ways in which inferential transforming of representations leads to misinformation.

Inference occurs in individual minds, but they also operate in institutional contexts that boost cognitive and emotional errors. The norms and values that influence behavior in educational, political, and commercial organizations can inspire misinformed inferences based on motivated reasoning, faulty causal reasoning, hasty generalization, and other error tendencies. The governments of China and the United States, for example, have sometimes encouraged the acceptance of misinformation about the origins of the novel coronavirus.

Memory

Breakdowns in the memory mechanisms for storing and retrieving can propagate misinformation in both individuals and organizations. It is a legitimate point that the information that people store in their neural memories is emotionally important because emotions indicate relevance to people's goals such as health and relationships. The unfortunate side effect of the impact of emotion on memory is that it makes people inclined to remember what they want to hear, such as that COVID-19 is easily treated, and also to remember what scares them, such as that the coronavirus came from international conspiracies. Like inference, memory can be influenced by motivations and fears. Human memory storage is also subject to decay and distortion because of confusion of old memories with new ones and loss of cognitive function through aging and dementia. But the major problem with individual human memories is that what gets stored is heavily shaped by emotional importance rather than by evaluations of accuracy.

This emotional bias on storing amounts to motivated remembering.[36] People do not remember something because it suits their goals, but motivation provides emotional importance to goal-relevant information despite its poor evidential status. Here are some COVID-19 examples:

- The goal of being healthy motivates people to downplay COVID-19 so they remember that most of the people who die are old.

- The goal of bonding with friends motivates people to agree with their antivaxxer buddies so they remember that vaccines occasionally have side effects.
- The goal of being unfettered motivates people not to wear masks so they remember that masks can cause breathing problems.

In such cases, people feel good about a conclusion so they are inclined to store it in long-term memory even if it is misleading or false.

Motivated remembering can lead to storage of misinformation, but motivated forgetting can lead to loss of real information. The Freudian idea of repression of unpleasant memories has sunk out of scientific discussion, but neural evidence for inhibitory processes in memory processes has been accumulating.[37] Here are some examples of how motivated forgetting can encourage the loss of real information about COVID-19:

- The goal of keeping calm and avoiding anxiety encourages people to forget that millions of people have died of COVID-19.
- The goal of being sociable encourages people to forget statistics about the prevalence of the coronavirus in their region.
- The goal of avoiding vaccines encourages people to forget the warnings by public health officials about the spread of COVID-19 among the unvaccinated.

Motivated forgetting does not directly produce misinformation, but it helps to suppress real information that might combat misinformation, thus generating motivated ignorance.

Memory retrieving is also subject to emotional biases. Bad retrieving encourages misinformation because it fails to recover evidence that might challenge false beliefs or other misrepresentations. For example, people may have heard about clinical studies of the effectiveness of vaccines against COVID-19, but when it comes time to decide whether to get a vaccine, they remember only negative claims that they got from social media. The biases and selectivity of memory retrieval discourage revision of beliefs and behaviors that could improve information and decision making. Even when people have learned important facts about the pandemic, they may fail to remember them at important decisions points, in accord with Ziva Kunda's explanation of motivated reasoning as based on selective memory (chapter 3). Hence, memory storing and retrieving that are essential to an individual's use of real information can encourage misinformation about COVID-19. Selective storing and retrieving can also be influenced by group identities, for example, when people who think of themselves as freedom-loving vaccination opponents easily forget what they hear about vaccine benefits.

Social storing and retrieving promote misinformation when they operate carelessly or maliciously. A British organization called the Center for Countering Digital Hate has identified twelve leading online antivaxxers that it calls the Disinformation Dozen.[38] These people store in websites and other publications misinformation about vaccines in general and about COVID-19 in particular. The top of the list is Joseph Mercola, who has a website and best-selling book *The Truth About COVID-19* that store and propagate many myths about the disease, for example, that COVID vaccines are more dangerous than the disease. Mercola and other misinformation spreaders who had been restricted on Facebook and YouTube migrated to the company Substack, a newsletter platform with scant boundaries.[39]

With respect to memory, the crucial problem with such sources is that they store information without proper evaluation of its accuracy. The people who create these sources have various motivations, such as selling their own alternative health products, which is what Mercola does on his websites. Other motivations can include achieving fame and influence as well as supporting politicians with similar views. Publishing misinformation in print and digital forms constitutes motivated remembering that puts many others at risk of retrieving falsehoods.

Motivated remembering and forgetting operate in institutions whose policies and norms run counter to truth. Storage and retrieval of COVID-19 misinformation is encouraged by Libertarian organizations that emphasize freedom over health and by alternative health companies that place their profits above evidence-based medicine.

Spread

Only a few decades ago, message sending was a laborious process requiring senders to write letters to be sent to correspondents, but technologies such as email, texting, Facebook, Twitter, Instagram, YouTube, Reddit, 4chan, and TikTok make it easy to reach large audiences with no prior evaluation of the quality of the message sent. For example, Mercola has more than 400,000 followers on Twitter, so his rants about the dangers of COVID-19 vaccines are communicated to a huge audience. The rapidity of social media sending gives it the potential to be an enormously valuable source of real information about medical topics, but careless sending on social media is the major source of misinformation about COVID-19. Facebook had internal documents about the spread of COVID-19 misinformation among its users but refused to share them with governments and academics.[40]

Misinformation about COVID-19 is often passed carelessly rather than because of motivating biases.[41] Before Trump was kicked off Twitter in 2021, he had 90 million followers on Twitter that allowed him to communicate misleading claims instantly. People who receive communications about a scientific study that used a controlled clinical trial to debunk a proposed treatment for COVID-19 may not understand why such trials are much stronger evidence than celebrity testimonials, so the content of the message is not comprehended or passed on.

Spread of misinformation about COVID-19 is facilitated by underregulated social media companies and also by institutions with their own agendas. Some political parties, religious groups, and commercial operations have adopted misleading views and values about COVID-19 that they fervently spread to their own membership and beyond.

Misinformation grows through interactions of the AIMS mechanisms. For COVID-19, acquisition by making stuff up about the origins of the disease feeds into motivated inferences about Chinese conspiracies, leading to poorly evaluated memories and irresponsible spread. People receiving the misinformation may then be inspired to make additional motivated inventions about Chinese government actions that interact with inference, memory, and spread. Hence the development of COVID-19 misinformation is encouraged by feedback loops among the relevant mechanisms.

REINFORMATION FOR COVID-19

Reinformation is the process of mending misinformation to regenerate real information. This process requires different techniques depending on whether the misinformation results from breakdowns in acquisition, inference, memory, or spread. Reinformation must battle the continued influence effect: misinformation can persist despite attempts to debunk it.[42]

Acquisition

For acquisition, the primary reinformation technique is to highlight the differences between making stuff up and gaining knowledge about the world by reliable methods of perception, systematic observations, experiments, and clinical trials. These methods have generated medical progress that gives billions of people lives much longer than those common a century ago, and have also produced technologies that expanded human opportunities for work, transportation, and entertainment. People need to be reminded of the instruments and experiments that provide reliable knowledge about COVID-19 and many other diseases.

From some philosophical perspectives, talk of the world and reality as independent of human thinking seems naïve, but the naïveté resides in exaggerating how much minds construct reality. The exaggeration is especially obvious with COVID-19, where no amount of psychological or social construction attempted by ineffective politicians has been able to dispose of a disease that has sickened hundreds of millions of people, of whom millions have died. In such medical cases and in many other problems such as climate change, reality bites back. Research by various companies and countries has yielded effective vaccines, but other efforts have led to failures revealed in clinical trials so that vaccine candidates produced by major companies such as Merck had to be abandoned.[43]

More specifically, public education is needed to help people appreciate the effectiveness of systematic observations based on reliable perceptions and instruments. People are unduly influenced by vivid anecdotes and can be swayed more by a story about a cousin who took ivermectin for COVID-19 and recovered than by hundreds of observations of people who got no benefit. Similarly, educators and journalists have an obligation to help people understand the difference between sloppy experiments that get dubious results and well-conducted clinical trials with large numbers of participants divided randomly into controlled groups. People cannot just be told to "trust the science" when they have no idea how and why science can be trustworthy. Decisions always require a combination of factual information and objective values such as saving lives and therefore require people to trust the values as well.

Once people have a rudimentary understanding of how scientific methods of information acquisition are superior to making stuff up, some of their misinformation may be corrected by the straightforward method of factual correction. For example, people who believe that ivermectin is an adequate treatment for

COVD-19 can be directed to studies that find no reliable support for its use. Studies have found that rebuttals of rejection of scientific claims can be effective.[44]

Factual correction may be aided by pointing out the motivations of people who spread misinformation about COVID-19. People have a hard time recognizing motivated reasoning in themselves but find it obvious in others. This difference should make it easy to understand why financial interests in alternative treatments or political interests in disease denial get in the way of solid conclusions about the prevalence, causes, and treatments of COVID-19.

Remedies for bad acquisition strategies can be directed at institutions as well as individuals. Laboratories, granting agencies, and commercial research groups will all do a better job of collecting and representing information about COVID-19 if their values and practices are aimed at truth and health. Early work on collecting information was hindered by lack of knowledge about the new diseases, for example, when the U.S. CDC failed its first attempts to produce a test for the virus. Overall, however, mainstream research on COVID-19 has been highly successful in acquiring real information about the disease, reducing the need for institutional modification, although we should remain concerned about political control of government health agencies in China and other countries. The best institutional modification to work toward is control or abolition of organizations that merely make stuff up about the causes and treatments of COVID-19, for example, the misnamed National Vaccine Information Center funded by Joseph Mercola.

Inference

What can be done with people who do not recognize that the pandemic is a serious threat to be mitigated through public health interventions such as wearing masks, keeping a safe distance, and avoiding large gatherings? Cognitive science provides guidance about how to deal with people whose beliefs and values get in the way of implementing the behaviors required to deal with COVID-19.

It would be naïve to suppose that merely presenting people with the relevant evidence is an adequate strategy. Scientists are trained to appreciate evidence-based reasoning and change their minds, for example, when WHO officials reversed their earlier recommendation about wearing masks. But ordinary people have no similar education in how to evaluate hypotheses based on rigorous evidence. They are more likely to rely on dubious sources of information such as friends and social media. In the absence of reliable evidence, they are especially prone to motivated reasoning that allows them to believe what makes them happy.

Few experimental studies have been done on how minds can be changed about COVID-19, but we can take some lessons from research on the equally controversial issues of vaccinations and climate change. A study at the University of California, Los Angeles (UCLA) found that people who are convinced that measles vaccinations are dangerous are not much affected by evidence, but their minds can be changed by vivid illustrations of the harsh effects of measles on children.[45] A scary picture is more effective than a line of argument. Perhaps people who are skeptical about the dangers of COVID-19 could be influenced by videos of victims gasping for breath. Figure 4.5 illustrates the use of multiple strategies to prevent coronavirus infection.

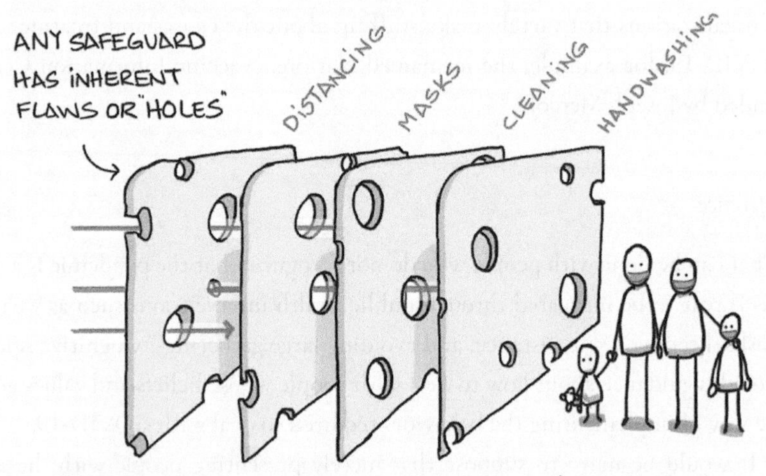

4.5 A model of how health interventions can combine to lessen disease risk.
Source: Jono Hey—sketchplanations.com, Creative Commons 4.0.

A less drastic intervention than the UCLA scare has succeeded in changing the minds of some climate change deniers. Science educators at the University of California, Berkeley found that short videos that clearly explain the underlying mechanisms for global warming caused by greenhouse gas emissions led to belief change in some individuals.[46] Perhaps an effective video could show how small droplets containing the coronavirus spread from infected people into the air and then into the noses of other people nearby, with resulting infection, coughing, and lung failure. Understanding the mechanism of infection as a vivid causal process might help people realize why masks can be helpful in blocking spread.

Much misinformation about COVID-19 arises from people's mistakes in evaluating their beliefs and from transforming beliefs into new ones without careful consideration of the evidence. The two main techniques for correcting such faulty inferences are critical thinking and motivational interviewing. The more familiar critical thinking technique operates in the two stages of error diagnosis and remedial reasoning. Error diagnosis identifies the faulty patterns of thinking that lead people to adopt the misinformation, while remedial reasoning proposes a logically legitimate way to replace misinformation with real information.

My analysis of inferences about mask wearing illustrates how two-step critical thinking can work. Figure 4.4 shows a faulty inference about masks based on the error tendency of motivated inference driven by emotional coherence, but figure 4.3 shows how to make a good inference to the best explanatory hypotheses using the mechanisms of explanatory coherence. Shifting from the misinformed thinking in figure 4.4 to the correct inference in figure 4.3 accomplishes reinformation.

Motivational interviewing provides an alternative way of shifting people's thinking, with emphasis on empathy rather than logic providing a more indirect and psychologically flexible way of changing people's minds. Here is how motivational interviewing could be used by an interviewer to deal with people reluctant to get vaccinated for COVID-19:[47]

1. Understand people's concerns about vaccines by asking them open-ended questions and empathizing with their concerns.
2. Be affirmative, reflective, and nonjudgmental about their vaccine concerns.
3. Identify discrepancies between people's current and desired behaviors such as staying healthy.
4. Summarize the issues and inform people while respecting their autonomy.

This method is not guaranteed to change people's minds, but its success with many problematic behaviors such as alcoholism suggests that it is worth trying as an antidote to misinformation about COVID-19.

Critical thinking suggests a more aggressive, logic-based method that would go like this:

1. Point out that prejudices against vaccines are based on bad evidence.
2. Describe the clinical trials that provide good evidence that the available vaccines are effective in preventing COVID-19.
3. Describe the huge costs of COVID-19 infection, including more than 5 million deaths worldwide.
4. Argue that these cost-benefit considerations make it rational to get vaccinated.

Whether this logical argument would convince as many people as motivational interviewing is an empirical question that can only be answered by experiments. I would like to see a study where skeptics about COVID-19 vaccines are randomly assigned to two groups and then addressed with either critical thinking or motivational interviewing. The resulting finding should start to reveal the comparative effectiveness of critical thinking and motivational interviewing as reinformation techniques. One reason to expect that the motivational interviewing approach might be more effective is that it works to establish trust: people are more easily convinced by those they trust. Skepticism about COVID-19 in the United States was shaped by people's distrust of Democratic politicians.[48]

The empathic understanding sought by motivational interviewing requires grasping the system of values and beliefs that makes people skeptical about COVID-19 treatments. The comparative emotional coherence of conflicting sets of values can be vividly depicted by cognitive-affective maps introduced in chapter 3.

Figures 4.6 and 4.7 are simplified maps of the values underlying disputes concerning mask wearing. Some countries such as China are committed to mask wearing as a technique for slowing the spread of disease, but masks remain controversial in the United States. Figure 4.6 shows the configuration of values that supports mask wearing as recommended by most public health officials. The concept of wearing masks gets strong positive emotional value because it fit well

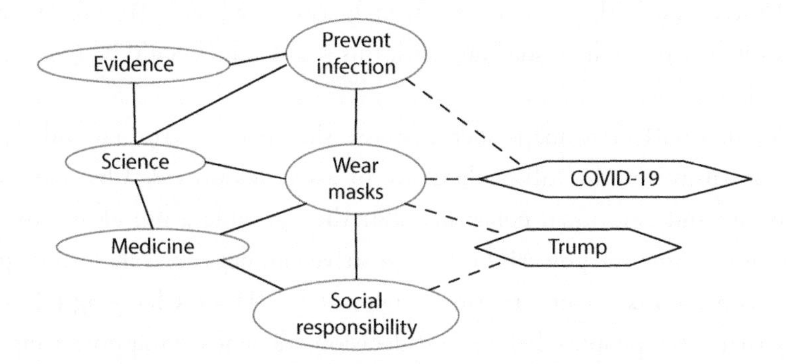

4.6 Cognitive-affective map of values supporting wearing masks. Ovals are emotionally positive; hexagons are negative. Solid lines indicate mutual support; dotted lines indicate incompatibility.

Source: Paul Thagard, "The Cognitive Science of Covid-19: Acceptance, Denial, and Belief Change," *Methods* 195 (2021): 92–102, p. 93, reprinted by permission of Elsevier.

with other values, such as keeping people healthy in the face of disease. Some of the links are based on causal connections, for example, that wearing masks prevents infection, but others are based on looser emotional associations, for example, that Trump made fun of masks.

In contrast, figure 4.7 shows a configuration of values of people who view public orders to wear masks as an infringement on their personal freedom. In

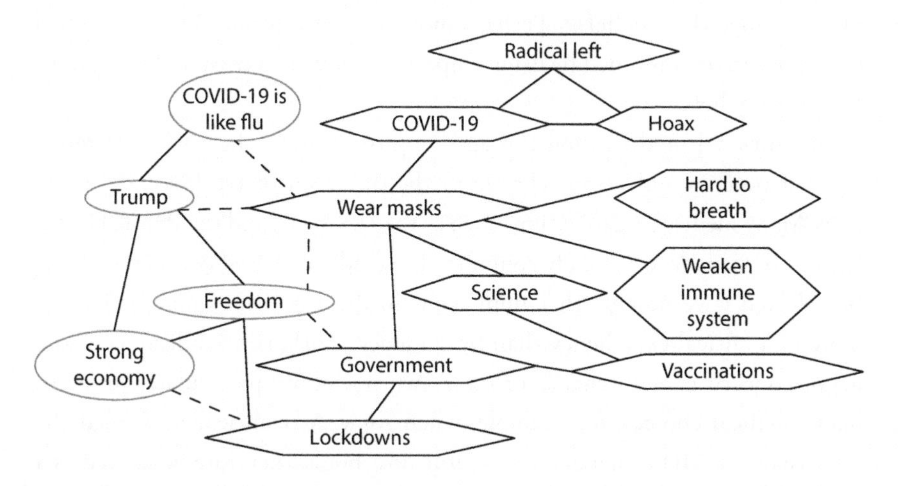

Figure 4.7 Cognitive-affective map of values opposed to wearing masks.

Source: Thagard, "The Cognitive Science of Covid-19," p. 95, reprinted by permission of Elsevier.

the United States, these values are strongly associated with Trump. The map makes it easy to see how someone with this set of values could be so opposed to mask wearing.

Cognitive-affective maps such as those shown in figures 4.6 and 4.7 do not substitute for the full explanation of attitudes provided by motivated reasoning and emotional coherence. But they provide a useful approximation for grasping the powerful role of positive and negative values in shaping how people think about situations such as COVID-19. Changing minds in ways that alter people's beliefs and decisions depends on appreciating the role of values and emotional coherence in how people think about matters important to them. Hence, cognitive-affective maps are a promising technique for encouraging the empathic understanding required for motivational interviewing.

I have suggested two techniques for mending misinformation resulting from faulty inferences. Critical thinking identifies errors that bias thinking and offers remedies in the form of more reliable forms of reasoning. Motivational interviewing uses empathy rather than logic to find out why people hold their emotionally powerful beliefs and values and uses gentle questioning to direct people away from misinformation. Cognitive-affective maps may prove to be a useful tool in an empathy-based approach to reinformation. Much experimental psychological research needs to be conducted concerning what persuasion techniques are most effective in convincing people to change their beliefs.[49] Perhaps motivational interviewing and critical thinking can be combined by using empathy to establish trust and using logic to change beliefs.

Institutional modification can also help to improve the quality of inferences if policies and norms change in the direction of truth-seeking practices. Many medical organizations such as the U.S. CDC and the Public Health Agency of Canada are already committed to evidence-based decision making. But elected governments that implement medical decisions may be swayed more by political expediency than by evidence. Authoritarian leaders whose major concern is maintenance of their own power are particularly prone to inane medical choices, for example, when some African leaders denied the seriousness of AIDS. Democratic institutions should therefore be viewed as a contributor to health.

Memory

How can memory be improved to support real information over misinformation? Social memory is potentially more malleable than individual memory, which operates with neural mechanisms inherited from millions of years of mammalian evolution. Turning off the inclination to remember what is emotionally salient looks impossible. Similarly, the inclination to believe what someone tells us may well be baked into the human brain.[50] This gullibility problem is partly a memory issue because to believe something requires storing it in memory along with some indication of approval.

Nevertheless, for COVID-19, several strategies can help people avoid storing misinformation. By far the most important is to encourage people to insert evaluating as a key part of the process of storing beliefs. Instead of hearing that ivermectin helps to cure COVID-19 and remembering the falsehood, people should hear it and question it before believing it and storing it in memory. Questioning invokes all the strategies already recommended with respect to acquisition and inference. Was the piece of information acquired by reliable processes such as clinical trials, or did it arise merely by making stuff up? Were the inferences that support it cases of explanatory inference that accounted for the full range of evidence and alternative hypotheses? How reliable and legitimate was the source of the information? The general process of critical thinking, combining error detection and remedial reasoning, should help to make memory storage more prone to real information than misinformation.

The biological process of memory retrieval from storage in individual brains is not open to modification, but the results of retrieving can be subject to the same critical thinking steps of error detection and remedial reasoning. When a person remembers claims about a cure for COVID-19, for example, there should be no automatic assumption that the claim is true. Retrieval is no guarantee of truth because what is stored in memory may have gotten there illegitimately. Hence, for individuals, the best way to lower the memory risks of misinformation are to supplement the basic biological processes of storing and retrieving with psychological processes of critical thinking. Just as inference can be improved by warning people of the dangers of motivated reasoning, so memory can be improved by warning people of the dangers of motivated remembering and forgetting.

We cannot change the basic way that individual brains store and retrieve memories, but social memory is much more manipulable by incorporating critical thinking. Unlike the automatic, unconscious processes of human memory and retrieval, placing a piece of information into a public source such as an article, book, website, or database is a conscious decision that can be interrupted by mechanisms of evaluating. Reputable scientific journals and book publishers use such interruptions, requiring review before the results of studies about the novel coronavirus or vaccines are accepted for publication and entered into the store of published articles. In contrast, personal websites such as Mercola's and social media threads have no such constraints: people can store whatever they want without any standards of evaluating.

Institutional modification can aim at reforming memory practices to support real information over misinformation. Change in this direction can occur not only in social media companies but also other organizations that control storing and retrieving, including other corporations, libraries, schools, and government agencies. The U.S. CDC, for example, is operated by people who are conscientious about the COVID-19 information that appears in its publications and websites. But other governments such as Brazil and private organizations such as Fox News spread misinformation about the disease, demonstrating the need for institutional change.

Everyone who places a piece of information on a website or social media site would ideally exercise due diligence using standards of evaluating and critical thinking. But many people who control such storing are motivated by personal goals such as sales and fame rather than by legitimate social values such as truth and justice. Then critical thinking is ineffective, but political action can operate because of the social nature of storing and retrieving. Publishers, websites, and social media can be subject to government action to reduce the amount of harmful misinformation that they store and retrieve. The main aim of such measures is to prevent media from spreading misinformation among large numbers of people, so I will discuss possible political solutions to social misinformation in the next section.

Spread

How can the spread of information about COVID-19 be controlled in ways that shift misinformation toward real information? The two most relevant techniques are critical thinking and political action. Critical thinking applies to sending and receiving of information because both require evaluating to detect errors and resolve them by remedial reasoning. Senders should learn to ask themselves about COVID-19 information that they are about to transmit whether they actually have good evidence for their recommendations. Senders need to be encouraged not to pass on information just because it is cool. A weak nudge to consider accuracy may not be enough, but thorough education about the dangers of medical misinformation might provide motivation to incorporate critical thinking into the process of sending.

Just as important, receivers of information should be encouraged to avoid the gullible practice of believing whatever they hear. When claims are made about COVID-19 vaccines and other treatments, people should learn to ask questions about the quality of the sources, including their expertise and motivations. Are the senders following good acquisition practices such as controlled clinical trials, or are they merely making stuff up? Do the senders have commercial motivations for pushing their own alternative-medicine products over evidence-based treatments?

Applying critical thinking to individual senders and receivers is unlikely to stop the flood of misinformation about COVID-19. Social media such as Facebook, YouTube, and Twitter have been major sources of falsehoods about the causes and treatments of the disease, and political scrutiny is required to control their irresponsible behavior. These media have taken some steps to restrict the spread of COVID-19 misinformation, such as YouTube taking down thousands of antivaccine videos.[51] But only government intervention can coerce them into doing all that is necessary to slow the spread of dangerous information. Freedom of speech is a crucial value for democratic societies, of course, but free speech has never been absolute and needs to be balanced against potential harms that it causes. Antivaccine propaganda has clearly caused harms because of the much greater incidence of COVID-19 among the unvaccinated, sometimes including death.

The political actions that can be taken against social media companies include encouragement, regulation, and antitrust legislation. In particular, Facebook has been investigated by the U.S. Congress and other governmental organizations for contributing to the spread of misinformation about COVID-19 and political conspiracies, including the U.S. Capitol riot of January 6, 2021. Facebook claims to be taking steps to control misinformation on its platform, but former employee Frances Haugen reported that Facebook is far more interested in enhancing its advertising revenue by engaging its readers with stories that are more provocative than accurate.[52] Facebook's own experiments showed that vaccine misinformation is reduced if posts are ranked based on trustworthiness rather than engagement, but following this policy produces less activity and hence less advertising revenue. Corporate greed triumphs over social needs.

Several measures have been proposed to make Facebook more socially responsible. Facebook could be required to change its business model so that it would get its revenue from client fees rather than advertising, eliminating its incentive to maximize advertising revenue. Another proposal is to change laws that shelter social media from legal liability for posts, as done by U.S. federal statute, Section 230. A third possibility is to eliminate Facebook's near-monopoly over the spread of some kinds of information by enforcing antitrust laws. All these measures would reduce the impact of Facebook in spreading misinformation, and similar restrictions might work for Twitter and YouTube. Such measures should be assessed according to whether restrictions on individual and corporate freedom are justified by the extent to which they provide benefits to people and prevent harm. The magnitude of the harm caused by COVID-19 misinformation justifies looking for ways to constrain social media, as chapter 8 discusses in more detail.

To sum up, two main techniques foster reinformation about COVID-19 with respect to the process of spread of misinformation. First, critical thinking can be applied to both sending and receiving to encourage individuals to apply high standards of evaluating information. Sending of information about the causes, treatment, and prevention of diseases should be screened to ensure that messages satisfy good evidential standards of both accuracy and relevance. Receivers of medical information should avoid passive and automatic acceptance of messages and instead adopt a policy of skeptical scrutiny. Efforts should be made to educate people about real information but also to encourage them to trust science as a way of getting useful knowledge.[53]

Second, political action is required to rein in particularly noxious vehicles of spread of misinformation that has become far more universal and rapid since digital social media became available after 2000. Facebook, YouTube, and Twitter originated in 2004, 2005, and 2006, respectively, and all have had benefits as well as costs for social communication. But along with traditional media such as Fox News, they have facilitated skepticism about COVID-19 that in the United States alone has cost thousands of lives. Slowing the spread of misinformation on social media requires legal changes that could affect their business models, protection from legal liability, and immunity from antitrust challenges. I hope that social media companies will voluntarily react to the threat of such controls by dramatically increasing their currently feeble attempts to curtail misinformation about COVID-19 and other pressing social issues.

The reinformation techniques I have described deal with COVID-19 falsehoods that are already in place, but preventive interventions can prepare people to recognize and resist misinformation. One study used computer games to inoculate people against being manipulated into accepting and transmitting COVID-19 misinformation, a kind of prebunking.[54]

OTHER MEDICAL MISINFORMATION

Medical misinformation has been rampant at least since ancient physicians told patients that their diseases were inflicted by the gods. Three ancient civilizations adopted theories of medicine according to which disease results from imbalances among crucial factors: for the Greeks, the humors of phlegm, blood, black bile, and yellow bile; for the Chinese, yin and yang; and for the Indians, the doshas of vata, kapha and pitta. These ideas still survive as forms of alternative medicine, but the lack of experimental validation shows that these balance metaphors are bogus and the traditions count as misinformation.[55] Historically these medical doctrines spread from physicians to patients but now they can be fostered by websites and social media.

More modern medicine has also been plagued by misinformation, often spread by individuals or companies whose interests run counter to truth. Here are some examples:

- Many companies tout dietary supplements such as echinacea whose efficacy has not been shown in clinical trials.
- Purdue Pharma introduced OxyContin as an effective opioid for controlling pain while misrepresenting the risk of addiction.[56]
- Countless bogus cancer treatments, ranging from alkaline diets to aromatherapy, to magnets, have been proposed.[57]
- The tobacco industry continued to claim that the evidence linking smoking to cancer was inconclusive long after their own studies had established causality.[58]

- The claim that vaccines cause autism was based on experiments that turned out to be fraudulent.
- Social media such as TikTok spread misinformation about abortion, including herbs claimed to induce abortion and false claims about the long-term effects of abortion.[59]
- When cases of mpox (monkeypox) spread in 2022, conspiracy theories about its origin occurred on TikTok and other media.[60]
- Dental misinformation includes false claims about the dangers of fluoridation that dramatically reduces tooth decay.[61]

In her book *Viral BS*, Seema Yasmin debunks dozens of medical myths and offers a useful Bullshit Detection Kit.[62] TikTok has become a potent source of medical misinformation being fought by only a few debunkers.[63] A fascinating exercise would be to analyze these important cases of medical misinformation based on how they result from breakdowns in information processes of acquisition, inference, memory, and spread.

Misinformation is also used to resist medical reforms such as the installation of a universal health-care system in the United States. Americans have been told that their health system provides more freedom and effectiveness than is found in countries such as Canada that have universal health care. But lack of universal health care cost hundreds of thousands of deaths in the United States during the COVID-19 pandemic, and Canada has had a pandemic death rate only about a third of the United States thanks to higher vaccination rates and more responsible mask wearing as well as universal health care.[64]

BEYOND MEDICAL MISINFORMATION

I have used COVID-19 as a dramatic, recent illustration of the development of medical information and misinformation. This case shows how real information can solidly and rapidly develop as the result of interactions among mechanisms for acquisition, inference, memory, and spread. Genetic sequencing of the SARS-CoV-2 virus in January 2020 led quickly to tests for the disease and within a year to successful vaccines. As of 2024, the disease has not been eradicated for various reasons, such as the dangerous infectiveness of the omicron variants and the failure to vaccinate people in the developing world. Even in rich countries in North

TABLE 4.1 Examples of information and misinformation about COVID-19

	Real information	Misinformation
Origins	COVID-19 probably originated in China through spread of a virus from animals or possibly escape from a laboratory in Wuhan, China.	COVID-19 originated in a U.S. army laboratory.
Causation	COVID-19 is caused by infection with SARS-CoV-2.	COVID-19 is caused by 5G cellphone towers.
Treatment	COVID-19 lung congestion is reduced by dexamethasone.	COVID-19 mortality is reduced by hydroxychloroquine.
Prevention	Masks reduce spread of COVID-19. Vaccines dramatically reduce occurrence of COVID-19.	Masks are ineffective. Vaccines are dangerous.

America and Europe, many people remain unvaccinated because of reluctance caused by misinformation.

COVID-19 serves to illustrate the sharp difference between real information and misinformation, as shown by the examples in table 4.1. The distinction between them is not just that the sentences in the misinformation column are false but also that the examples have different features associated with the mechanisms that produce them.

Identifying mechanisms and their breakdowns provides explanations as well as descriptions of fundamental aspects of information and misinformation. The eight classes of mechanisms in the AIMS theory explain the origin and transmission of information, while breakdowns in these mechanisms explain the spread of misinformation. COVID-19 illustrates the operation of these mechanisms and breakdowns, showing how substantial medical progress has occurred in understanding and treating the disease but also how medicine has been impeded by persistent misinformation. Chapter 5 describes how understanding climate change has similarly been marked by progressive information and obstructive misinformation.

CHAPTER 5

STORMS

Climate Change and Scientific Misinformation

COVID-19 has been a worldwide disaster with more than 6 million documented deaths and millions more uncounted. But climate change looms as a much greater disaster with an estimated 50 million deaths from malnutrition, disease, and heat stress from 2030 to 2050, according to the World Health Organization (WHO).[1] In my book *Mind-Society*, I marked World War I as the all-time most irrational human enterprise, because the leaders of five countries stumbled into a conflict that cost millions of lives and led to World War II, which killed millions more.[2] But the peak of human irrationality is looming today with failures to prevent further climate change by controlling greenhouse gases.

COVID-19 was a surprising threat that combined infectiousness and lethality, but countries dealt with it immediately through intense public health measures and novel medical treatments. In contrast, although the dangers of climate change have been known for decades, even countries with well-intentioned leaders have dithered rather than enact the strong measures required to avert catastrophe. Global warming also increases the chances of future pandemics through environmental changes that expose people to more cross-species viral transmission.[3]

Real information about climate change is abundant courtesy of the United Nations Intergovernmental Panel on Climate Change (IPCC), whose most recent reports provide a wealth of observational, inferential, and valuational information. Its impact has been deflected unfortunately by misinformation spread by climate change deniers with industrial and political motives. Climate change reinformation requires identifying these distortions and dealing with them using a combination of cognitive, emotional, and political strategies. We can again rely on the AIMS theory that explains how real information and misinformation are based on *a*cquisition, *i*nference, *m*emory, and *s*pread.

REAL INFORMATION ABOUT CLIMATE CHANGE

In 1988, the United Nations created the IPCC, which now has 195 member countries.[4] In 2021, it published a report on the physical science basis for climate change with almost four thousand pages, bringing together the latest advances in evidence from "paleoclimate, observations, process understanding, and global and regional climate simulations."[5] This report argues that climate science shows that global warming is increasing, that human influence is the main cause of this increase, but that time still suffices for actions that interrupt the path to disaster. In 2022, the IPCC released two additional reports dealing with effects of climate change on human lives and with measures that can be taken to mitigate these effects.

Acquisition

For individual humans, much information comes from the senses receiving signals from the world, such as light from the sun. Scientific observations concerning climate are superior to ordinary perception in two respects. First, they employ instruments such as thermometers to interact with the world to produce data that are more quantitatively exact than can be provided by bodily senses. Second, the observations are much more systematic because they cover conditions comprehensively at relevant places and times.

Here are some of the instruments used by climate scientists to accumulate data about the state of the world about quantities such as temperature:

Temperature: Thermometers measure heat and cold at the Earth's surface, in the upper atmosphere, and in bodies of water.
Precipitation: Rain gauges measure the accumulation of rain and snow.
Gas levels: Gas sensors use electrochemical and other mechanisms to measure the amount of greenhouse gases such as carbon dioxide and methane.
Sea level: Tide gauges use sensors to measure daily high and low tides; satellite altimeters measure the distance of sea level from space using radar.

Ocean chemistry: pH sensors measure the extent of ocean acidification; oxygen meters measure the extent of oxygenation.

Glaciers: Physical and electronic devices measure changes in the mass of glaciers.

These instruments enable climate scientists to gather information about the world by interacting with it.

According to the AIMS theory of information, acquisition is the initiating process that operates with the mechanisms of collecting and representing. The instruments for measuring temperature, precipitation, gases, sea level, ocean chemistry, and glaciers collect data that constitute evidence according to the requirements of reliability, intersubjectivity, repeatability, robustness, and causal correlation with the world. For example, thermometers are reliable in accurately measuring temperature, intersubjective in being usable by all people, repeatable in yielding stable measurements, robust in that different kinds of thermometers (e.g., analog and digital) provide similar results, and causally connected with the world because heat and cold in the world cause changes in the thermometer.

Collecting data using such instruments generates representations in computer databases for simple entries such as that the temperature in Waterloo, Ontario, at 8:13 a.m. on January 19, 2022 is -9°C. These entries can be transformed into sentences such as the one you just read and also into mental representations such as my belief that the temperature at that time was -9.

Summaries of information collected by instruments can use words, but they can more vividly use pictorial representations such as graphs and maps. Maps are available that summarize temperature increases in different parts of the world over the past fifty years. The IPCC 2021 report contains countless maps and graphs that represent changes and distributions of climate measurements. Many more can be found at the IPCC's Interactive Atlas that includes images that use motion for dramatic depiction of actual and possible changes.[6]

Single instrument readings are useless for determining patterns of climate change and correlations with greenhouse gases. Instrumental observations are systematic in tracking measured values over both time and space. For example, temperature changes have taken place over the past fifty years in many different parts of the world. Global warming that occurred over short periods of time only or only in a few parts of the world would have little significance for human welfare. Systematic observations are the basis for IPCC generalizations such as: "Widespread and rapid changes in the atmosphere, ocean, cryosphere and biosphere have occurred."[7] Observations concerning temperature changes and greenhouse gas accumulations generate correlations that provide the basis for causal inferences.

In medical contexts such as COVID-19, the most causally useful observational information is based on controlled experiments, including clinical trials. Experiments about climate change are rarely possible, however, because of the difficulty of manipulating variables, such as the amount of carbon dioxide in the air. The effects of attempted reduction of greenhouse gas emissions will not be evident for decades. The mainstay techniques of epidemiological experiments—random assignment to conditions and double blinding—are impossible to carry out for climate change. Nevertheless, much can be learned about climate by systematic observations using instruments. Summarizing these observations requires inferences consisting of generalizations about data, but explaining and predicting climate change requires more daring inferences.

Inference

The accumulated observational evidence is sufficient to warrant inferences about the causes of global warming and possible ways of dealing with it. The investigation of climate change requires inferences that go beyond observation to answer these questions: What are the causes of climate change? How will climate change develop in the future? What can be done to prevent future disastrous effects of climate change?

The IPCC report boldly asserts: "It is unequivocal that human influence has warmed the atmosphere, ocean and land."[8] Here "unequivocal" means clear, certain, and beyond doubt. This conclusion is important because it implies that human influence can change and thereby slow or reverse the rate of warming. Other ways of stating the conclusion include saying that global warming is caused by increases in greenhouse gas emissions and that climate change is anthropogenic. Influences and causes are observable by human perception even aided by instruments, so they must be inferred. Nevertheless, the consensus among climate scientists that global warming is anthropogenic has approached 100 percent.[9]

The logic of the IPCC's causal inference is not obvious from their text, but the evidence they present is impressive. Here are some of the relevant pieces of evidence:

- Since 1750, humans have caused increases in greenhouse gas concentrations. This causal conclusion is based on observations that human activities such as that manufacturing, mining, and transportation generate carbon dioxide and other gases.
- Global surface temperature has been increasing since 1850.
- The same period has also brought increases in precipitation, sea level, and glacier retreat, and the rate of increase is rising.
- Weather extremes such as heat waves, floods, droughts, and cyclones are increasing around the world.

Climate models consisting of mathematical equations implemented in computer programs simulate how greenhouse gases affect the balance between energy incoming from the sun and energy outgoing from the Earth. How do these models combine with evidence about global warming and human activities to justify the conclusion that humans are responsible for climate change?

Chapter 4 described how causal inferences in medicine are based on explanatory coherence. We know about increases in climate quantities such as temperature over the same time that greenhouse gases have been increasing as the result of human activity. We can then conclude that the best explanation of global warming and other climate changes is human activity, as shown in figure 5.1. The

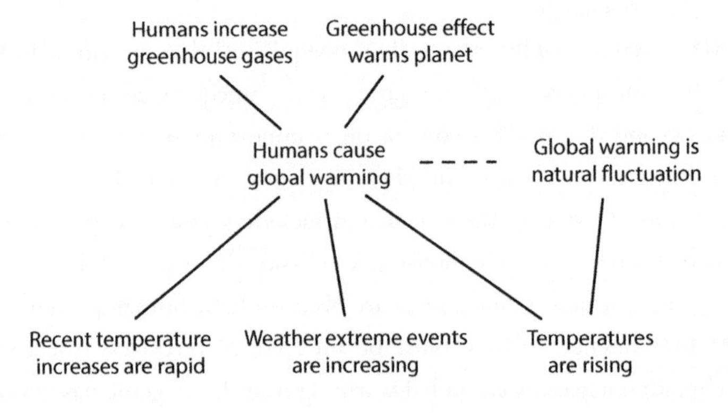

5.1 The explanatory coherence of the hypothesis that humans cause climate change. Solid lines indicate explanations; the dotted line indicates incompatibility.

main alternative hypothesis is that global warming and other climate changes are just natural fluctuations, which fails to explain why recent temperature and other increases are now undergoing more rapid increases as the amount of greenhouse gases in the atmosphere increases.

The hypothesis that humans are causing climate change is coherent with the available evidence for three reasons.[10] First, human activities explain the observed facts of climate change, including rapidly increasing temperature and extreme weather events. Second, how human activities produce these effects is explained by the greenhouse effect generated by humans. Third, alternative explanations of climate changes such as random fluctuation are too weak to cover all the available evidence. Therefore, the IPCC and climate scientists in general are justified in concluding that humans are causing climate change. This causal inference generates real information about the world.

Climate science is concerned not only with explaining past changes but also with predicting future changes that will affect human lives. Predictions are made by climate models that use computer simulations to infer the likely outcomes of possible changes in variables that drive climate change. The 2021 IPCC report considers five scenarios for the results of different levels of greenhouse gas emissions for temperature and other effects. The validity of these predictions depends on the accuracy of the inputs to the model and the quality of the representations of physical mechanisms represented by the equations in the model. Computer models for climate change play the same valuable roles that I described for COVID-19 models: making hypotheses explicit and exact, revealing consequences of the hypotheses for comparison with the world, experimenting with variations, and identifying unexpected interactions.

If the inputs are approximately correct and the mechanisms are adequately captured by the equations in the model, then the model yields real information about the world. For example, the IPCC estimates that high amounts of carbon dioxide emissions over the next sixty years could lead to average temperature increases of more than 4°C. Modeling enables the IPCC to assert: "Global surface temperature will continue to increase until at least the mid-century under all emissions scenarios considered. Global warming of 1.5°C and 2°C will be exceeded during the twenty-first century unless deep reductions in CO_2 and other greenhouse gas emissions occur in the coming decades."[11] An assessment of fourteen climate-tipping elements provides evidence for urgent action to mitigate climate change.[12]

For prediction, climate models function like the COVID-19 models described in chapter 4. Mathematical equations describe the causal relations between variables, and computer programs calculate the effects of current and expected inputs. Then COVID-19 models make predictions about future levels of cases, hospitalizations, and death in the same way that climate models make predictions about future levels of temperature, precipitation, and extreme weather events. These inferences cannot generate as much confidence as the causal explanations of past cases of disease or climate change, but they nevertheless provide provisional extrapolations of future occurrences. Useful predictions depend on the models capturing the causal relations in the world that determine the future course of disease or climate. Improved models also provide increasing confidence that previously unusual extreme weather events such as heat waves, droughts, and hurricanes are caused by global warming.

Chapter 4 emphasized that medical decisions cannot be based on scientific information alone but require integrating that information with value judgments. Similarly, climate decisions depend on values concerning undesirable effects on humans, where the values contribute to decisions when combined with scientific information about the causes and effects of global warming. At a ridiculous extreme, the determination that climate change will soon render the Earth unhabitable would not require any action if we placed a high value on human extinction.

The IPCC report displays acute awareness of the importance of values: "values—fundamental attitudes about what is important, good, and right—play critical roles in all human endeavours, including climate science."[13] The values specifically mentioned include life, subsistence, stability, and equitable distribution of costs and benefits. Values influence decisions about how to deal with climate change, but they also affect decisions about how to conduct research and how to communicate the results to the general public. Scientists need to try to avoid both false alarms and missed warnings.

We can grant that values are emotional attitudes but still see them as potentially objective given a rich theory of emotions. Emotions are not just bodily reactions but also require cognitive appraisals of the significance of a situation for a person's goals.[14] For example, my reaction to the prospect that 50 million people will die because of climate change is visceral but it is also cognitive because my goals include the flourishing of the human species.

To count as objective values, the relevant goals must reflect the needs of humans, as argued in chapter 4. The values mentioned by IPCC—life, subsistence, and equality—are directly connected with human needs so our emotional attachment to them can be viewed as real information. Emotions are information, but whether they are real information or misinformation depends on how well the appraisals they include reflect human needs, both biological and psychological. For example, the mob that stormed the U.S. Capitol in January 2021 was informed by anger that reflected their view that Donald Trump's election had been stolen, but their anger was misinformation because he had been defeated in a fair election.

The real information compiled by IPCC in its various reports required decades of work by thousands of climate scientists. This labor partially reflects their professional interests, but it also reflects deep concern with the future of humanity. This concern is based on a constellation of values, which is illustrated by the cognitive-affective map in figure 5.2. This map shows how stopping global warming is a positive value because it fits with saving lives and avoiding harm to people, along with other values of sustaining environments and supporting equality. Equality is a value for climate change management because global warming disproportionately brings suffering to poor people who have fewer resources to deal with the dangers of global warming. Rich people can turn on air conditioners, move to cooler climates, and buy food and water, whereas poor people are stuck with weather extremes or compelled to become refugees.

The right side of figure 5.2 shows why global warming has negative value because of its harmful effects, which include heat waves, floods, droughts, famine, tornadoes, and cyclones (hurricanes and typhoons). Ample evidence supports the causal connections between global warming and harmful effects, for

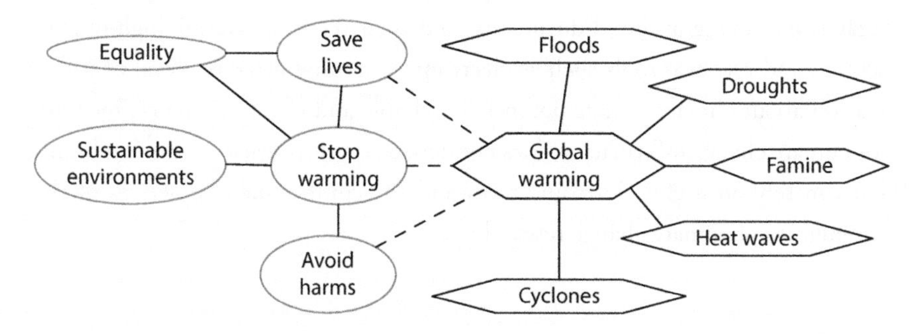

5.2 Cognitive-affective map of the values of people who are concerned about stopping global warming. Ovals indicate desired values; hexagons indicate disliked values. Solid lines indicate emotional associations; the dotted lines indicate emotional incompatibility.

example, between global warming and heat waves that cause much suffering and loss of life. Because suffering and loss of life run counter to human needs, the values shown in figure 5.2 are objective and therefore constitute real information about why climate change is bad. In contrast, misinformation spread by climate change deniers emanates from illegitimate values such as corporate profits, political power, and unconstrained freedom.

Memory

Public memory for real information about climate change includes the IPCC reports, online databases such as the Interactive Atlas, and an enormous number of articles and books. Google Scholar generates more than 4 million results from a search for "climate change." Much information about climate change occurs in tightly refereed scientific journals such as *Science* and *Nature*. Empirical and modeling results are summarized in the IPCC reports, which undergo the scrutiny of their hundreds of authors. Hence, the storage of scientific climate change research is a stellar example of how evaluation should precede the placing of information into memory.

Retrieving climate change information from memory storage is complicated by the vast amount of text and images available. Even finding pertinent information in the 2021 IPCC physical sciences report is difficult when it has thousands of pages. Fortunately, the twelve long chapters of the main report are preceded by a concise summary for policy makers and a more detailed technical summary, which allow readers to glean the most important conclusions of the IPCC working group.

Evaluation of retrieved information can take into account various factors, such as the source of found documents and their recency. Overall, high-quality sources and retrieval tools such as electronic searching make retrieval of information about climate change potentially reliable and effective. Hence, memory for climate change information does not depend on the vagaries of human brains but can rely on useful electronic tools and the commendable practices of the community of climate change researchers.

Spread

These practices also effectively generate the spread of real information about climate change. Sending of information is constrained by the careful reviews performed for IPCC documents and articles in scientific journals. Communication also takes place by more casual methods such as email, Twitter, websites, and newspapers, but the underlying constraints on spread of scientific information help to ensure that regular evaluation limits the communication of mistakes.

The scientific training of climate researchers prepares them to scrutinize information that is sent to them so that reception is also subject to careful evaluation. Scientists know that some journals are more rigorously reviewed than others and that some groups of researchers produce more reliable results than others. Hence the receiving of information by researchers follows high standards for the spread of information.

Outside the scientific community, the spread of information is more haphazard. Ordinary people are not cognitively equipped to read dense IPCC reports or scientific journals. Some popular sources with competent journalists are responsible fortunately in reporting important findings based on high-quality research. Magazines such as *Scientific American* and *Science News* popularize research findings but usually do so with high standards of accurate communication. Credible newspapers such as the *Guardian*, *New York Times*, and *Washington Post* are generally reliable in selecting and reporting on scientific research about climate change. Hence, the spread of climate change information outside universities and research institutes can be based on sound principles of sending and receiving. In contrast, the spread of misinformation about climate change is based on different practices involving unconstrained social media, websites for special interests, and biased reporting.

MISINFORMATION ABOUT CLIMATE CHANGE

Despite the overwhelming scientific consensus about climate change, powerful groups continue to deny that humans are causing global warming. The two interconnected groups that most forcibly reject climate science are companies that profit from fossil fuels and right-wing politicians who are often funded by them. Opposition to climate change research grew in the 1980s in response to growing reports that dealing with climate change would require serious cuts to greenhouse gas emissions. Climate change denial has tried to use misinformation to undermine both the observational information concerning the occurrence of global warming and the inferential information that the human activity that produces greenhouse gases is the main cause of global warming. The AIMS theory of information explains climate change denial as resulting from breakdowns in processes of acquisition, inference, memory, and spread.

Climate change misinformation is generated and spread by various groups and individuals, but I will focus on the role of one American foundation, the Heartland Institute, which has vigorously opposed the regulation of greenhouse gas emissions since 1993.[15] In contrast to the IPCC and politicians who take it seriously, the Heartland Institute maintains that climate changes are small and can be attributed to natural fluctuations rather than human greenhouse gas emissions. Therefore, government actions to reduce greenhouse gas emissions are unnecessary. It dismisses the IPCC and their allies as "climate change alarmists." It organized reports from the self-styled Nongovernmental International Panel on Climate Change that challenges the conclusions of the IPCC.

In the United States, another major force for climate change denial is the Republican Party, which in 2019 had more than one hundred members of Congress who were publicly skeptical about anthropogenic climate change.[16] Republican politician Trump has dismissed global warming as an "expensive hoax."[17]

Acquisition

Whereas the IPCC uses reliable instruments to collect massive amounts of real information about changes in temperature and other climate variables, the Heartland Institute and other climate change deniers collect no new evidence. Instead, they challenge interpretations of available data accepted by the IPCC and climate scientists in general, dismissing the IPCC as a political rather than a scientific entity.[18] The IPCC is criticized for being biased toward government action and therefore prone to exaggerate the extent of global warming and sea level increases. The Heartland Institute does not deny the occurrence of small climate changes but analyzes them as random and minor enough not to require explanation or intervention.

The book *Merchants of Doubt* by Naomi Oreskes and Erik Conway describes how climate change deniers use the same strategy employed by tobacco companies to forestall government limitations on sales of cigarettes.[19] Tobacco companies funded groups such as the Heartland Institute to cast doubt on evidence-based claims that cigarettes cause cancer. The strategy did not require proving that cigarettes do not cause cancer but instead claimed that there was too much uncertainty about the evidence to be confident about the strong causal claim that would justify government action to reduce cigarette sales. Oil companies such as ExxonMobil also funded groups like the Heartland Institute to raise doubts about climate change data. Uncertainty justifies inactivity. Manufactured uncertainty is also used by Republican legislators to justify their antipathy toward government intervention. If we cannot be sure that global warming is occurring and that human industrial activity is responsible for it, then we are off the hook for controlling that activity. Skepticism amounts to motivated ignorance.

Climate change denial displays breakdowns in the information mechanisms of acquisition, which ordinarily leads from collecting by interaction with the world to representing using words and images. Deniers avoid novel interactions with the world but block representations of observations that conflict with their goals. They frequently indulge in making stuff up, for example, claiming that global warming would be good because fewer people will freeze to death.

The transition from collecting to representing requires generalization from data, for example, the IPCC conclusion that global warming is increasingly occurring. The kinds of representations used by climate change denials are the same as those used by climate change scientists—sentences and images such as graphs—but the representations serve to convey doubt and uncertainty rather than action-justifying confidence.

Inference

On the surface, the arguments of the Heartland Institute appear scientific. Science operates by collecting evidence and looking for causal explanations of that evidence. Instead of jumping to the first causal story that comes to mind, scientists appreciate the responsibility to consider alternative explanations. Thus, climate change scientists should not automatically infer that global warming is caused by human production of greenhouse gases without considering other possible explanations. Climate change deniers such as the Heartland Institute claim that the alternative to anthropogenic warming is that natural fluctuations occur in climate variables.

The IPCC gives reasons for rejecting this interpretation, for example, the strong relation between increases in greenhouse gases and increases in global temperatures. This relation is analogous to the dose-response relationship that is used as one of the standards that justify causal inferences concerning diseases. For example, the fact that the more cigarettes people smoke, the more likely they are to get cancer is one of the reasons that scientists since the 1960s have agreed that smoking causes cancer. Almost all climate scientists and most political leaders accept the IPCC conclusion that the hypothesis of anthropogenic climate change is far more plausible than natural fluctuation.

So why are climate change deniers so vehement that climate change is not significantly caused by human activity? Perhaps they are simply providing an honest alternative interpretation and explanation of complex data. Or perhaps their conclusions may be motivated less by concern for truth and human welfare than by nonscientific interests such as personal gain, the interests of fossil fuel companies, and conservative dislike for government interaction.

The 2008 Heartland Institute report disavows personal gain as a motive of the writers of the report. They say they were not motivated by financial interest because no grants were provided for writing the report, nor are they motivated by political aims because they were not tied to any government organizations or party candidates. The ties of the editor of the report Fred Singer, however, display financial and political interests.[20]

Singer served as a consultant for petroleum companies such as ExxonMobil and Shell Oil, and he also received payments from organizations that got funding from those companies, including the Heartland Institute and the Science and Environmental Policy Project. He also worked with the right-wing Frontiers of Freedom Foundation, which espouses property rights and economic freedom. Hence, Singer's motivations for his conclusions went far beyond the scientific. Another prominent scientist who attacks the standard view on climate change, Willie Soon, has benefitted from more than $1 million in support from fossil fuel companies.[21] Republican politicians have their campaigns funded by oil companies and foundations supported by them.[22]

The interests of scientific and political climate change deniers clearly suffer from conflicts of interest. Scientists are supposed to be motivated by truth, explanatory power, and social benefits. Politicians should be motivated by the needs of their constituents. Conflicts of interest occur in situations where the actual motivations of decision makers such as personal gain are different from the motivations that their social responsibilities require for them. Conflicts of interest frequently arise because the interactions between cognition and emotion in human brains make us prone to emotion-driven motivated inference.[23] Such

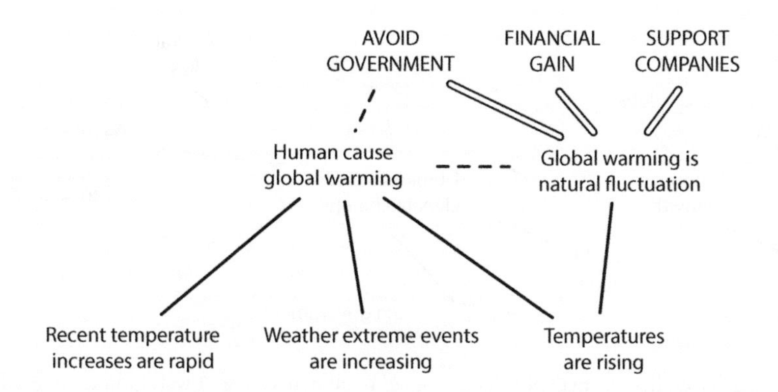

5.3 Motivated reasoning based on goals (shown in capital letters) supports the fluctuation hypothesis. Double lines indicate motivations, solid single lines indicate explanatory coherence, and dotted lines indicate incompatibility.

thinking enables people to convince themselves that they are doing right when they are actually putting their personal interests ahead of their social responsibilities. This self-deception fits the personal goals pattern of motivated reasoning.

How motivated reasoning works in climate change denial is shown in figure 5.3. The hypothesis of climate fluctuations has less explanatory power than the hypothesis that human greenhouse gases cause climate change, but it is nevertheless preferred because it fits strongly with goals such as personal finances, support for fossil fuel companies, and avoidance of government intervention. Emotional coherence with goals triumphs over explanatory coherence with evidence. The motivated reasoning for politicians is similar but with more emphasis on getting reelected thanks to support from fossil fuel companies and conservative donors. A more complete treatment would also show skepticism for the evidence about rising temperatures.

Besides evidence for warming and the hypothesis that warming is caused by humans, climate change deniers are skeptical about predictions of future changes. A crucial part of the IPCC case for government intervention is that global warming will continue unless strong actions are taken. Because climate change deniers are strongly motivated to avoid government intervention, they dislike such predictions and therefore challenge the models that generate them. Models are criticized for not capturing past changes and for making unjustified extrapolations about future changes. As already described for medicine and

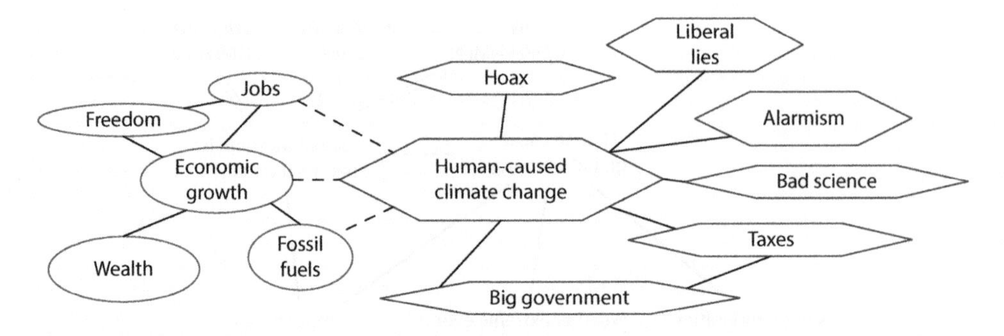

5.4 Cognitive-affective map of values opposed to climate change. Ovals indicate desired values; hexagons indicate disliked values. Solid lines indicate emotional associations; dotted lines indicate emotional incompatibility. The online Supplementary Material contains a more detailed map.

climate, modeling does indeed have problems of identifying the relevant causal factors and the complex relationships among them. Thus, models can be viewed only as approximations of likely but not guaranteed events. Nevertheless, models can be appreciated as best available estimates, which is much better than dismissing them because they yield unpalatable conclusions.

For climate change, the unpalatability of anthropogenic global warming is more a matter of values than evidence, with values such as personal freedom, corporate free enterprise, and small government playing the dominant roles. Figure 5.4 shows a cognitive-affective map of the economic values employed by climate change deniers.[24] Human-caused climate change gets a negative value (emotional attitude) because it is regarded as incompatible with a constellation of treasured political and economic values that include freedom, wealth, and private property.

Values are an indispensable part of decision making, but applying them has two dangerous pitfalls. First, good decisions depend on values that are legitimately based on universal human needs. The values deployed by the IPCC and government leaders shown in figure 5.2 include avoiding harm, saving lives, and promoting equality, which help to satisfy the needs of the huge majority of people. In contrast, the values of climate change deniers are skewed toward the needs of a small minority of rich and powerful people.

Second, values should not swamp the use of evidence to make correct inferences about the current state and eventual future of the world. We might want to make the appealing conclusion that climate change is so dangerous for human needs that it must not be real, but this motivated inference runs counter to the evidence-based inference that climate change is a serious problem. Motivated reasoning based on the goals of powerful individuals interferes with accomplishing the long-term objective goals of humanity.

Motivated reasoning about climate change can also fit the group identity pattern. No one self-identifies as a climate change denier, but other identities have contributed to the spread of misinformation about global warming. In the United States, people who think of themselves as Republicans and Trump supporters, including many members of Congress, are prone to reject the scientific consensus about climate change. The group identity pattern of motivated climate change denial amounts to something like the following statement: "I'm a Republican so I'm not the sort of person who succumbs to Democratic lies about the science of climate change."

Another motivated reasoning pattern applied to climate change is motivated ignorance. Some U.S. states have had policies and laws that prevent people from even talking about climate change.[25] Motivated ignorance may be connected with the specific emotions pattern of motivated reasoning if people react to fear about the drastic consequences of climate change by refusing to think about it.[26] The shift to collecting real information about global warming and adopting measures to control it would benefit from a shift from blind anxiety to hope.[27]

In some individuals, climate change denial comes in a package with skepticism about COVID-19 and opposition to vaccines. The coherence among these attitudes is emotional rather than empirical because evidence supports a different package. Climate change threatens to increase the occurrence of pandemics as environmental disruption increases human exposure to animal viruses.[28]

Memory

Memory is important for saving information for future use. The maintenance of information depends on methods for storing and retrieving that include evaluations of what is worth storing and what is safely retrieved. The memory of individuals is unavoidably haphazard because people have limited conscious control over what they put into their memories and what they remember afterward. But social processes can ensure better standards of storing and retrieving, and scientific research on climate change displays such standards. Information intended for an IPCC report or a good scientific journal is scrutinized by multiple qualified researchers. Thus, people who retrieve information from these sources can have some confidence that they are avoiding misinformation.

In contrast, memory processes used by climate change deniers are rarely scrutinized. They rarely publish in peer-reviewed journals because they cannot meet their scientific standards. Deniers such as the Heartland Institute can post whatever they want on their websites, and almost all climate-denying books receive no prior review, especially the growing number of self-published books.[29] Hence, misinformation about climate change violates the evaluative standards of storing and retrieving real information.

Spread

Memory and spread are connected because one of the ways to spread information is to store it in a location that someone else can access, as happens with magazines and websites. But spread can also work more immediately when one person communicates with others by conversation, public speaking, or media such as radio and television. Climate change scientists speak at scientific conferences and also at more public occasions.

Climate change misinformation can spread much more aggressively than scientific information because of communication of inflammatory messages via many media. For example, a study of YouTube videos accessed by climate change queries found that the majority espoused climate change denial rather than scientific views.[30] Climate change denial is also rampant on Facebook, Twitter, and Instagram.[31] Social media companies have pledged to restrict spread of climate change misinformation but have not been very effective.[32] As I discussed in relation to misinformation about COVID-19, these social media companies continue to show more concern with driving advertising than in spreading truths. In 2022, Twitter announced that it would restrict misleading ads about climate change, but it said nothing about the deluge of misinformation in its tweets.[33]

The right-wing television channel Fox News has personalities who interview climate change deniers and spread their views while deriding scientific alternatives. Scientists and politicians who take climate change seriously are dismissed as "alarmists." As with the Heartland Institute, Fox News is more concerned with propagating an ideology of personal freedom and business priorities than in pursuing general human interests.

According to the AIMS theory of information, the spread of real information results from episodes of sending and receiving that are constrained by evaluations. Before sending, messages should be evaluated for their truth and relevance to human needs. Upon receiving a message, the message should be evaluated on the same criteria before being believed, passed on, or stored in locations where it can be retrieved by other people. Without this crucial evaluative component, misinformation is readily spread.

Climate change deniers engage in motivated sending where the aim is to propagate biased claims, and their followers engage in motivated receiving to believe claims that fit with their prejudices. A 2021 study found that Facebook posts denying climate change were increasing substantially, with thousands of posts generating more than 1 million views.[34] The resulting cascades of misinformation can even be automated; bots on Twitter are commonly used to replicate messages opposed to climate change action, for example, in 2017 when President Trump announced withdrawal from the Paris Agreement on climate change.[35] As occurred for COVID-19 misinformation, a small number of widely accessed websites such as Breitbart are responsible for a large percentage of Facebook posts on climate change denial.[36]

The spread of misinformation about climate change involves all three kinds of information—observational, inferential, and valuational. Skeptics reject the observation that global warming is occurring at a rapidly increasing rate and the inference that this increase is due to human emissions of greenhouse gases. Along with these factual claims, climate change deniers also aim to spread their preferred values of personal freedom, restricted government, and untrammeled corporations. The spread of such values occurs by repetition and visible approval, for example, when television commentators advocate freedom. Value spread also works by derision, when commentators sneer and make fun of opposing views at more centrist media, as when Fox News commentators mock CNN. The spread of values often operates by emotional contagion where positive or negative views are passed on to receptive audiences.

REINFORMATION ABOUT CLIMATE CHANGE

For dealing with climate change threats, the foremost policies are reducing greenhouse gas emissions and finding other ways to reduce carbon dioxide in the atmosphere, for example, by carbon capture and geoengineering. Applying these strategies requires changing minds about climate change so that political leaders and voters support the required actions. Mental change requires fighting rampant misinformation about climate change by replacing it with real information, using eight reinformation techniques: detecting misinformation, identifying sources, recognizing motives, factual correction, critical thinking, motivational interviewing, institutional modification, and political action.

Detecting Misinformation

Before misinformation can be fixed, it must be identified using several methods. The first is incompatibility with real information that is backed by systematic observations, instruments, experiments, and valid inferences. If a climate change denier proclaims that a particular cold winter month shows that global warming is not happening, the claim can be identified as misinformation because it conflicts with decades of data that show global warming is occurring as a general trend that is compatible with short-term cooling occurrences.

The second method for identifying a claim as misinformation is contradiction of statements by recognized authorities. No expert is infallible, but many climate change scientists have decades of widely appreciated research. For example, James E. Hanson is a much-published and much-cited climatologist who has been warning about the risks of global warming since the 1980s. Views about climate that contradict his are likely to be misinformation.

The third method for detecting misinformation is to be skeptical of sources known to be undependable. For example, the *Washington Post* counted more than thirty thousand false or misleading claims made by Trump during his four years as president.[37] Similarly, Joe Rogan is an enormously popular podcast host who has a long history of extreme views about issues such as COVID-19 vaccinations, and the guests he invites to discuss climate change are nonexperts such as Randall Carlson and Jordan Peterson. Anything that Rogan proclaims has a good chance of being misinformation.

The final method for identifying misinformation relies on novel computational methods such as machine learning. An international group used computers to analyze 255,449 documents about climate change denial from conservative think tanks and blogs, generating a model that detects specific claims contrary to science.[38] This analysis generated a classification of denial into five classes of assertions: global warming is not happening, human greenhouse gases are not causing global warming, climate impacts are not bad, climate solutions will not work, and climate science is unreliable.

These four methods show that detecting misinformation can be based on more than an intuitive dislike of some claim. Contradiction of real information, disagreement with authorities, and utterance by certifiably unreliable sources are all signs of misinformation that warrants correction.

Identifying Sources and Recognizing Motives

All communication is helped by understanding the background and motives of the audience. Teaching, for example, is aided by a detailed knowledge of what students already know and what they want to know. Similarly, convincing people who are misinformed that they should change their minds depends heavily on understanding the beliefs and desires that drive those minds.

In evaluating and correcting people who spread misinformation, we need to identify the spreaders and their motives. Such identification does not commit the fallacy that philosopher's call arguing ad hominem, which is attacking the person without addressing the issue at hand. Motives are legitimately relevant to handling misinformation because of the six patterns of motivated reasoning presented in chapter 3. To evaluate the information sent, we need to know the motives for sending it.

Normally, when people say things, it is fine to believe them. Social life would be intolerable if people had to doubt everything said by family members, friends, and acquaintances. People operate with a default assumption that people say things because they believe them and the beliefs are mostly true.[39] However, if an utterance contradicts what we already believe, contradicts an accepted authority, or comes from a suspicious source, then we legitimately examine the claim much more carefully.

Identifying utterers' motivations is an important part of this scrutiny. First, do they actually believe what they are saying? If not and the utterance is false, then the utterance is a lie and counts as disinformation. Oil companies such as ExxonMobil already had internal evidence that global warming was a problem by the 1980s but continued efforts to downplay it.[40] If people are lying to us, their testimony gives us reasons to disbelieve rather than believe them.

Second, if the utterers believe what they are saying, we should ask whether their belief is based on a strong examination of the evidence or on motivated reasoning from personal goals. To answer this question, we need to know what these goals are. For example, we know that climate change denier Fred Singer received financial benefits from oil companies and conservative foundations for espousing climate change skepticism, which also fits with his right-wing politics. Given the abundance of evidence that global warming occurs as the result of human greenhouse gas emissions, we can conclude that climate change denial emanates from motivated reasoning rather than inference to hypotheses that best explain the evidence.

The third reason for scrutinizing the motivations of adherents to misinformation is that it affects the choice of how best to revise it. To transition believers from disinformation to real information, we need to understand their overall general motivations and their particular motivations for a specific belief. Different recommendations apply to different motives operating in misinformed persons. Here are some tentative recommendations for reinformation strategies that should be tested for practical efficacy:

- If the persons are generally motivated toward accuracy and truth and if their beliefs are not distorted by personal motivations, then attempt to reinform them by factual correction.
- If the persons are generally motivated toward accuracy and truth, but their beliefs are distorted by personal motivations and reasoning errors, then attempt to reinform them by critical thinking.
- If the general motivations of the persons are unknown, and their belief is embedded in a network of personal values and group identities, then attempt to reinform them by motivational interviewing.
- If the persons are entirely motivated by personal and political goals and if they are impervious to factual correction, critical thinking, and motivational interviewing, then do not waste time trying to reinform. Instead, look to political action for ways to reduce their ability to spread misinformation to others.

Thus, identifying the motivations of climate change deniers can help us decide whether to try to deal with them using factual correction, critical thinking, motivational interviewing, or political action.

Factual Correction

Some adherents to climate change skepticism are just innocents who were exposed to misinformation despite a general desire to have true beliefs and no particular ax to grind against government interventions. For example, someone who thinks that extreme weather events are no more common than they used to be can be presented with easily assimilated facts about the increasing occurrence of heat waves, floods, and other extreme weather events. Factual correction should help well-meaning individuals to appreciate the large evidence base that supports global warming.

Factual correction is also helpful for dealing with the theoretical background to climate change. One of the reasons why climate change scientists believe that human activity is the main causes of global warming is the well-understood mechanism by which greenhouse gases prevent the sun's energy from being reflected back to space. Berkeley researchers found that almost no Americans understand the basic global warming mechanism, but that short explanations of it increased climate change acceptance across the liberal-conservative perspective.[41]

Critical Thinking

On my two-step view, critical thinking requires both error detection and remediable reasoning. For climate change, the main error to be detected is motivated reasoning, which we saw was common in deniers who ignore the mounds of evidence reported by the IPCC for conclusions that fit with their goals of personal gain, profits for fossil fuel companies, and freedom from government interference.

John Cook has identified other thinking errors commonly committed by climate change deniers: fake experts, impossible expectations, cherry-picking, and conspiracy theories.[42] Fake experts with minimal climate credentials are funded by oil companies or conservative think tanks to spread doubt about global warming and its human causes, an instance of the fallacy that logicians call false authority. Impossible expectations include the demand that anthropogenic climate change be shown to have the deductive certainty found only in mathematics rather than the inductive confidence that can reasonably be expected of scientific theories with practical significance. Cherry-picking is the selective choice of data that cast global warming into doubt, for example, focusing on short periods of cool temperatures.

Chapter 4 mentioned the operation of conspiracy theories in skepticism about COVID-19, and conspiracy theories are similarly common among climate change deniers. The fact that 97 percent of all climate change researchers accept anthropogenic climate change can be dismissed with the accusation that they are all part of a progovernment conspiracy to meddle with the economy.[43] On this view, the IPCC is just a group of conspirators trying to increase their own funding and government power. No evidence about conspirators, communications, or plans supports this conspiracy theory, which is fueled purely by making stuff up and motivated reasoning. Another environmental conspiracy theory popular on YouTube concerns chemtrails, which are condensation trails left by aircraft supposedly spreading chemicals for nefarious goals such as solar radiation management or human population control.[44] Chapter 6 provides a detailed analysis of conspiracy theories that indicates how to evaluate and debunk them.

Besides motivated reasoning, other error tendencies are relevant to explaining climate change. Confirmation bias is the tendency to seek evidence for views already held, as climate change deniers commonly do, although it is hard to separate the purely cognitive process of seeking confirmation from the emotional process of motivated reasoning. The fallacy of hasty generalization is going from a small and unrepresentative sample to a general conclusion, which climate change deniers do when they cherry-pick data to deny the general trend toward global warming.

Identifying inferential errors is complemented in critical thinking by remedial reasoning, which requires identifying the logic appropriate for the task at hand. The reasoning tools relevant to good judgments about climate change include observational generalization, causal inference to the best explanatory hypotheses, practical decision making, and probabilistic calculation.

Observational generalization is taking vast amounts of data concerning temperature and aspects of climate and reaching overall conclusions such as that global warming is increasing. The standards for this transformation of data into phenomena include ensuring that all relevant evidence is considered and that alternative explanations of data are ruled out, including fraud, incompetence, and random occurrence.

For causal conclusions such as that climate change is caused by humans, the main reasoning is inference to the best explanation, which requires considering all the relevant evidence and only accepting hypotheses that are part of the best explanation of that evidence, in comparison to alternative explanations.

The hypothesis that climate change is primarily caused by human production of greenhouse gases should be accepted because it provides a better explanation of many kinds of evidence such as global warming, as I showed earlier in this chapter.

Climate change requires not only inferences about past and future conditions but also reasoning about what to do. Governments and individuals face hard choices about actions intended to stop or at least reduce global warming that can have dire consequences. Decisions must be made on issues such as emissions limits, carbon taxing, and geoengineering. Such decision making can be construed as inference to the best plan, which chooses a set of actions because they are better at accomplishing desired goals than alternative actions.[45] Just as inference to the best explanation can fail when it does not consider all the evidence and alternative hypotheses, inference to the best plan can fail when it does not consider all the relevant goals and alternative actions. Climate change deniers assail governments for underestimating the economic costs of actions such as reducing the use of coal, and their alternative set of actions is simple: do nothing because climate change is not a threat. Taking climate change seriously requires much more complex reasoning about how to deal with it.

Some normatively appropriate reasoning requires use of probability and statistics. The IPCC seems to be using probabilities when it states:

> The following terms have been used to indicate the assessed likelihood of an outcome or a result: virtually certain 99–100 percent probability, very likely 90–100 percent, likely 66–100 percent, about as likely as not 33–66 percent, unlikely 0–33 percent, very unlikely 0–10 percent, exceptionally unlikely 0–1 percent. Additional terms (extremely likely 95–100 percent, more likely than not >50–100 percent, and extremely unlikely 0–5 percent) may also be used when appropriate.[46]

Unfortunately, the IPCC never says what it means by probability, which is clearest when it applies to frequency, for example, when the probability of a coin flip turning up heads is .5 because in the long run a fair coin will turn up heads 50 percent of the time. In contrast, the probability of global warming and anthropogenic climate change are not frequencies but rather must be construed as degrees of belief, which are hard to interpret and compute.[47] For example, using Bayes's theorem to calculate the probability of the claim that climate change is caused by human greenhouse gas emissions requires knowing countless conditional probabilities such as the indeterminate probability of global warming given greenhouse gas emissions.

Similarly, decision making about how to tackle global warming is not easily construed with the economist's favorite tool of maximizing expected utility using probabilistic computations. The probabilities and utilities are largely unknown, for example, the probability that a carbon tax will reduce emissions to a specific extent. Thus, decision making about climate change is better interpreted as a qualitative matter of inference to the best plan than as a speciously quantitative process of maximizing utility.

Critical thinking can also be used as a preventive measure by methods that constitute prebunking or inoculation.[48] People can be protected from misinformation by warnings about impending flawed argumentation such as the use of flawed experts while being told about the strong consensus among climate authorities concerning the occurrence and causes of global warming. From the perspective of the AIMS theory of information, inoculation works by priming people to do more evaluating when they are receiving messages. Simple interventions that prime people to use critical thinking can limit the influence of fake news on Facebook.[49]

In sum, good reasoning patterns can supplant the thinking errors commonly used by climate change deniers. Making remedial reasoning work with hard-core deniers is unlikely, but it might help with people who have more open minds.

Motivational Interviewing

If logical reasoning fails to budge the minds of climate change deniers, another strategy is motivational interviewing, which has been tried in the climate domain.[50] An approach closer to psychotherapy than logic might operate as follows:

- Understand people's concerns about climate by asking them open-ended questions and empathizing with their fears and insecurities about weather events and government actions.
- Be affirmative, reflective, and nonjudgmental about their concerns.
- Identify discrepancies between people's current and desired behaviors such as dealing with increasingly common extreme weather events.
- Summarize the issues and inform people while respecting their autonomy.

This approach is unlikely to work with entrenched climate change deniers enmeshed in their motivated inferences, but it might help with moderate deniers who have concerns about personal freedoms threatened by the restrictions that governmental action concerning climate change would require.

Institutional Modification

The generation and spread of real information about climate change has benefitted enormously from productive institutions that include laboratories, universities, government agencies, and the IPCC. Misinformation has unfortunately gained from the activities of other institutions that include fossil fuel companies, foundations, political parties, traditional media, and social media. How can institutions be modified to promote reinformation? At least five kinds of institutional change might contribute to the reduction of misinformation about climate change and other problems: (1) creating new institutions dedicated to real information; (2) eliminating institutions that spread misinformation; (3) changing the composition of the institution to have more responsible members; (4) changing the explicit policies of the institution to discourage disinformation; and (5) changing the implicit values, norms, and practices that enable institutions to spread misinformation.

We already have the IPCC as the major institution for acquiring and spreading real information about climate change, but local institutions are also needed. Consider the institutions generated by the Swedish activist Greta Thunberg. The first institution she influenced was her own family when she convinced her parents to make better lifestyle choices. She started lobbying the Swedish parliament and joined with fellow students to start a new organization, School Strike for Climate, which spread across 150 countries.[51] Similar new grassroots organizations are needed to agitate for better information about climate change and political action to manage it.

Candidates for institutional elimination include the Heartland Institute and the website wattsupwiththat.com, which bills itself as "the world's most viewed site on global warming and climate change." With contributors that have included Fred Singer, this website is devoted to denying the climate emergency and attacking the IPCC. Lucky for them, democratic countries have strong protections for free speech, so I am not suggesting that the Heartland Institute and Watts Up with That be shut down. Such legal elimination should be reserved for provably violent organizations such as Al Qaida and the neo-Nazi Atomwaffen Division. Ideally, exposure of the agenda and falsehoods of climate change–denying organizations can lead to their gradual fading as the climate crisis worsens.

Prospects are few for changing the membership of the Heartland Institute or the American Petroleum Institute, which lobbies for large oil and gas companies. But other, more flexible organizations might be modifiable by adding participants. For example, universities can become more responsive to climate change issues by hiring faculty and other researchers dedicated to collecting real information about climate problems and solutions.

Many institutions could benefit from explicit changes to their official policies about information. After Elon Musk bought Twitter in 2022, he shifted its policies from attempts to control extreme views toward free speech absolutism, which allowed reinstatement of banned members such as Trump, Kanye West, and Peterson. Twitter (renamed X in 2023) immediately saw increased far right communications such as those espousing anti-Semitism. Since 2020, Twitter had a policy against coronavirus misinformation that led to the suspension of more than eleven thousand accounts and removal of more than 100,000 pieces of content. This policy was ended in November 2022.[52] Even Musk, however, has to pay attention to the exodus of advertisers from Twitter and may have to modify his policies. Advertisers and users can lobby other media companies such as Facebook and Google to strengthen their policies against misinformation.

The effects of institutions on information and generally on the behavior of their members are not simply the result of official policies. More subtle influences come from unwritten values, norms, and practices that shape the routines operating inside the institution. For example, a police department may not have any explicit policy in favor of stopping and arresting people of color, but the attitudes, expectations, and communications of mostly white officers can produce an implicit bias in that direction. A government agency dealing with climate issues may operate with implicit bias for misinformation if its political leadership takes a strong stance against climate transparency.

Political Action

The techniques of factual correction, critical thinking, motivational interviewing, and institutional modification will not always suffice to stop climate change deniers from propagating misinformation. Political action against climate change denial may require infringement on the valuable right to freedom of speech. That right is never universal because people do not have the right to harm others by defaming them as criminals or threatening violence. Autonomy is an important ethical principle, but so are benefiting people, avoiding harm, and maintaining justice by treating people equally.[53] Climate change denial causes harm by preventing decisions that can save millions of lives. Freedom is important to human flourishing but so is having a habitable planet.

Political action can be used to reduce the spread of dangerous climate change ideas by controlling the excesses of social media. I have described how Facebook, Twitter, and YouTube have not only allowed the spread of misinformation but have actually encouraged it. The encouragement comes from the use of algorithms that reward messages that are inflammatory rather than accurate. Social media have taken a few steps to reduce the amount of misinformation they spread about climate change, but more steps could be taken, as outlined in chapter 8.

Political action is required to encourage governments to introduce such measures against spread of dangerous misinformation. Even more important, political action is required to elect representatives who are aware of the dangers of misinformation and who are sufficiently courageous to enact laws to help control it. Opposition to such elections and legislation comes forcefully from interests who benefit from unlimited use of fossil fuels, including people who are rich, powerful, and politically connected. Nevertheless, anyone convinced of the disastrous prospects of climate change over the next decades should be politically active and help install governments that will deal seriously with global warming and the misinformation that encourages it.

OTHER SCIENTIFIC MISINFORMATION

Besides climatology, misinformation afflicts other areas of science. The medical misinformation discussed in chapter 4 is also scientific information because modern medicine is part of science. Other popular sources of misinformation connected with science include proponents of creationism, astrology, extrasensory perception, and flat-earth claims. Most dangerous of all is the general denial that rejects the whole scientific approach to knowledge. People urged to "do your own research" think that they can do so by consulting a few random YouTube videos, like the man depicted in the cartoon in figure 5.5.

"Honey, come look! I've found some information all the world's top scientists and doctors missed."

5.5 Cartoon about scientific information.
Source: Jon Adams, reprinted by permission.

General Science Denial

Modern science has ancient roots in Babylon, Greece, and Baghdad. The field of science began to flourish in sixteenth-century Europe through the work of scholars such as Nicolaus Copernicus, Galileo Galilei, and Isaac Newton. From physics, it expanded to chemistry in the eighteenth century through work by Antoine Lavoisier, John Dalton, and others. Biology became thoroughly scientific in the nineteenth century through advances such as Robert Hooke's discovery that organisms consist of cells, Charles Darwin's theory of evolution by natural selection, and Louis Pasteur's germ theory of disease. Subsequently, psychology and other sciences have since developed to share the common methodology that includes systematic observations using instruments, careful experiments, and inference to theories that best explain the empirical evidence.

Alternatives to science include religion, mysticism, personal intuition, and political orthodoxy, all of which have been favored over science when they conflict. Some people contend that religious texts such as the Bible and the Qu'ran are better authorities than science about ultimate reality or that the world is inherently incomprehensible and better viewed mystically than scientifically. Some think that their own intuitive feelings are inherently valid and dominate any amount of scientific evidence. Extreme political views on both the left and the right have assailed science when it conflicted with their ideologies, for example with Joseph Stalin's advocacy of Trofim Lysenko's biology and Adolf Hitler's rejection of Albert Einstein's physics.

The two key questions for the defense of science are: how does science differ from other potential ways of knowing, and how is it better at revealing reality? Simple answers to the first question, that science is unique in being verifiable or falsifiable, have not stood up to scrutiny; and doubts have been raised about whether science can actually be demarcated from other enterprises.[54] I think that demarcation works not by giving a strict definition of science but by contrasting its typical features with those of a pseudoscience such as astrology:[55]

1. Science explains using mechanisms, whereas pseudoscience lacks mechanistic explanations.
2. Science uses correlation thinking, which applies statistical methods to find patterns in nature, whereas pseudoscience uses dogmatic assertions or resemblance thinking, which infers that things are causally related merely because they are similar.
3. Practitioners of science care about evaluating theories in relation to alternative ones, whereas practitioners of pseudoscience are oblivious to alternative theories.
4. Science uses simple theories that have broad explanatory power, whereas pseudoscience uses theories that require extra hypotheses for particular explanations.
5. Science progresses over time by developing new theories that explain newly discovered facts, whereas pseudoscience is stagnant in doctrine and applications.

Scientific progress is marked by consensus about fundamental theories and methods. Controversies continue in physics, chemistry, and biology, but scientists in these fields generally agree about the importance of experiments and accept the same general theories such as relativity theory, quantum theory, genetics, and evolution by natural selection.

These features give science major advantages over other alleged ways of knowing such as religion, mysticism, personal intuition, and political dogmas. Besides a remarkable degree of progress and consensus, science has made an astonishing contribution to technological successes such as flight, electronic machines, antibiotics, and vaccines. These applications attest to the ability of science not only to gain knowledge by interacting with the world but also to apply knowledge to the world through technology based on the science. Science is not just a different way of knowing the world: it is demonstrably better than available alternatives such as creationism and astrology.

Creationism

Before Darwin, most scientists agreed with ordinary people that biological species were designed and created by God. Creationism carried with it misinformation such as that creation took place less than seven thousand years ago. To justify teaching creation in U.S. public schools, an attempt was made to concoct "scientific creationism" or "creation science," but this enterprise is easily shown to be starkly different from scientific methods of experiment and theory evaluation.[56] Creationism explains by spirituality rather than mechanisms, is dogmatic rather than statistical, ignores alternative theories of life's origin, is not simple in that it requires many assumptions about God's intentions, and has not substantially progressed since the Bible was written. Creationism can therefore be rejected as pseudoscientific misinformation.

Astrology

Astrology was once as much a part of ancient science as astronomy, but today it is easily reclassified as a pseudoscience because of its lack of progress in finding evidence for the effects of celestial objects on human lives. Nevertheless, astrology survives in newspaper horoscopes and occult practitioners who exploit people's needs to understand and predict their lives. As explanation of human behavior, astrology has long been superseded by psychology, which uses scientific methods of systematic observation, controlled experiments, and evidence-constrained theorizing about mental mechanisms. When people think they are learning about themselves because they were born under the sign of Taurus or the influence of Mars, they are actually victims of misinformation.[57]

Extrasensory Perception

Extrasensory perception (ESP) is the alleged ability of human minds to receive information without using the usual senses of vision, hearing, touch, smell, and taste. People are ascribed paranormal, psychic abilities such as telepathy (reading minds), clairvoyance (seeing the future), remote viewing, and levitation. In the early days of psychology, ESP was taken seriously by William James and other scientists, but the current consensus in psychology is that the evidence for ESP has been discredited by faulty experiments and fraud. For example, when people are convinced by psychics that they can communicate with deceased love ones, they are succumbing to motivated reasoning based on their desire to hold on to the dead rather than on good evidence.[58] Another problem with ESP as science is that, aside from some mumbling about quantum effects, no mechanism explains how telepathy and other paranormal sensing could take place.

Flat-Earthers

The most amusing current group of science deniers are the flat-earthers who deny that the Earth is a sphere. The popularity of this view was initially spurred by YouTube videos and now abounds in social media communications and conferences.[59] Flat-earthers complain that they are derided as primitive even by anti-vaxxers and climate change deniers. One striking aspect of the flat-earthers is that they need to dismiss vast amounts of scientific evidence about the shape of the Earth, for example, pictures collected by satellites and astronauts. Thus, they adhere to dramatic conspiracy theories about the lengths gone to by government agencies such as NASA to maintain the belief that the Earth is round. Flat-earthers are also prone to other conspiracy theories such as those about COVID-19 and Jewish plots.

The different forms of pseudoscience discussed here are too far on the fringes to be much of a threat to science, but they must sometimes be combatted in everyday life. The reinformation strategy of factual correction should suffice, for example, in pointing out that scientific psychology offers much better explanations of human behavior than mythical astrological effects. Flat-earthers are too silly to bother with, but creationism and ESP sometimes become prominent enough in social and educational contexts to require debunking. Critical thinking provides the tools for identifying the motivated reasoning of their proponents, such as religious orthodoxy, and for offering remedial reasoning that shows their weaknesses in the face of evidence and alternative scientific theories.

Lee McIntyre has usefully identified five mistakes that are common to science deniers concerning the shape of the Earth, climate change, COVID-19, and other issues: reliance on conspiracy theories, cherry-picking evidence, reliance on fake experts, setting impossible expectations for science, and using illogical reasoning.[60] My reinformation tool kit provides ways of dealing with all of these.

BEYOND SCIENTIFIC MISINFORMATION

Science serves as a model of how people can acquire real information through interaction with the world using observations, instruments, experiments, and theories that explain evidence. Sciences such as physics, chemistry, biology, and climatology have amassed astonishing amounts of knowledge that reliably guide human actions. Nevertheless, science is sometimes challenged by misinformation in pursuits such as creationism, astrology, parapsychology, and climate change denial.

Over four decades, climate science has amassed a wealth of real information about global warming, showing that it is already a serious problem that will get worse in the coming decades. The main cause of global warming and other threatening phenomena such as rising sea levels has been identified as human production of greenhouse gases, particularly by burning fossil fuels. Some governments are attempting to prevent climate disasters by reducing greenhouse gas emissions through limitations on use of coal and oil. Large oil companies and agricultural organizations that previously fought recognition of climate change

TABLE 5.1 Examples of information and misinformation about climate change

	Real information	Misinformation
Global warming	The Earth has been warming steadily.	Earth's warming is sporadic.
Causation	Global warming is caused by human-created greenhouse gas emissions.	Global warming is random fluctuation.
Prediction	Continued warming will produce drastic climate effects.	Climate events will occur as they always have.
Action	Fossil fuel use must drop sharply to prevent disastrous global warming.	No political and economic changes are necessary.

are responding more often to changes in popular opinion by claiming concern, but doubts have been raised about their sincerity.[61] Major oil companies such as ExxonMobil and Shell have adopted a clever and hypocritical strategy of presenting a publicly appealing face on climate change. They each claim to be making their own contribution to fighting climate change through approaching carbon neutrality, but they continue to fund the American Petroleum Institute, which lobbies against substantial government action.[62]

Climate change denial rejects IPCC conclusions about the dangers of climate change with mounds of misinformation about warming trends and alternative explanations for them. Deniers operate on the fringes of science, but they receive attention from people with much to gain from the success of fossil fuel companies, including politicians funded by them and by their owners. Another source of support for climate change denial is conservatives who dislike the government intervention in the economy that is required to rein in the production of greenhouse gases.

Perhaps some climate change deniers could be influenced by factual correction, critical thinking, and motivational interviewing. But two forms of political action also serve to control the spread of misinformation about climate change: lobbying governments to control social media, and working to elect politicians ready to act on the extreme risks of uncontrolled climate change. Modification of media and educational institutions can also tilt them toward real information. The contrast between real information and misinformation about climate change is summarized in table 5.1.

Global warming is now the most serious current threat to human survival, outranking pandemics, nuclear war, and the robot apocalypse in which humans bow to artificial intelligence. Climate disasters can be averted by developing increasingly affordable renewable energy sources such as solar power and by making massive changes in behavior that generates greenhouse gases. A crucial part of this transformation is replacing misinformation by real information. This replacement requires dealing with conspiracy theories, the concern of chapter 6.

CHAPTER 6

PLOTS

Conspiracy Theories and Political Misinformation

The most outrageous misinformation consists of conspiracy theories, such as claims that the Holocaust is a hoax concocted by Jews, and that the 9/11 destruction of American buildings was directed by the U.S. government. We saw conspiracy theories contributing to misinformation about COVID-19 and climate change, for example, that the virus was originally designed as a bioweapon and that the Intergovernmental Panel on Climate Change (IPCC) conspires to expand government and limit fossil fuel companies. Dozens of other conspiracy theories are currently in wide circulation, ranging from the relatively harmless claim of flat-earthers that NASA invents evidence that the Earth is round to the politically toxic QAnon story that Democratic politicians and government officials run a pedophile ring.[1]

Some conspiracies are real: Roman senators plotted to assassinate Julius Caesar, and the Oath Keepers, a U.S. militia group, planned the attack on the U.S. Capitol on January 6, 2021. How can we distinguish real information about real conspiracies from misinformation about fake conspiracies? To go beyond the obvious difference that the real conspiracy claims are true and the fake ones are false, we can examine them from the perspective of the AIMS theory of information and misinformation. Real information about conspiracies uses the AIMS processes of *a*cquisition, *i*nference, *m*emory, and *s*pread, to explain how agents communicate with each other and develop plans that turn into actions.

A plausible conspiracy explanation provides evidence for the motives and interactions of the conspiring agents. In contrast, bogus conspiracy theories arise because of severe breakdowns in the mechanisms that support acquisition, inference, memory, and spread. For example, acquisition of real information about real conspiracies is supported by reliable perceptions, but misinformation about fake conspiracies arises from making stuff up using motivated invention.

I illustrate the applicability of the AIMS theory to political conspiracies by considering two important examples of real conspiracies: the assassination of Julius Caesar and the Oath Keepers' attack on the U.S. Capitol. In both cases, we can spell out the cognitive and emotional mechanisms that drove the conspiracy and produced plans and actions. In contrast, examination of fake conspiracies reveals their origin in different mental operations such as motivated reasoning, as I show for QAnon and the theory of white replacement. Most conspiracy theories are scary, but some are amusing, as is the one shown in figure 6.1.

Conspiracy theorists are disturbingly immune to evidence against their passionate views, but we can nevertheless consider strategies for converting

Conspiracy theory of relativity

6.1 Amusing conspiracy theory: conspiracy theory of relativity.

Source: Andertoons, reprinted by permission.

misinformation into real information. After identifying falsehoods and motives, we can try to employ factual correction, critical thinking, and motivational interviewing to penetrate conspiracy mindsets. Failure of these attempts supports the use of political action to forestall the dangerous effects of conspiracy theories.

Political activities in governments illustrate kinds of misinformation besides conspiracy theories. Deception and propaganda are used by leaders to further their ends, and democratic government requires strategies for detecting and correcting such misinformation. Additional political misinformation provides ways of justifying inequality, as chapter 7 shows.

REAL INFORMATION ABOUT REAL CONSPIRACIES

Understanding how real conspiracies work requires specifying the mental mechanisms operating in the conspiring agents and the social mechanisms by which they interact.[2] Recall from chapter 1 that a mechanism is a combination of connected parts whose interactions produce regular changes. In the mind of each agent, the parts are mental representations, such as the belief that Caesar was assassinated, and the interactions are processes, such as the inference that Brutus led a plot against him. A conspiracy is a social mechanism where the parts are the participating agents and the interactions are the communications among them that produce plans and actions. I illustrate how conspiracies work with a simple, fictional example and then go into more detail with two important historical cases.

Communicating Agents

To get the basic idea of how conspiracies work, I return to the characters of Pat and Sam. A conspiracy involves secret planning with other people to do something bad, so let us suppose that Pat and Sam are short of money and decide to defraud a neighbor, Chris. Pat and Sam tell Chris that Pat has a rare disease that requires treatment with a medicine that can only be bought online for thousands of dollars.

In this case, Pat and Sam are the two agents in the conspiracy, and understanding their actions requires attention to their mental operations, including their motives, goals, beliefs, and emotions. A motive is a reason for doing something, and the prime motive of Pat and Sam combines (1) their goal to get money, (2) their belief that Chris can be tricked into providing it, and (3) their emotions such as greed and contempt for Chris's gullibility.

We could further break down the mental processes of Pat and Sam to include their concepts, which are wordlike mental representations such as *money* and *gullible*. Their beliefs include rules, which are mental representations with an if-then structure, for example, the assumption that *if we get money from a gullible person, then we can spend it*. Pat and Sam may also be working with analogies, for example, planning this fraud based on a previously successful case where they tricked someone else. The mental representations operating here need not all be verbal because Pat and Sam may be thinking with images, such as a mental picture of Chris giving them money or the imagined satisfaction of carrying out the scam.

Pat and Sam also engage in inferences to construct their plans. Inferences involving rules are an important part of planning as people think: if we trick Chris, then we get the money we want, so we should trick Chris. Sometimes when plans conflict, Pat and Sam need to do an inference to the best plan, for example, deciding that they could get more money by defrauding Chris than by approaching a different victim. Pat and Sam may also employ analogical inference when they reuse a fraud plan that worked in a previous case. If the plan works again, then Pat and Sam could generalize it into a rule: *if you want to defraud someone, then use the plan*. Additional inferences could be generated by unforeseen events, for example, if they begin to suspect that Chris might be becoming suspicious of them.

A conspiracy is a social process as well as a mental one, so we must analyze the communicative interactions between conspirers such as Pat and Sam. In their case, most of the communication would be verbal, that is, done by talking, but they could also have pictorial communications by sharing images, such as diagrams of the plan to trick Chris. Additional nonverbal communication could come from emotional contagion when Pat and Sam are talking excitedly with each other about a joint plan. Pat and Sam may also increase their emotional understanding of each other by mutual empathy, where they put themselves in each other's shoes and imagine how they are feeling about their collaboration.

The communication between Pat and Sam could also employ technologies such as telephone, email, and text messaging. Pat and Sam would be careless to record their conspiring against Chris in emails or texts, but they may be unaware

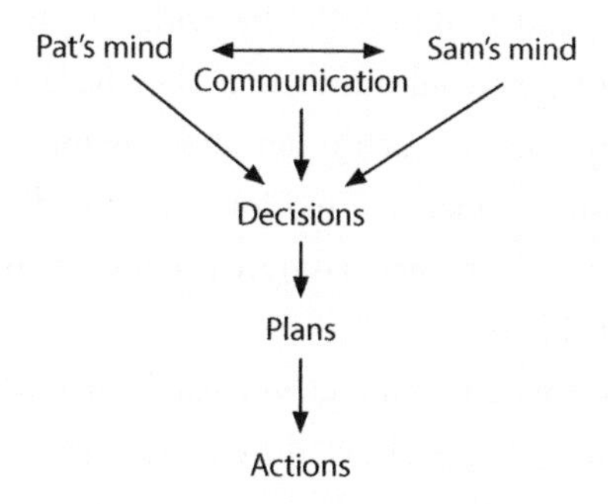

6.2 The structure of a conspiracy using communications, agents, plans, and actions (CAPA). Arrows indicate causal relations.

of the insecurity of these messages. Telephones can be tapped but usually leave no permanent record of joint plans. By verbal, nonverbal, and digital communication, Pat and Sam can develop the plans that lead to their joint action against Chris, using decisions that choose between alternative plans, such as different ways of manipulating Chris. The structure and processes of the conspiracy are shown in figure 6.2, which depicts the interactions of minds leading to joint plans that produce harmful actions. Call this the CAPA schema of conspiracies, where communications among *a*gents lead to *p*lans and *a*ctions.

This account of the mental and social processes in conspiracies shows how real information can result from mechanisms of acquisition, inference, memory, and spread. Chris or the police could acquire information about Pat and Sam's conspiracy by perception of their communications, for example, by Chris accidentally overhearing Pat and Sam plotting. Instruments, such as microphones and cameras, could help with observations by enhancing ordinary perception. Instruments could also help to observe Pat and Sam by tapping their phones or hacking into their electronic communications. Acquisition requires representing as well as collecting data, and Chris or investigators could represent what they learn about the conspiracy by words or by diagrams, such as arrows connecting pictures of the conspirators. Acquisition of information about conspiracies usually uses perceptions and systematic observations, but experiments are rare unless Chris thinks of some clever way to manipulate Pat and Sam to see how they react.

Inference is also crucial for understanding a conspiracy because the mental states of conspirers are not directly observable. The utterances, messages, and other behaviors of Pat and Sam provide the evidential basis for inferring their intentions. For example, we could infer that they have a financial motive for fraud from their complaints to each other about being broke, as well as from bank accounts and other evidence. Emotions are not directly observable, but behaviors such as cursing their financial state provide the basis for inferring that they are angry about being poor.

Inferences about the plans of Pat and Sam can be made from other behaviors, for example, if they buy tickets to leave the country. Inferences are also required to connect mental states to plans, and to connect plans to actions, for example, when the goal of getting money leads to the fraudulent lies. Such inferences are legitimate if they are made by inference to the best explanation that ties together all the available evidence while thoroughly considering alternative hypotheses. Then explanatory coherence justifies the inference that Pat and Sam are conspiring to defraud Chris.

Memory is also important for real information about conspiracies because the results of acquisition and inference must be stored and retrieved with care. Chris as an individual might be haphazard about keeping track of information about Pat and Sam, but police investigators are more systematic about storing what they learn in notebooks and computer files. They should be careful about assessing the reliability of what is stored and cautious about retrieving information for future purposes. For example, if rumors about what Pat and Sam are doing come only from Quinn, who is known to dislike them, then this report should be flagged as possibly misleading.

The development of real information about the conspiracy depends on effective methods of spread. Like conspiracies, the investigation of conspiracies is usually a group process that involves multiple agents. Police investigators typically work as teams whose members share information and collaborate to generate plausible accounts of agents, motives, plans, and actions that constitute a conspiracy. An effective team sends and receives information guided by evaluation concerning the value and credibility of the messages. It usually takes a team of investigators to capture a team of conspirators. The team is often part of a larger institutions, such as a police department or a newspaper.

My fictional story of Pat and Sam is merely illustration, not evidence, for the CAPA schema of conspiracy and the AIMS theory of gaining information about conspiracies. Evidence consists of real cases that are illuminated by applying this account of the structure of conspiracies and real information about them.

The Assassination of Julius Caesar

In 44 BC, the Roman dictator Julius Caesar was murdered by a group of around twenty senators. The distinguished historian Barry Strauss draws on ancient sources to describe their conspiracy in *The Death of Caesar*.[3] The conspiracy began with two senators, Brutus and Cassius, who recruited the others, including Decimus, who became an additional leader. The agents in the conspiracy consisted of the group of senators who planned and carried out the assassination. Their institutional context was the Roman Senate.

The motives of the members of the Roman nobility who killed Caesar are summarized by Strauss. "They believed that they were carrying out their sworn duty to defend the Republic. By attacking Caesar, the assassins believed, they were covering themselves with glory. They did it out of conviction, they did it out of self-interest, they did it out of hatred, they did it out of jealousy, and they did it out of honor."[4] The senators had become convinced that Caesar had arrogantly decided not to share power with the Senate but instead to operate as king. The main goal of the senators was to prevent Caesar from becoming all powerful, and they believed that the only way to stop him was by killing him. This goal and belief combined with the emotions of hatred, jealousy, and desire for glory and honor to motivate the senators to plan the assassination.

The senators communicated by meeting in small groups, and their planning required important decisions. Some senators wanted to kill Antony also to prevent him from avenging Caesar's death, but Brutus argued that they should only kill the tyrant Caesar. The senators decided to kill Caesar at a senate meeting where he would not be protected by friends. To share responsibility, the senators agreed to stab Caesar simultaneously with their daggers. These plans led to the effective joint action of Caesar's execution, so the assassination of Caesar fits my CAPA schema of a conspiracy as consisting of communicating agents, plans, and actions. Communication by conversations among numerous conspiring agents produced decisions that generated plans that led to the murderous action.

Writing more than two thousand years after the conspiracy, Straus had to rely on ancient sources, particularly Nicolaus of Damascus, Suetonius, Plutarch, Appian, and Cassius Dio. Of these, only Nicolaus was alive at the time of the assassination and could draw on contemporary witnesses, but the others drew on their own sources. Strauss's five sources agree overall about events but disagree about some details. Strauss's account of events also drew on his general knowledge about Roman society. We cannot be certain that he got the conspiracy exactly right, but his judicious account nevertheless amounts to acquisition of information about the conspiracy by collecting and representing what the historical sources provide.

Strauss's conclusions about the key motives of the conspiring agents require inferences that attribute goals, beliefs, and emotions to the participants. Some relevant information comes from the writings of contemporaries such as Cicero, who was sympathetic to the assassins. Memory of the components of the conspiracy has been passed down to us through generations of historians, although it is lamentable that many early writings have not survived. Spread of information about the conspiracy has occurred through repeated acts of memory, that is, through storing by writing things down and retrieving by reading the writings of generations of historians. Errors may have been introduced, for example, when Shakespeare's magnificent play *Julius Caesar* gets details wrong about the location of the murder and Caesar's last words, which were probably not "Et tu, Brute." Nevertheless, we can conclude that Strauss and other historians have given us real information about a real conspiracy.

The Attack on the Capitol by the Oath Keepers

A mob of over two thousand protestors attacked the U.S. Capitol building in Washington, DC, on January 6, 2021. The whole mob was not a conspiracy because they were there for a variety of reasons connected with supporting Donald Trump, without an overall organization. Hundreds of them have been charged with a variety of offenses, but eleven members of a particular group, the Oath Keepers, were charged with "seditious conspiracy," which consists of conspiring against the authority of the state.

By May 2022, three members of the Oath Keepers had pleaded guilty to seditious conspiracy.[5] In November 2022, a jury convicted two additional members, including its leader Stewart Rhodes, of seditious conspiracy, and found three other members guilty of other charges.[6] Three other members were found guilty of seditious conspiracy in January 2023.[7]

The Oath Keepers are an American antigovernment militia who have been involved in various disputes and protests since 2009.[8] Their alleged motives and plans for action are summarized in the U.S. Justice Department's indictment:

> The seditious conspiracy indictment alleges that, following the Nov. 3, 2020, presidential election, Rhodes conspired with his co-defendants and others to oppose by force the execution of the laws governing the transfer of presidential power by Jan. 20, 2021. Beginning in late December 2020, via encrypted and private communications applications, Rhodes and various co-conspirators coordinated and planned to travel to Washington, D.C., on or around Jan. 6, 2021, the date of the certification of the electoral college vote, the indictment alleges. Rhodes and several co-conspirators made plans to bring weapons to the area to support the operation. The co-conspirators then traveled across the country to the Washington, D.C., metropolitan area in early January 2021.
>
> According to the seditious conspiracy indictment, the defendants conspired through a variety of manners and means, including: organizing into teams that were prepared and willing to use force and to transport firearms and ammunition into Washington, D.C.; recruiting members and affiliates to participate in the conspiracy; organizing trainings to teach and learn paramilitary combat tactics; bringing and contributing paramilitary gear, weapons and supplies—including knives, batons, camouflaged combat uniforms, tactical vests with plates, helmets, eye protection and radio equipment—to the Capitol grounds; breaching and attempting to take control of the Capitol grounds and building on Jan. 6, 2021, in an effort to prevent, hinder and delay the certification of the electoral college vote; using force against law enforcement officers while inside the Capitol on Jan. 6, 2021; continuing to plot, after Jan. 6, 2021, to oppose by force the lawful transfer of presidential power, and using websites, social media, text messaging and encrypted messaging applications to communicate with co-conspirators and others.[9]

The Oath Keepers were motivated to attack the U.S. Capitol by the desire to oppose the presidential transfer of power, and they mostly communicated with each other electronically. They constructed plans to recruit personnel and gather weapons, and plotted to use force at the Capitol to accomplish their goal of stopping the transfer of power. Video evidence shows two groups of Oath Keepers entering the Capitol. In June 2022, members of another extremist group, the Proud Boys, were also charged with seditious conspiracy on the basis of similar evidence.[10] In October 2022, a former leader, Jeremy Bertino, pleaded guilty to seditious conspiracy.[11]

The guilty pleas and convictions show that the Oath Keepers fit the CAPA schema of conspiracy, with communicating agents, plans, and actions. The Federal Bureau of Investigation (FBI) and Department of Justice produced real information through methods compatible with the AIMS methodology. Rather than just making stuff up, these institutions collected information by systematic observations of the actions and communications of the Oath Keepers. Inference is required for conclusions about the motives of the Oath Keepers on January 6, but these are plausible given their past proclamations and communications. Analysis of their electronic messages provides evidence about their developing plans, and video evidence shows the execution of the plans in violent actions. The FBI and Department of Justice built up an extensive memory of the acquired information about the case, and spread of information took place among the investigators. Spread increased to the public when the trial occurred, governed by legal constraints about what kinds of evidence can be presented in court.

A pressing question is whether Trump was also part of a seditious conspiracy to retain power. Answers are being sought by a congressional committee[12] and by the Department of Justice directed by a special counsel.[13] Both institutions are using AIMS methods to show that Trump's behavior and other evidence reveal that he conspired to overthrow the election. The key question is whether his public pronouncements and tweets amounted to communications that made him part of the conspiracy to block a legal election. Trump did not directly interact with members of the Oath Keepers and Proud Boys, but they avidly followed his Twitter messages and speeches. Trump's allies Roger Stone and Michael Flynn were in direct contact with the groups charged with seditious conspiracy. Another question is whether the $250 million raised by Trump to fight what he claimed was a stolen election amounted to fraudulent deception; at the least, money raising gave Trump another motivation to continue what has become known as the Big Lie.

On August 1, 2023, Trump was indicted in Federal District Court on charges of conspiracy to defraud the United States, conspiracy to obstruct an official government proceeding, and conspiracy to deprive people of a civil right. The details of the indictment confirm that the case fits the CAPA schema of conspiracies and that the investigators have been operating with proper AIMS methods.

In chapters 4 and 5 on COVID-19 and climate change, respectively, I used the scientific community as a good example of how spread of information can be controlled to support real information. The legal system provides another example because countries have stringent rules about what kinds of evidence can and cannot be presented in court. Evidence must be relevant to the legal issue at hand, but it can be excluded if it is substantially more prejudicial than probative. Other excluded evidence concerns coerced confessions and information acquired because of privileged relations, such as attorney-client and doctor-patient roles. Legal systems provide constraints on the spread of information in the form of limitations on what kinds of evidence can be presented in court. Chapter 8 compares how the legal system regulates evidence with similar constraints in science and journalism.

The Oath Keepers attack on the Capitol exemplifies both the CAPA schema for conspiracy and the AIMS theory of information. Processes of acquisition, inference, memory, and spread produce real information about the agents, communications, plans, and actions of a conspiring group. Many other examples of real conspiracies could be analyzed to show that real information results from using AIMS processes to learn about a conspiracy. For example, the 1973 overthrow of the democratically elected government of Chile involved the Chilean military working with the U.S. government, both motivated by their desire to remove the left-wing president Salvador Allende. Communications by meetings and telephone generated plans to take over the government, leading to actions that included massive imprisonment and executions.[14]

The AIMS theory shows how to acquire real information about real conspiracies, but breakdowns can lead to misinformation about actual conspiracies. The coordinated attacks on the United States of September 11, 2001, were the results of a real conspiracy by al-Qaeda terrorists to hijack four airplanes and crash them into buildings. Ample evidence supports the existence and operation of this conspiracy, but conspiracy theorists have proposed numerous alternatives, such as that the collapse of the World Trade Center was the result of controlled demolitions planned by the U.S. government to justify invasions of Afghanistan and Iraq. Here, allegations of fake conspiracies interfere with recognition of a real conspiracy. Fake conspiracies are more commonly alleged in the absence of any conspiracy at all.

MISINFORMATION ABOUT FAKE CONSPIRACIES

The AIMS theory explains how bogus conspiracy theories qualify as misinformation. Some conspiracy theories are true, for example, the one about Caesar's assassination, but false conspiracy theories are mistaken about the supposed agents, communications, plans, and actions. These mistakes result from breakdowns in the processes of acquisition, inference, memory, and spread. Conspiracy theories frequently arise from making stuff up rather than acquisition using observation; from motivated reasoning rather than evidence-based inference; and from spread by social media rather than constrained vehicles such as scientific publications, legal proceedings, and responsible journalism. I illustrate how this works by considering two notorious conspiracy theories: QAnon and the Great Replacement theory of white genocide.

QAnon as Conspiracy Theory

QAnon originated in 2017 with a post on the website 4chan by a character Q who claimed to be a high government official with a Q-level security clearance.[15] Q's proclamations built on earlier accusations that Hillary Clinton and other Democrats were involved in a pedophile ring but added the prediction that a "storm" led by Trump would clear out the evildoing. Q's postings were picked up by other 4chan users and quickly spread to other social media, including Twitter, Reddit, and YouTube, and were also advocated by prominent personalities such as Sean Hannity, Roseanne Barr, and Alex Jones. By 2020, Facebook had thousands of QAnon theme groups with millions of members and followers.[16] A 2021 American poll found that 29 percent of Republicans believed that Trump had been secretly fighting a group of child sex traffickers that included prominent Democrats.[17] Millions of Americans as well as thousands of people in Great Britain and other countries have bought into the QAnon conspiracy theory.

The alleged conspiracy fits the CAPA schema. The accused agents are Democrats such as Hillary Clinton and other elites such as Hollywood actors, who are presumed to have communicated with each other to make plans to traffic children for sexual exploitation, resulting in despicable actions. The QAnon conspiracy theory sometimes merges with COVID-19 conspiracies, for example, with the claim that the pandemic was a way of covering up pedophile activity.[18]

QAnon as Misinformation

The astonishing success of QAnon cries out for explanation, which the AIMS theory provides. Real information results from appropriate processes of acquisition, inference, memory, and spread, whereas misinformation results from failures of these processes, which occurred luridly with QAnon.

As the Caesar and Oath Keepers examples illustrate, acquisition of real information about conspiracies requires collecting and representing through perception and systematic observation. None of that has occurred with QAnon because no one has presented observational evidence that Democrats collaborate to traffic children for sexual abuse. Pedophile rings do exist, and the tragic evidence of their operation can be acquired through witnesses and cracking encrypted video files on computers. No such evidence has been brought forward for the pedophile activities claimed by QAnon. Instead, the claims about a Democrat trafficking ring and Trump's moves against it have arisen purely by making stuff up, a gross violation of the norms of acquisition. These claims are generated by motivated invention in which evidence-independent imagination generates emotionally appealing conjectures.

Many other tragic cases of sexual abuse of children have been revealed, with acts committed by perpetrators that include priests, teachers, and scout leaders; but evidence for such abuse comes convincingly from testimony by the abused. For example, thousands of cases of sexual assault occurred in the Canadian residential schools for Indigenous children, as revealed by subsequent testimony.[19] In contrast, no victims have come forward to substantiate the accusations of QAnon. Trump has refused to repudiate QAnon, whose support he values, but he has not supported the claim that he was trying to combat a pedophile ring. QAnon predictions about Trump's actions have turned out to be false without undermining support for QAnon.

The QAnon conspiracy theory requires inferences about the motivations of the alleged pedophiles. Around 1 percent of adults are sexually attracted to children, but not all of them act on that attraction and only a small portion form pedophile rings to traffic victims.[20] QAnon has provided absolutely no evidence that the Clintons and other Democrats are motivated by pedophilia. The inferences made by QAnon are clearly not based on evidence, and I will examine shortly how they arise from motivated reasoning and other biases.

Testimony is often a reliable inferential source of information. If a trusted friend tells you that a politician is corrupt, you can infer that the politician is corrupt because your friend probably said it on the basis of good evidence. But such testimonial inferences do not apply to Q because Q's identity is unknown, Q's sources are unknown, and Q's predictions have often turned out to be wrong. When Q says something that is repeated on social media, the saying and repeating provide absolutely no reason for believing it.

Memory processes for QAnon also violate appropriate standards for storing and retrieving information. Storage for Q's pronouncements and extensions of them consists of the vast number of messages on social media, which are placed with no concern for standards of relevance or accuracy. Similarly, when advocates of QAnon retrieve such messages, they do so with attitudes of adulation and faith rather than critical examination.

The mechanisms for spread of information about QAnon are similarly immune from critical evaluation. Sending occurs by posting to social media without any of the critical scrutiny found in scientific peer review, legal standards of evidence, or ethical journalism. People post to social media on the basis of considerations of interest rather than truth, so spread is utterly unconstrained, although some social media sites have tried to cut back on QAnon activity. Spread of QAnon misinformation results from the same information incontinence that operates with COVID-19 and climate change. QAnon is insufficiently organized to qualify as an institution, but spread of its conspiracy theories has benefitted from organizations that include 4chan, 8kun, and several social media companies.

QAnon illustrates how misinformation need not originate by making stuff up but can rely on prejudiced conspiracy theories that circulate in popular culture. QAnon's anti-Semitism, evident in its attacks on George Soros and the Rothschilds, picks up on ancient attitudes. In a similar way, QAnon did not originate theories about pedophile rings but intensified their communication.

So the AIMS theory of information provides a plausible account of QAnon misinformation as resulting from defective acquisition, inference, memory, and spread. Why then do so many people believe in QAnon?

QAnon Adherence

Psychologists have identified characteristics that dispose people toward conspiracy theories, including financial insecurity and suspicious personality.[21] QAnon supporters are inclined to adopt other conspiracy theories such as that climate change is a hoax, but we need to look at specific mental mechanisms that lead people to adhere to QAnon. I propose that a main factor is motivated reasoning mingled with identity considerations, as described in chapter 3. Other emotional distortions including fear-driven inference also contribute, as do deference to bogus authorities and religious patterns of thinking.

Motivated reasoning, when people reach conclusions that align with their personal goals rather than evidence, is common in politics.[22] The motivations of QAnon believers are clear from their political affiliations as Republicans and supporters of Trump. They accordingly want to believe that Democrats are evil and capable of pedophilia. Testimony by Q and supporters provides a whiff of evidence that supports the motivated reasoning about pedophilia. In addition, the desire to see Trump as a strong and virtuous candidate provides support for the belief that he was fighting the pedophile ring.

These motivated conclusions fit the personal goals and group identity patterns of motivated inference. QAnon supporters strongly see themselves on the side of Republicans and Trump, and view Democrats as the evil opposition. Thus, they are motivated to accept views that align with their own side, with a strong sense of themselves as Republicans and Trump supporters, which are identities based on the self-phenomena described in chapter 3. These phenomena include concepts such as *conservative*, evaluations such as *self-respect*, and motivations such as *stability*. Motivated collective cognitions serve to defend the in-group, justify the power aspirations of the in-group, and manage anxiety by explaining unexpected events.[23]

Motivating self-concepts are captured by the cognitive-affective map in figure 6.3, which shows how supporting QAnon fits with other values that are central to the identities of some conservatives. The result is strong emotional coherence with QAnon based simultaneously on what the supporters value and how they like to think of themselves, including strong opposition to Democrats and pedophiles. The resulting picture has little connection with facts but a strong emotional resonance.

The cognitive-affective map in figure 6.3 depicts what QAnon supporters like and dislike, but it does not show specific emotions. Such intense conservatives may not just dislike Democrats, elites, and pedophiles but actively hate them, and they may feel angry toward them for their alleged evil actions. Some of them not only liked Trump but even adored him for his near-messianic actions. Fear can also be a powerful motivator, and it would be natural to be afraid of Satanic pedophiles. Pride is a positive emotion, but it can motivate people to think worse

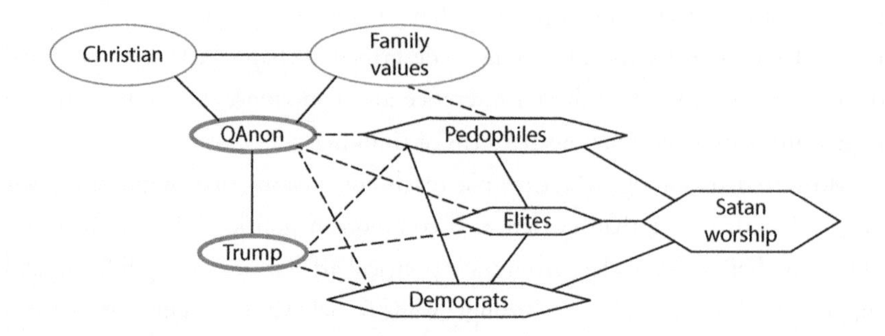

6.3 Cognitive-affective map for QAnon supporters. Ovals indicate positive values; hexagons indicate negative ones. Solid lines show emotional support; dotted lines indicate incompatibility. Not shown are additional links, such as incompatibility between Christianity and Satan worship.

of people who do not belong to the group identity that inspires pride, such as white supremacy. Hence, motivated reasoning about QAnon claims sometimes fits the emotions pattern.

The positive emotional values associated with QAnon and Trump explain why their supporters are inclined to make motivated inferences that ignore evidence, but a terrifying complex of negative values shows the potential operation also of fear-driven inferences that support scary beliefs that no one would want to have. Motivated inference supports the association of Democrats with Satanic pedophiles, but the basic existence of Satanic pedophiles gets support from fear-driven inference. Satanic pedophilia is so scary that it attracts attention and thereby gets credibility that far exceeds the evidence for it. Another unpleasant but strongly motivating emotion is disgust, to which conservatives are often sensitive.[24] Being intensely disgusted about Satanic pedophilia in Washington is not pleasant but it attracts attention that can be misinterpreted as plausibility.

Another source of adherence to QAnon is advocacy by false authorities. We depend on other people for most of our information, so testimony-based inferences are often legitimate. Most of our beliefs come from inferring that a claim is true because it comes from a trusted friend or journalistic source. But false authority is a fallacy where the source cannot be trusted to be right about the issue at hand, for example, when we take seriously a celebrity endorsement of a medical treatment. QAnon got rolling because nonexperts were eager to build a support network around Q's vague pronouncements, and people who were already following these nonexperts on 4chan and social media were ready to believe them because the view fit with their personal goals. Adherence then results from a lamentable combination of motivated reasoning and belief in bogus authorities.

Puzzlement about the gullibility of QAnon supporters reduces with realization that many of them are religious and therefore already accustomed to believing much on faith.[25] As a child, I was raised Catholic and dutifully recited the Apostles Creed:

> I believe in God, the Father Almighty, Creator of Heaven and earth; and in Jesus Christ, His only Son Our Lord,
>
> Who was conceived by the Holy Spirit, born of the Virgin Mary, suffered under Pontius Pilate, was crucified, died, and was buried.
>
> He descended into Hell; the third day He rose again from the dead;
>
> He ascended into Heaven, and sits at the right hand of God, the Father almighty; from thence He shall come to judge the living and the dead.
>
> I believe in the Holy Spirit, the holy Catholic Church, the communion of saints, the forgiveness of sins, the resurrection of the body and life everlasting.

I believed these assertions, even though only a few of them, such as the existence of Jesus Christ, had any evidential basis. My belief arose because of the influence of my parents and religious teachers, whom I had no reason to doubt.

None of the books and articles on misinformation that I have read discuss religion as an important source, perhaps because of reluctance to upset religious people or to violate legitimate freedom of religion. But religions encourage misinformation in three ways. First, they generate claims that people are expected to believe based purely on faith; because religions contradict each other, these claims cannot all be true.[26] Second, the operation of faith discourages uses of more reliable methods such as observation and evidence-based inference. Third, religions have leaders accepted as authorities who get to propagate dogmas that people erroneously accept as trustworthy testimony. Hence, religion not only spreads religious misinformation: it encourages ways of thinking, such as uncritical acceptance of false authorities, that help the spread of other kinds of misinformation.

Another source of advocacy of QAnon ideas is thematic merchandise such as the QAnon T-shirts worn by some rioters at the U.S. Capitol on January 6, 2021.

After that event, Amazon took down numerous QAnon merchandisers who previously had been selling items that included coffee mugs, bumper stickers, and books. Similarly, misinformation merchandise contributed to COVID-19 denial because of the large stake that natural remedy purveyors have in making money from alternative medicine. The Freedom Convoy truckers protest that shut down Ottawa for weeks in February 2022 was a real conspiracy against the Canadian government that challenged democracy while spreading misinformation about vaccines. The protest quickly spawned a flood of merchandise such as hats and flags that people could use to express their allegiances and identities. The Infowars website of Alex Jones promulgates multiple conspiracy theories while making money from sales of questionable nutritional supplements.[27] Examination of the motivations of misinformation spreaders should include how conspiracy theories and other lies can be used to make money for conspiracy entrepreneurs.[28]

In sum, adherence to the QAnon conspiracy theory results from many sources, including motivated reasoning, fear-driven inference, false authorities, religious backgrounds, entrepreneurial activities, and misleading institutions. For each individual, several of these influences can intermingle to produce the result that QAnon makes sense to that person, where making sense combines cognitive and emotional coherence. Conspiracy theories can therefore be appealing without any factual basis.

I have shown how the QAnon conspiracy illustrates four kinds of motivated reasoning that fit personal goals, group identity, emotions, and invention patterns. The other two patterns also apply. Motivated perception about crowds can operate when they are estimated to be larger than they really are or when they are claimed to be more peaceful than videos reveal. Motivated ignorance was apparent after the January 6 riot when Republicans strongly resisted calls for congressional and judicial investigation of its causes. The congressional inquiries into the January 6 uprising were broadcast by major American news networks except for Trump-supporting Fox News, another instance of motivated ignorance. Hence, the QAnon case supports recognizing motivated reasoning as a major source of belief in fake conspiracy theories.

Promulgaters of conspiracy theories are sometimes remarkably candid about what they are doing. Trump said in a speech: "If you say it enough and keep saying it, they'll start to believe you."[29] His supporter Steve Bannon proclaimed: "The real opposition is the media. And the way to deal with them is to flood the zone with shit."[30] Repetition and a torrent of misinformation can overwhelm evidence and evaluation.

The Great Replacement Theory

European and North American countries are undergoing demographic changes because of low birth rates and immigration of nonwhites. This trend is a sociological fact, not a conspiracy theory, but conspiracy theory enters the picture with claims that elites plot with immigrants to accelerate these changes. The French author Renaud Camus called the trend toward increasing numbers of Muslim immigrants in his country the Great Replacement (*le Grand Remplacement*).[31]

Right-wing politicians and journalists in other countries have expressed similar fears. For example, U.S. Republican members of Congress such as Paul Gosar and the Fox News host Tucker Carlson maintained that President Joseph Biden and other Democrats want to increase immigration from the third world to reduce the political power of "people whose ancestors live here."[32] Replacement theories have motivated terrorist killings in Norway, New Zealand, and the United States.

Replacement theory has ancestors in other conspiracy theories that viewed a privileged group as threatened by minorities, for example, the anti-Semitic claims discussed in chapter 7 about Jewish plots to run the world. Current replacement theories have different targets depending on perceived threats in different countries, with French advocates concerned mostly about Muslims and American advocates concerned mostly with Hispanic immigrants. What they have in common is the claim that leftist politicians work with nonwhites to reduce the power of traditional white majorities.

These claims fit the CAPA schema for conspiracies because they allege that agents communicate with each other to make plans to carry out actions. The agents are the elites and immigrants who are motivated to reduce white power. Their forms of communication are not specified but meetings and other contacts can easily be imagined. The agents are accused of making plans to increase immigration and nonwhite population growth, and white women are criticized for not having enough children. Actions resulting from the conspiracy are supposed to include relaxation in immigration rules, for example, in the U.S. treatment of illegal immigrants from Mexico.

As the Julius Caesar and 9/11 examples illustrate, real information about conspiracies requires evidence that agents actually communicated with each other to produce plans aimed at actions. In contrast, replacement theories are like QAnon in presenting no evidence for the relevant agents, communications, and plans. Demographic changes are undoubtedly occurring in Europe and North America, but there is no evidence that they result from antiwhite conspiracies.

Canada has one of the world's highest rates of immigration, with foreign-born people making up around a fifth of the current population. The major sources are India, China, the Philippines, and Nigeria, so the proportion of white people is declining.[33] But even conservative parties in Canada recognize immigration as a major source of Canada's growth and prosperity, which otherwise would decline because of a low birth rate. Canada stringently admits immigrants on the basis of education, work skills, family connections, and refugee status. No conjectures about communicating plotters and antiwhite plans are needed to explain the dramatic changes in Canadian society, and replacement conspiracies are similarly implausible in France, the United States, Austria, and other countries where they are common. I was appalled by a 2022 survey that found that 37 percent of Canadians agree with a form of the replacement theory.[34] Nearly half of U.S. Republicans also support it.[35]

The popularity of replacement theories comes partly from uncritical operations of memory and spread because people hear about them from social media and right-wing news sources. But the basic question is, Why do some individuals find the idea that a conspiracy is working to replace white people appealing? As for QAnon, candidate psychological processes for adoption of replacement theories include motivated reasoning, fear-driven inference, other emotional inferences, and false authorities. I have not noticed much entrepreneurial activity favoring replacement claims, although the Infowars website hawks nutritional supplements side-by-side with rants about immigration.

What are the main motivations that promote adoption of replacement theories? The dominant racist goal of their advocates is to maintain white supremacy, in keeping with the belief that the white race is different from and superior to other races. This goal often meshes with the religious goal to maintain dominance of Christianity in the face of threats from Islam and other religions. Because immigration is viewed as a threat to the white race and its Christian religion, the goal is to block actions such as increasing immigration. Advocates of the Great Replacement theory are not motivated to believe in replacement because they do not want that outcome. But they are motivated by dislike of elites and minorities to believe that elites and minorities are conspiring to produce the replacement.

The group identity pattern of motivated reasoning operates powerfully to support the replacement theory because its adherents strongly identify themselves as white and Christian and want to see these groups retain social and political dominance. Group identity spills over into the personal goal pattern of motivated reasoning because individuals want to feel good about themselves and to be confident that they will not be replaced in their jobs and communities. Specific emotions also contribute to motivated reasoning when fear and hatred of immigrants and nonwhites drive people to believe that violence should be used to stop the replacement. When you hate someone, you tend to want and expect that bad things will happen to them.

Other patterns of motivated reasoning can also support the replacement theory. Motivated invention is behind generation of the claim that elites, politicians, and minorities form a conspiracy. Motivated perception may operate when people claim to observe large increases in criminal activities. Failure to recognize the substantial contributions that immigrants make to their new countries is supported by motivated ignorance.

Belief in replacement as a statistical fact results partly from actual demographic changes but also from fear-driven inference. The scariness of nonwhite ascendancy generates so much attention that gullible people find themselves focusing on it and thus believing that it is true. Fear drives attention, which drives belief. Anxiety has long been a force behind white genocide theories expounded by Nazis and neo-Nazis, who blamed Jews for plots to make the white race extinct through immigration, interracial sex, low birth rates, and abortion. Additional negative emotions directed toward nonwhites include hatred, anger, and contempt, all of which intensify the perceived threat and increase the motivation to stop the downward slide in the numbers of whites.

The internal coherence of this appalling white-supremacy worldview is displayed in the cognitive-affective map in figure 6.4, which depicts whites as superior in intelligence, morality, and culture to nonwhites. Among racist whites, this configuration provides a sense of identity that is threatened by immigration. In Europe, some white nationalist movements are called "identitarian" (*identitaire, Identitäre*) because of the desire to maintain what they take to be traditional identities. White supremacists see immigration and higher immigrant birth rates as challenges to the self-representations and self-evaluations crucial to the emotional significance of personal identities. White supremacists not only see themselves as white and Christian; they also place a strong positive evaluation on

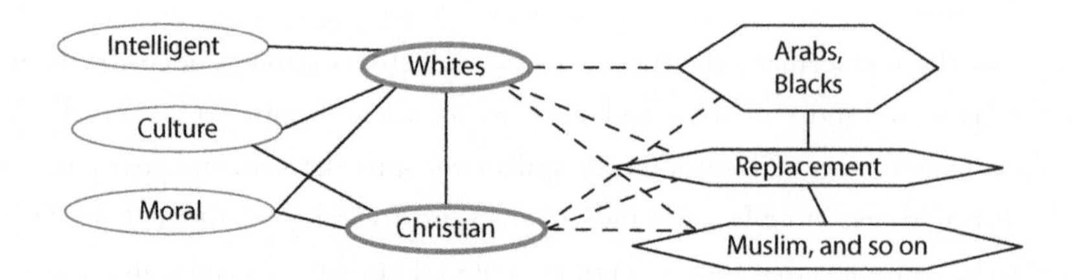

6.4 Cognitive-affective map of white supremacist advocacy of replacement theory, using the same conventions as those shown in figure 6.3.

these characteristics and have no desire to change. Replacement takes on a strong negative value because of association with nonwhites and its conflict with values of being white and Christian.

Another contributor to acceptance of replacement conspiracy theories is false authorities. People are more likely to take the conspiracy seriously if they hear from celebrities such as Camus and Carlson that replacement of whites by nonwhites is a serious threat encouraged by elites. When people learn about replacement theories from their favorite Facebook groups or Twitter feeds, they may mistake the messengers for people who know what they are talking about. Institutions that have contributed to replacement theories include political parties and media companies.

Replacement conspiracy theories are like the QAnon theory in compensating for their utter lack of observational evidence by appeal based on motivated reasoning, fear-driven inference, other emotional reactions, and false authorities that encourage the spread of inflammatory claims. A full treatment would show that dozens of other conspiracy theories currently in circulation are subject to the same analysis. Disabusing believers of these conspiracy theories is difficult because of the powerful cognitive, emotional, and social mechanisms that support them. Nevertheless, we can consider how reinformation strategies might work for debunking conspiracies.

REINFORMATION ABOUT FAKE CONSPIRACIES

For QAnon, Great Replacement, and other conspiracy theories, the reinformation problem is structurally the same. Claims are made that fit the CAPA schema with allegations about communicating agents who plan to carry out frightening actions. Challenging these claims can use the eight techniques recommended for misinformation about COVID-19 and climate change.

Detect Misinformation

The indispensable first step in converting misinformation about conspiracies into real information is to recognize false claims, which requires distinguishing fake conspiracies from real ones. The CAPA schema points to the crucial questions about the reality of the alleged plans and actions of the suspected communicating agents.

Proving definitely that no conspiring agents are operating is difficult because it requires establishing the universal negative that no plotters exist. This would logically require showing for all relevant people that they are not conspiring, through a general inquiry that eliminates the conspiracy. For example, conclusively refuting QAnon would require showing that all Democrats are not Satanic pedophiles. Refutation is easier when a specific accused is mentioned such as Hillary Clinton or George Soros, where we can point out that they have been public figures under journalist scrutiny for decades with no observed connections to pedophiles. But both QAnon and replacement theories imagine unspecified conspiring agents whose existence cannot be exclusively refuted.

Even without such refutation, however, we can have good reasons for thinking that a proposed conspiracy theory is false because evidence is totally lacking that agents are communicating to make plans. The logical burden of proof is not on skeptics who deny the existence of the conspiracy but rather on the conspiracy theorists, who should be able to provide evidence about the existence and activities of a group of agents. Evidence could take various forms such as eyewitness testimony, documents, videos, or audio recordings. In my book *Natural Philosophy*, I proposed an *existence procedure* according to which we should believe in the existence of entities if and only if they are proposed by hypotheses that are most coherent with all the available evidence.[36] Similarly, let me propose here the *nonexistence principle*:

> *IF* (1) evidence for the existence of something is lacking, (2) serious attempts have been made to find evidence for it, and (3) it is the sort of thing for which evidence should be findable, *THEN* we can reasonably conclude that it does not exist.

For instance, we can reasonably conclude that unicorns do not exist because people would love to observe them but no one ever has.

One response to such skepticism is to invoke the principle that absence of evidence is not evidence of absence, but this principle fails when evidence for something has been pursued but not found. For example, nineteenth-century scientists tried to find evidence of the existence of an ether through which light waves were supposed to travel, but the failure of their attempts led naturally to the conclusion that no such ether exists. The QAnon and replacement theorists have not even attempted to find evidence for the existence of the conspiracies they conjecture, so we need a slightly different principle: if people who have made claims about existence have made no attempts to find evidence, then the claims are probably false. Contrast their approach with my Caesar and January 6 cases, where investigators have worked hard to find evidence of agents engaged in conspiracies.

A particularly easy way to spot misinformation about conspiracies is when people hold contradictory versions, for example, people who believe both that Princess Diana was murdered and that she faked her own death. The online *Conspiracy Theory Handbook* warns of other traits of conspiracy theories: overriding suspicion of official accounts, nefarious intent ascribed to conspirators, persecuted victims, immunity to evidence, and the tendency to reinterpret random occurrences as intentional.[37] All of these can be used as suggestions of fake conspiracies, although real conspiracies can also have victims who suffer from nefarious intent, for example, the 9/11 terrorist attacks.

Misinformation can be harder to spot in what Russell Muirhead and Nancy Rosenbaum all the "new conspiracism."[38] They describe how Trump as president remained vague about who conspired to rig elections, supporting his claims only by repetition and the catchphrase that "a lot of people are saying." Nevertheless, many of his Republican followers remain convinced that the 2020 election was rigged despite numerous judicial inquiries that found no wrongdoing. Not having specific claims about agents, communications, plans, and actions makes identification of fake conspiracies even more difficult, but the claims can be challenged when translated into allegations about bad election practices in particular constituencies.

Identify Sources and Recognize Motives

The sources of conspiracy theories are sometimes obvious, as in Trump's vociferous claims about a stolen election and Renaud Camus's publications about the great replacement. But sometimes the originators of conspiracy theories are anonymous such as Q and the authors of the anti-Semitic *Protocols of the Elders of Zion*.

Either way, the motives of conspiracy theorists are usually evident from their rhetoric. QAnon advocates are pro-Trump and anti-Democrat. Replacement theorists are prowhite and anti-immigrant. The writing and speaking of right-wing conspiracy theorists usually make it apparent how the suggestion of conspiracy supports their conservative political goals. For example, the claims of replacement theorists support their goals to restrict immigration.

Motives by themselves do not disqualify claims as misinformation because sometimes truths are arrived at by people with evil intent, for example, when Nazi rocket scientists learned much science while trying to build weapons. Nevertheless, identifying motives is valuable for explaining the adoption of beliefs lacking in evidence. Such beliefs can be supported by motivated reasoning, fear-driven inference, and other emotional effects. Different motives require different reinformation strategies. If beliefs are strongly motivated by strong personal goals, especially ones based on a strong sense of identity, then factual correction and critical thinking strategies are likely to fail. The more empathic strategy of motivational interviewing may be more useful in dealing with a complex of emotions. Psychotherapy might have more chance of disrupting the racist hatred of replacement theorists than cold logic. Identifying motives is not a direct strategy for changing minds but serves as a useful preliminary for actual change.

Factual Correction

Straightforward factual correction is unlikely to work with conspiracy theories that propose mysterious agents and communications because we cannot exhaustively show that such conspiracies do not exist. But some conspiracies are refutable because they make assertions that are easily shown to be contradicted by well-established evidence. For example, the Pizzagate conspiracy theory that helped to inspire QAnon claimed that a pedophile ring was operating in the basement of a specific Washington pizza restaurant that turned out not to have a basement. Claims that the Holocaust was made up by Jews are easily refuted by thousands of eyewitness testimonies and photographs. The vague lie that an election was rigged is hard to refute, but specific allegations about voting irregularities such as corrupted voting machines can be disproven, as happened in dozens of cases after the 2020 U.S. election. Thus, some aspects of conspiracy theories can be reinformed by merely getting the facts right.

Critical Thinking

Correcting conspiracy theories usually requires a more complex approach consisting of error detection plus remedial reasoning. I presented motivated reasoning, fear-driven inference, and false authorities as the most important error tendencies that support the conspiracy theories I analyzed, but other psychological biases and logical fallacies can also contribute. Here are some examples, with pointers to the kinds of reasoning that can provide a remedy for the misinformation generated.[39]

The *clustering illusion* is the tendency to see nonexistent patterns in random events. Conspiracy theorists find patterns in unconnected events, such as when replacement theorists assign a common cause to low birth rates among whites and high birth rates among nonwhites. The remedy for such thinking is better statistical inference and causal analysis.

Vividness is the tendency for information that is particularly salient or emotionally charged to be given undue influence, for example, when replacement theorists take a large family of Muslims as a sign of an overall plan. The remedy is to consider all evidence independent of its emotional impact.

Confirmation bias is the tendency of people to seek information that supports their views and to ignore information that contradicts them. Conspiracy theories look only for evidence of the conspiracy and ignore alternative explanations. The remedy is to follow the standards of explanatory inference that require considering all the relevant evidence and alternative hypotheses.

Overconfidence is the tendency to be too sure of assumptions and opinions, as conspiracy theorists usually are despite scant evidence. The remedy is to make confidence proportional to the amount of evidence and the quality of inferences drawn from it.

False cause is the tendency to infer that two events are causally related just because one happened after the other. Conspiracy theorists, for example, might see a meeting between world leaders as a cause of a migration crisis that takes place around the same time. The remedy for bad causal inference is to consider alternative explanations, including chance.

Pointing out these thinking errors and their remedies might work with conspiracy theorists who like to view themselves as critical thinkers. But an alternative could be directed more toward emotion than cognition.

Motivational Interviewing

The *Conspiracy Theory Handbook* recommends talking with conspiracy theorists using empathy rather than ridicule to encourage their open-mindedness.[40] This suggestion is compatible with the technique of motivational interviewing, which advocates this procedure:

- Understand people's concerns about a conspiracy by asking them open-ended questions and empathizing with their fears and insecurities about political threats.
- Be affirmative, reflective, and nonjudgmental about their concerns.
- Identify discrepancies between people's current and desired behaviors such as being fair and dealing with immigrants.
- Summarize the political issues and inform people while respecting their autonomy.

This approach can try to grasp what makes conspiracy theories appealing and generate opportunities for subtle alterations of emotional commitments. Drawing a value map might help to understand the background assumptions and goals of the conspiracy theorist.

Motivational interviewing can also take into account that endorsement of conspiracy theories is more common among people who feel powerless and vulnerable.[41] Empathic conversation can try to identify the origins of these feelings, for example, in economic insecurity, and to understand how blaming elites might help people to deal with threats such as unemployment. The goal is to edge people away from emotions such as fear, anger, and hate toward greater confidence and hope. The *Conspiracy Theory Handbook* also recommends approaching conspiracy theorists via trusted messengers, former members of their extremist community who might be taken more seriously.

Another part of motivational interviewing could be getting at the identity issues that provide much of the impetus for belief in conspiracies. What goes into the self-representations and self-evaluations of people who strongly think of themselves as white and Christian and therefore threatened by immigrants and Muslims? Changing people's identities is hard, but there might be ways of moving them away from conspiracy-laden fears, for example, by encouraging the more egalitarian aspects of Christianity, as in "we are all God's children."

Institutional Modification

Some institutions contribute to the generation and spread of false conspiracy theories. QAnon and replacement theory have benefitted from websites such as 4chan, 8kun, and Infowars. Both theories have been spread by posters on social media platforms such as Twitter and Facebook. Right-wing political parties in France, Hungary, and other countries have adopted and promulgated replacement theories in support of advocacy of white supremacy.

For most of these institutions, abolition is probably the best hope because changing their policies, values, and norms is unlikely to work. Social media companies can be lobbied to encourage them to limit the transmission of conspiracy theories more strictly.

On a more positive note, other institutions can be used to discourage conspiracy theories that are based on making stuff up. Educational organizations such as universities can highlight the differences between bogus conspiracy theories and real information about real conspiracies. Multilingual fact-checking services can push back against misinformation that flourishes beyond English.[42] Police departments can investigate groups who use conspiracy theories to organize antidemocratic movements. Legal institutions can prosecute criminals who use conspiracy theories to foment violence. Lawsuits can punish harmful claims such as Alex Jones's allegations that the 2012 Sandy Hook school massacre was a hoax; he now owes hundreds of millions of dollars in damages to maligned parents.[43] In extreme situations, military forces may be required to control seditions based on false conspiracies.

Political Action

The psychological approaches of critical thinking and motivational interviewing, and the social approach of institutional modification, can be supplemented by two sorts of political action. First, election activity is crucial to keep conspiracy theorists from influencing those in power. Conspiracy theorists encourage violence and oppression of groups they think are behind the perceived dangers. Political leadership and police control are sometimes needed to prevent harm to the vulnerable.

Second, political activity needs to be directed against the rampant spread of conspiracy theories on social media, which have been lax in allowing the proliferation of false but dangerous views. U.S. policymakers have been looking at how to change the law that shields tech companies from lawsuits over their posts, and another tactic is to create a new digital safety bureau that regulates digital platforms.[44] Such changes are necessary to slow down the uncontrolled spread of toxic ideas on social media. New threats arise continuously from new technology such as the Telegram platform, which is heavily used by far-right groups, jihadists, and child pornographers.[45] Other new platforms awash in misinformation include Trump's Truth Social, Gab, and Parler.[46] QAnon accounts banned from social media have migrated to Truth Social.[47] Chapter 8 reviews ways of combating misinformation that is spread on the internet and social media.

One other possible strategy for combating conspiracy theories is cognitive infiltration of extremist groups.[48] Government agents could enter online networks covertly to undermine conspiracy theories, but the use of deception to overcome misinformation is ethically problematic. Lying is not always wrong because sometimes it prevents harm, but truth should be a dominant value, as chapter 9 contends.

OTHER POLITICAL MISINFORMATION

Politics is rife with other kinds of misinformation besides conspiracy theories, notably deception and propaganda. Deceptions are carried out by partisan operators and by governments for military and other purposes. Deceptions are disinformation because the propagators know that what they are saying is false, in contrast to conspiracy theories which are usually believed by those who spread them. Propaganda is communication intended to influence an audience and it may be true or false, so it is not always misinformation, but propaganda often consists of flagrant disinformation.

Deception

Deceptions can be carried out by particular politicians or by whole governments. After Trump lost the 2020 U.S. election, he continued to maintain that the election was stolen by fraudulent votes, a response that has been labeled the Big Lie. This response is not obviously a conspiracy theory because Trump never points to a group of agents who communicated to pull off the stolen election, although others blame the stolen election on George Soros. From Trump's perspective, the Big Lie may not even count as deception or disinformation because Trump may actually believe he won the election through a combination of motivated reasoning and utter disbelief that he could have lost to "sleepy" Joe Biden. However, numerous Republicans want Trump's support, so they repeat the Big Lie even though they know that his defeat was legal. They are thus guilty of deception and disinformation. Other nonconspiracy political lies include the claim that former president Barack Obama was not born in the United States.

Politicians have often used lies to pursue their interests.[49] Joseph Stalin concocted lies about treasonous activities to carry out a massive purge of officials in the late 1930s. Richard Nixon cascaded lies while trying to survive the Watergate scandal of 1972. In the 1950s, Joe McCarthy and Herbert Hoover exaggerated the extent of communist activity in the United States to increase their own power. Vladimir Putin prepared for Russia's invasion of Ukraine in 2022 with a series of lies about Ukraine's history, which amounted to propaganda as well as misinformation and disinformation.

The most striking governmental use of deception is for military purposes.[50] In World War II, the Allies carried out a systematic deception to convince the Germans that the D-Day invasion would take place at Calais rather than the beaches of Normandy. This deception shows how disinformation can be morally justifiable because the Allies' trick delayed Adolf Hitler from responding to the actual invasion, saving many lives and facilitating the defeat of an evil government. In wartime, governments can also be justified in concealing some news from the population to maintain morale that is crucial for the war effort. With less ethical justification, the Soviet Union and the United States waged disinformation campaigns against each other for decades.[51]

Another complicated war deception is a "false flag" operation, where one side pretends to be the other to deflect responsibility. For example, the Germans started World War II by invading Poland, but they faked an attack by Poland against Germany. Even more complicated are conspiracy theories about false flag operations such as the ridiculous claims that the 9/11 attack was actually a false flag operation by the United States or that the January 6, 2021, attack on the U.S. Capitol was a false flag operation intended to undermine Trump. These cases are conspiracy theories about fake deceptions.

Such lies and deceptions fit with the AIMS theory of information, misinformation, and reinformation. The deceptive claims do not result from collecting information by reliable means but instead from using motivated invention in the service of making stuff up. Their spread exemplifies motivated sending in which one side transmits information to trick the other, resulting in misinformation. Reinformation occurs most easily by factual correction in which evidence reveals the falsity of the deception, for example, when the faking of the alleged 1939 Polish attack on Germany was revealed.

New methods are becoming available for detecting deception. Researchers used linguistic analysis of Trump's tweets to develop a model of the types of words he used when he was being deceptive. This linguistic analysis was shown to be 77 percent accurate at identifying his deceptions.[52]

Propaganda

In contrast to specific lies and deceptions, propaganda is the more systematic attempt to change the attitudes and behaviors of a group by means of targeted communications.[53] These communications can include press reports, radio and TV announcements, movies, visual arts such as flags, photography, music, literature, and events such as rallies and concerts. Propaganda increasingly uses internet social media such as Facebook.

Besides political groups, propaganda can serve religious, economic, moral, and social interests. Propaganda is often a mixture of real information and misinformation, for example, when a country advertises itself by a combination of facts and fantasies. All countries engage in propaganda to some extent, but Nazi Germany took it particularly seriously, with Joseph Goebbels serving as minister of propaganda for twelve years.

Acquisition of real information may contribute to propaganda, as when a country brags accurately about its degree of prosperity. But the goal of propaganda is not accuracy but rather change in attitudes and behavior by whatever means are available. Propaganda does not aim to educate its targeted group but rather to manipulate it by emotional associations and appeals. For example, the 2022 Russian invasion of Ukraine was accompanied by claims that Ukraine was dominated by neo-Nazis (see chapter 8).

People can be manipulated by propaganda when it spurs them to engage in motivated and fear-driven inferences rather than ones based on evidence. For example, if Russians are motivated to believe that the invasion was justified, then they can more easily believe Ukrainians are neo-Nazis, even though Nazis have little influence in Ukraine.

The AIMS theory of information is useful for understanding propaganda because it highlights the breakdowns in proper information mechanisms that make propaganda effective. Propagandists prefer making stuff up to reliable methods of acquisition and inference. Memory and spread operate in propaganda without the constraints provided by evaluations of accuracy. Propagandists are content to store and send whatever information produces the desired changes in the target population, regardless of whether it is false or misleading. Thus, propaganda has often led to negative social effects such as tyranny, racism, and other forms of inequality.

The AIMS account of reinformation provides hints about how propaganda can be combatted by the same techniques used against conspiracy theories. The targets of these strategies are not the generators of propaganda, who already know that much of what they are saying is false, but instead the people at whom the propaganda is aimed. Propaganda recipients can be inoculated or deprogrammed by identifying misinformation and the motives of the propaganda purveyors, and by practicing factual correction, critical thinking, motivational interviewing, and political action. I will illustrate this with reference to the Russian propaganda campaign that targeted Ukrainians and others during the 2022 invasion. Chapter 8 contains a more complete analysis of misinformation in that war.

The first step in tackling propaganda is to identify misinformation in it, for example, the false Russian claims that two eastern Ukraine republics are independent countries. Identification comes as the result of noticing contradictions with known facts, such as the history of Ukraine. Propaganda can also be suspected when the source has a history of spreading misinformation, as in the Russian trolls on Twitter who supported QAnon.[54] Second, propaganda is more easily spotted when we can detect the motives of its spreaders, for example, the desire of Putin and other Russians to reestablish the territories of the Soviet Union and the old Russian Empire.

If people are in danger of succumbing to propaganda misinformation, factual correction is the first line of defense when claims are clearly wrong. For example, it is easy to establish the extent to which the government of Ukraine is influenced by neo-Nazis by noticing that the most right-wing party, Svoboda, won only one seat in the Ukraine parliament in 2019.

A more complicated way to help victims of propaganda can resort to critical thinking, pointing out to them that they have been manipulated into succumbing to standard thinking errors. People should be able to spot how propaganda has exploited their desires and fears, for example, by exploiting the desire to avoid war and thus convince people that Ukraine could easily dispense with the contested Donbas region. Propaganda may also exploit other error tendencies such as false cause to draw people into bad inferences, for example, believing the propaganda lie that the Russian invasion was prompted by Ukrainian killing of Russians.

Propaganda victims could be helped more delicately by motivational interviewing that empathically helps them to identify ways in which their beliefs and values have been illegitimately altered by propaganda. Perhaps people could draw a value map of the emotion-laden concepts that propaganda has conveyed to them. Nudging rather than arguing might then help people to recognize and cast off the effects of propaganda.

Institutional modification can also fight propaganda. The most extreme move is to shut down dangerous propaganda distributors such as Nazi parties. Propaganda can be countered by reinformation organizations, such as responsible news media and government agencies. People in countries that spread propaganda can be encouraged to criticize lying leaders.

Political action can be a useful response to propaganda by impeding organized attempts to control opinion. Blocking propaganda may require limitations on free speech, such as wartime censorship practices, or just establishing expectations on social media that they restrict activities of individuals and groups who are spreading propaganda. In 2022, media platforms including Facebook, Apple, Twitter, Google, Instagram, TikTok, and YouTube took steps to lessen Russian influence.[55]

BEYOND POLITICAL MISINFORMATION

Misinformation is a scourge, but we can be heartened that propaganda and deception are susceptible to the same reinformation strategies as conspiracy

TABLE 6.1 Examples of information and misinformation about conspiracies

	Real information	Misinformation
Agents	Roman senators Oath Keepers	Democratic Party pedophiles Leaders plot replacement
Communications	Verbal and electronic messages	Alleged conversations
Plans	Kill Caesar Insurrection	Abuse children Overwhelm whites
Actions	Kill Caesar Attack the U.S. Capitol	Sex trafficking Unlimited immigration

theories. Political misinformation is open to diagnosis and treatment using the same methods that worked for medical and scientific misinformation. In all these areas, we can use the AIMS mechanisms to distinguish between real information and misinformation and to replace lies with facts.[56]

This chapter showed the difference between real information about real conspiracies and misinformation about fake conspiracies. The structure of real conspiracies is captured by the CAPA schema, which looks for the agents, communications, plans, and actions that constitute a conspiracy. Evidence for these elements and inferences about the motives of the agents provided reasons to believe that the conspiracy is real. Table 6.1 summarizes real information and misinformation about the conspiracies I have analyzed.

Fake conspiracies such as QAnon and the Great Replacement theory are marked by lack of evidence about agents and their activities. The popularity of conspiracy theories is not explained by their employment of legitimate information methods but by defective mechanisms of acquisition, inference, memory, and spread. Observational collecting is displaced by making stuff up, and evidence-based inference is displaced by motivated reasoning, fear-driven inference, and other error tendencies. Memory and spread are encouraged by emotional appeals rather than limited by objective evaluations.

QAnon, Great Replacement, and other right-wing conspiracy theories aim to block moves toward equality for downtrodden social groups. Chapter 7 looks more comprehensively at ways in which misinformation has been used to support social inequality.

CHAPTER 7

EVILS

Inequality and Social Misinformation

Inequality kills. Abundant evidence shows that unequal societies suffer from higher rates of death, disease, unhappiness, and political discord. Yet economic inequality has increased steadily in recent decades, and social discrimination based on race, sex, gender, class, caste, and religion remains a large problem worldwide.

The maintenance of inequality depends on social forces such as the power of those in command, but this chapter focuses on the role of misinformation. Since the eighteenth century, enlightened views have maintained that all humans are equal, but movements for the elimination of inequalities have been thwarted by arguments that inequality is unavoidable and therefore legitimate. The legitimation of inequality has depended on many kinds of historical and social misinformation. Restoration of real information about the origins and ongoing causes of inequality can contribute to adoption of just policies that move societies toward greater satisfaction of human needs. The AIMS theory of information and misinformation explains inequality effects as resulting from *a*cquisition, *i*nference, *m*emory, and *s*pread.

Inequality interacts with the other main problems discussed in this book. COVID-19 cases and deaths have disproportionately affected people who are poor or subject to racial discrimination. Educational inequalities rose because of pandemic restrictions that kept children out of school. Climate change is increasingly affecting the lives of disadvantaged people who lack the resources to move or to adapt in other ways, and unequal societies have elites whose conspicuous consumption uses massive amounts of unsustainable resources, for example, by using private jets and thus lots of fossil fuel. Political conspiracies often target racial or religious minorities and serve to direct power toward groups eager to maintain inequality. Hence, dealing with misinformation about inequality complements dealing with medical, scientific, and political misinformation.

Problems of inequality concern misinformation about facts and about values. Factual misinformation is used to support inequality using arguments designed to show that equality is neither feasible nor desirable. Chapter 3 argued that values can also be subjects of misinformation, and I will describe lines of thought that lead to the undervaluing of equality with respect to other values such as freedom.

After outlining the nature of equality and inequality, I review evidence for the negative effects of inequality on human lives. History reveals many ways in which misinformation has been used to support inequality. Ancient myths from religious and philosophical sources provided early defenses of inequality. Later doctrines used to buttress inequality include the divine right of kings, philosophical stories about the origins of civilization, social Darwinism, fascism, economic mythology, pseudoscientific racism, and mistakes about sex and gender. Combatting these doctrines requires reinformation that replaces falsehoods about the human condition with evidence-based understanding of how equality can be enhanced rather than dismissed.

EQUALITY AND INEQUALITY

The concept of equality has been central to political discussions since the eighteenth-century revolutions in America and France, although concerns about equality go back to the Old Testament.[1] But what is equality? Dictionary definitions such as "the state of being equal, especially in status, rights, and opportunities" are not informative or useful, even with consultation of the definition of "equal" as "being the same in quantity, size, degree, or value."

We can characterize the concept of *equality* using examples, features, and explanations.[2] First, the standard examples of equality include historical cases where people have been treated the same regardless of sex, color, religion, sexual orientation, and so on. Another class of standard examples of equality includes particular instances, such as equal rights to free speech, equality of opportunity, and equality of legal treatment. No country is a stellar exemplar of equality, but some such as Denmark and Canada do relatively well, for example, when Canada in 2017 outlawed discrimination based on gender identity and expression.

Second, the typical features of equality include people getting the same treatment, fairness, and recognition that their humanity is not diminished by characteristics such as being of a particular sex, race, religion, sexual orientation, and status in general. Social equality does not require that people be identical with respect to physical features and talents, only that they are equal in opportunities to satisfy the needs required to flourish as human beings.[3] Such requirements cover biological needs (food, water, air, shelter, health care) and also psychological needs for autonomy (freedom from control), relatedness (connection to other people), and competence (opportunity for accomplishments).[4] Human individuals differ from one another across numerous dimensions, including height, weight, attractiveness, intelligence, and energy. Moral equality does not require that people be exactly the same in every way but rather that society should treat them equally with respect to ethically relevant features.

The concept of equality has an important explanatory role, helping to answer questions such as why people are dissatisfied and agitating for better treatment. The concept of equality also has a strong moral dimension, explaining why oppression and discrimination are viewed as wrong.

Inequality is not just the absence of equality but can be characterized by standard examples, typical features, and explanatory roles. First, the exemplars of inequality include enslavement of people based on their race, discrimination against women with respect to voting and jobs, oppression of minority religious groups, and mistreatment of gay and lesbian people. Economic inequality is also prominent, both in absolute terms (hundreds of millions of people in the world with almost no income) and relative terms (the disproportionate share of wealth of the top 1 percent). Specific examples of inequality abound, such as slavery in the United States before emancipation, Nazi murder of Jews, and abject poverty in Africa.

Second, inequality's typical features include violations of human rights, unfairness, lack of recognition, and inadequate access to health care, education, and safety. Third, the concept helps to explain why American and French revolutionaries were so resentful of tyranny and why subordinated people and their allies today become activists trying to overcome social oppression of particular groups.

Opposing political views place different priorities on equality, with left-wing views such as social democracy and liberalism (in the American sense) ranking it high, and right-wing views such as libertarianism ranking it far below individual freedom. The cognitive-affective maps in figures 7.1 and 7.2 show the contrasting configurations of values in these ideological approaches.

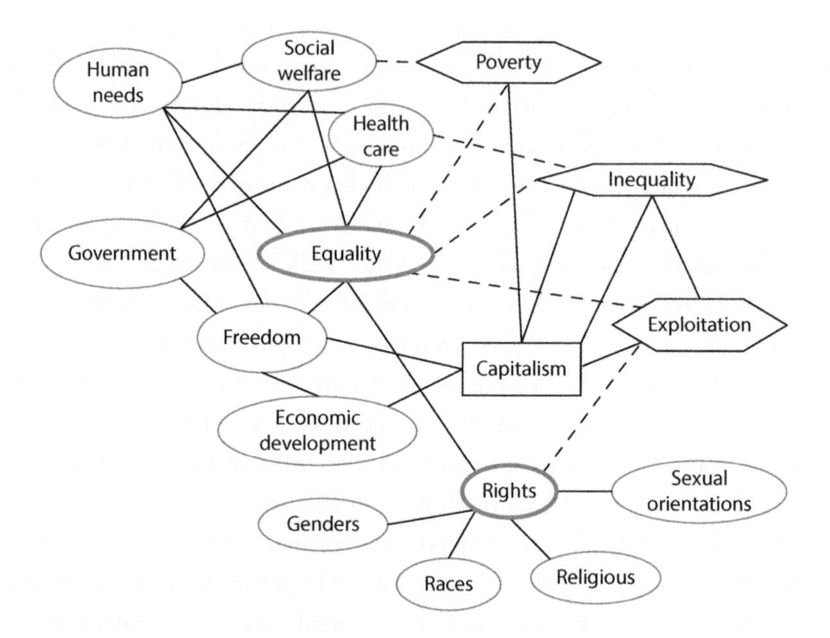

7.1 Cognitive-affective map of social democratic (liberal) values. Ovals are emotionally positive, hexagons are negative, and the rectangle is neutral. Solid lines indicate mutual support; dotted lines indicate incompatibility.

Source: Paul Thagard, "Social Equality: Cognitive Modeling Based on Emotional Coherence Explains Attitude Change." *Policy Insights from Behavioral and Brain Sciences* 5, no. 2 (2018): 247–56, p. 251. Reprinted by permission of Sage.

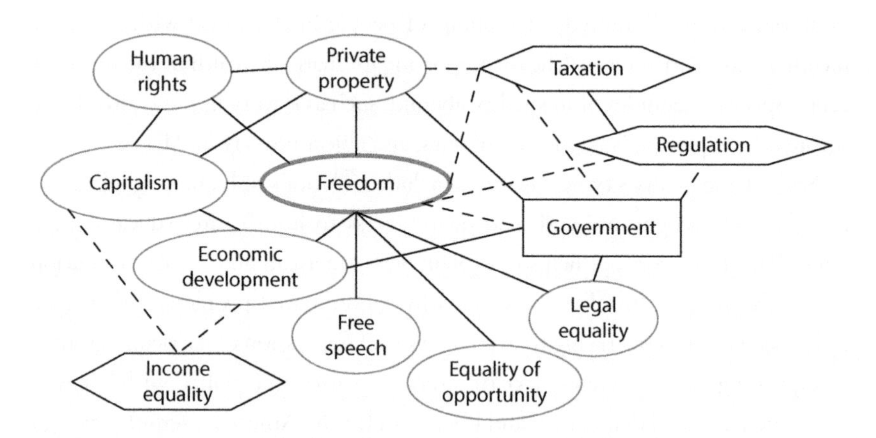

7.2 Cognitive-affective map of libertarianism. Mapping conventions are the same as for figure 7.1.

Source: Thagard, "Social Equality, " p. 251. Reprinted by permission of Sage.

WHY INEQUALITY MATTERS

The U.S. Declaration of Independence proclaimed in 1776: "We hold these truths to be self-evident, that all men are created equal, that they are endowed by their Creator with certain unalienable Rights, that among these are Life, Liberty and the pursuit of Happiness." Obviously, these truths are *not* self-evident because the equality of men (and women) has frequently been denied. The main writers of the Declaration, Thomas Jefferson and Benjamin Franklin, were slave owners, so their commitment to equality was confined to white men. It did not extend to women, who were not allowed to vote in the United States until 1920. Right-wing arguments contend that people are so inherently unequal that attempts to impose equality will inevitably make societies worse.[5] So evidence-based arguments must be given for the desirability of equality and the corresponding undesirability of inequality.

Economists and other social scientists have documented harmful effects of inequality that include: decreased happiness; increased rates of crime, ill health, obesity, and teenage pregnancy; lack of social cohesion; lack of trust; limited human development; and economic instability.[6] Health and social problems are more common in more unequal countries such as the United States, the United Kingdom, and Portugal than in more equal countries such as Japan and Sweden.

Inequality between countries contributes to illegal immigration, which is stressful both for the immigrants and for citizens whose precarious financial lives are threatened by low-priced immigrant labor. Additional reasons for valuing equality over inequality include opportunity to satisfy needs, availability of legal representation, and access to health care. Inequality is also a source of stress and anxiety when people at the bottom of the society suffer from insecurity and lack of respect.[7] Respecting others requires taking their needs seriously, which is difficult for some people when others are drastically inferior in income, wealth, and cultural resources.

Inequality concerning income and wealth undercuts equality of opportunity, which depends heavily on access to education. Poor people tend to have much less access to educational resources because they live in bad neighborhoods with inferior schools. One of the reasons that the United States has less upward social mobility than other wealthy countries is that university education is much more expensive. So even with respect to what is supposedly a narrow right to equality of opportunity, income inequality is a serious concern.

Equality before the law is severely challenged when people do not have access to good legal representation. Wealthy people can hire expensive lawyers to ensure that they are more likely to win legal cases rather than losing them when dealing with the legal system.

People need good health to function fully as human beings, and the negative impact of inequality on health has been well documented.[8] The impact of inequality is partly that poor people cannot afford medical treatments. In addition, people who are low in social hierarchies tend to have less control over their lives, which leads to more stress and resulting diseases and unhealthy behaviors.

Social initiatives can deal with these negative effects of inequality. Taxes on income and wealth can be used to reduce the huge gaps between rich and poor, with the revenue used to support social programs that ensure that all people can meet their biological and psychological needs. The goal is not to make the rich have less money to spend on their wants but rather to guarantee the well-being of people on the lower rungs of society who need help with food, shelter, and health care as well as with autonomy, relatedness, and competence. An international wealth tax could redistribute resources across countries as well as within them.[9]

One innovative way of lessening equality is to use taxes to provide all members of the society with a basic income, which ensures that people can take care of their vital biological needs without bureaucratic tests and interference.[10] Support for such programs comes not only from the left but sometimes also from conservatives who see basic income as a more efficient and less controlling alternative to traditional welfare operations.

A more established way to overcome social as well as economic inequality is to pass laws that prohibit discrimination based on factors such as sex, religion, race, ethnicity, sexual orientation, and gender identity. Laws against such mistreatments are morally justified because discrimination prevents people from satisfying their needs for autonomy, relatedness, and competence.

REAL INFORMATION ABOUT INEQUALITY

I have described how acquisition of real information derives from perceptions, instruments, systematic observations, controlled experiments, and randomized clinical trials. Controlled experiments are rare in social situations, although there are a few examples, such as implementations of basic income plans.[11] Occasionally, natural experiments allow important observations, such as when David Card won the 2021 Nobel Memorial Prize in Economic Sciences for showing that raising the minimum wage in New Jersey did not produce loss of employment.[12]

Clinical trials with double blinding and randomized assignment are impossible for social phenomena such as equality and inequality. Equality and inequality are too relational and abstract to be subject to direct perception, and there are no instruments that directly measure them. Thus, the main method for acquiring real information about inequality is systematic observation of people's income, expenditures, and wealth.

Information about income is readily available from government statistics concerning income taxes. These data are not perfectly accurate because people may misreport income to evade taxes, but nevertheless they provide a good approximation for income amounts in many countries. Supplemental information can be collected by questionnaires such as the Canadian Income Survey, which "gathers information on labour market activity, school attendance, disability, unmet health care needs, support payments, child care expenses, inter-household transfers, personal income, food security, and characteristics and costs of housing."[13]

The standard way of calculating inequality in income distributions is the Gini coefficient developed by the Italian statistician Corrado Gini. A mathematical formula generates a measure of the degree of income inequality in a society ranging from 0 (complete equality) to 1 (complete inequality). For example, relatively egalitarian European countries such as Denmark tend to have Gini coefficients around .3, less egalitarian democracies such as the United States have higher coefficients around .4, and much poorer and unequal countries such as Brazil have still higher coefficients around .5. The Gini coefficient provides little information about how income is distributed and does not reveal increases in the relative wealth of the top 1 percent.[14]

Information about expenditures is collected by consumer surveys such as those conducted by the U.S. Bureau of the Census.[15] Given information about how much individuals spend, mathematical techniques such as a version of the Gini coefficient can be used to measure the degree of inequality of expenditures.

In contrast to income, wealth measures the total assets of an individual, including homes, automobiles, personal valuables, businesses, savings, and investments. Information about wealth can be collected by surveys and by publicly available information such as ownership of houses, stocks, and companies. Gini coefficients for wealth can be generated that display huge inequalities, but more salient gaps are indicated by specific facts such as that, in 2017, the world's eight richest people had as much wealth as the poorest 50 percent of the population, more than 3 billion people.[16]

The *World Inequality Report 2022* provides an excellent survey of current observations, such as that the bottom 50 percent of the world has only 2 percent of the total global wealth, while the top 10 percent has 76 percent of the total.[17] Because wealth discrepancies tend to grow faster than income discrepancies, wealth inequality is likely to increase. Between 1995 and 2021, the share of wealth owned by the top 0.1 percent rose from 7 percent to 11 percent. From 1945 to 1980, inequality shrank in many parts of the world because of government policies that included increasing social benefits and high tax rates. But inequality has increased steadily since then because of policies inspired by leaders such as Ronald Reagan and Margaret Thatcher.

The real information in the *World Inequality Report* comes from the World Inequality Database, which provides abundant information represented in tables, graphs, maps, and databases.[18] Tables, graphs, and maps are also used in the *World Inequality Report* along with two hundred pages of text. The World Inequality Database combines information from the world's countries compiled by household surveys, administrative data, rich lists, and national accounts, all of which count as systematic observations. The COVID-19 pandemic dramatically increased the gap between the top of the wealth distribution and the rest of the population; global billionaire wealth increased by more than 50 percent between 2019 and 2021.

Instead of the mathematically obscure Gini coefficient, inequality is depicted more vividly by the *World Inequality Report* as differences between the share of total income or wealth of the top 10 percent and the bottom 50 percent. For example, page 187 of the report has a graph that shows the change in relative income between Canada's top 10 percent and bottom 50 percent between 1920 and 2019. Inequality has substantially increased since 1980, which would not be so concerning if those in the bottom half had all their economic needs met. But many of them suffer from lack of food, housing and health care not covered by Canada's government health system, which lacks dental and other services.

Thanks to institutions such as universities and government agencies, we have an enormous amount of real information available concerning inequalities of income and wealth. Statistical data and mathematical analysis tell us much about shortfalls in equality, but they do not directly address ethical questions of why inequality is bad or practical questions of how equality can be increased.

MISINFORMATION ABOUT INEQUALITY

Given the demonstrated harms of inequality and the availability of ways of reducing it, the question arises: Why is inequality increasing rather than decreasing? A major part of the answer is that the wealthy who benefit most from inequality exercise large amounts of political power, even in democracies. Political parties are heavily dependent on donations from people whose interests discourage redistribution of income and wealth. Another part of the answer concerns misinformation. Many people, even those who would benefit from greater equality, have been pumped with false beliefs and defective values that serve to justify inequality.

Here is a general argument intended to show that equality is not a fundamental moral value:

> Inequality is natural because people vary with respect to biological features such as height, strength, energy, and intelligence. People cannot be leveled down to the same abilities, so we should expect that some people will be more successful in accumulating wealth. Interfering with this accumulation will infringe on people's rights to fundamental freedoms that include self-control and the right to property. Inequality matters only for a narrow set of rights, such as free speech, equal opportunity, and equality in legal treatment. History shows that restricting such freedoms with the aim of a more equal distribution of wealth produces totalitarian regimes such as the Soviet Union. Moreover, such controlled societies have a dismal record of achieving the economic growth that benefits everyone: a rising tide lifts all boats.[19]

In sum, freedom is a more fundamental value than equality.

This line of reasoning has blatant flaws. Opposition to inequality does not suppose that people are perfectly equal in all respects such as physical abilities.

Equality does not demand complete leveling of income and wealth, only that all people have sufficient resources to ensure that their vital human needs are met. Countries such as Sweden and Canada have developed social programs like national health systems that help to meet human needs while maintaining high amounts of freedom. Hence, freedom and equality are compatible, and a reasonable balance can be achieved without undue coercion.

Nevertheless, many areas of social misinformation stand in the way of reducing inequality; they operate in religion, history, philosophy, social science, and popular culture. In response, we can pursue reinformation by various techniques, especially critical thinking.

Ancient Myths

Myths are forms of misinformation that are culturally established by traditionally accepted stories. The Bible's creation story establishes inequality between men and women because the first man Adam was created before the woman Eve, who was formed from his rib. Not only was Eve derivative, her subordination to Adam was specified by God when he said: "Your desire shall be for your husband, and he shall rule over you."[20] The New Testament uses a peculiar analogy to confirm woman's inferiority: "Wives, submit to your own husbands, as to the Lord. For the husband is the head of the wife even as Christ is the head of the church, his body, and is himself its Savior. Now as the church submits to Christ, so also wives should submit in everything to their husbands."[21] Hence, the religious myth of the creation of Adam and Eve serves to justify inequality between men and women.

In the *Republic*, Plato has Socrates recommend the myth that all citizens have metals in their souls that explain their unequal place in society. Rulers have gold, their helpers have silver, while farmers and craftspeople have iron and brass.[22] Socrates recognizes that this story has no truth, so it qualifies as disinformation. But he recommends it anyway because the lie would have a good effect in making people more inclined to conform to their social roles. Plato's myth of the metals was an early use of misinformation to justify inequality as a natural condition.

Aristotle used an analogy, probably the worst in the history of philosophy, to justify the unequal status of slaves, who constituted about one-third of Athenian society. He said that non-Greek slaves were naturally inferior, just like women, and so deserving of a subordinate position.[23] Misinformed claims about natural inferiority are a common way of justifying inequality, and they are still used today, for example, with respect to IQ.

Ancient myths are easily subject to reinformation by factual correction. There was never any evidence for the story of Adam and Eve, and not even Plato believed his myth of the metals. Aristotle had no evidence that slaves and women were naturally inferior, and he should have been aware that people ended up as slaves because they were part of conquered populations, not because of any personal failings. In my book *Balance*, I evaluate analogies as strong, weak, bogus, and toxic, and Aristotle's analogy obviously falls into the toxic category.[24]

Theological Justifications

The Bible's Book of Genesis gave a theological justification for the dominance of men over women. Similarly, the Book of Samuel anoints Saul and David as kings of Israel, providing a religious basis for the dominance of monarchs. Other cultures, including the Roman, Byzantine, and Chinese empires, viewed monarchs as divinely established. In Medieval Europe, people accepted that the monarch was empowered by God, but opposition to the divine right of kings developed in the sixteenth century.[25] In 1689, John Locke criticized the view that the power of the monarch descended from God's gift to Adam and maintained that monarchy was justified only by social consent.[26] Christianity has largely abandoned the idea that the power of kings and rulers comes from God, but the Islamic State formed in Iraq in 1999 reclaimed religious authority.

The divine right of kings extended justification to the dominance of nobles whose authority derived from the monarch. The difference between the wealth of ordinary people and the wealth of rulers was appropriate because God had ordained it. If whatever happens more generally is God's will, then discrepancies in income, wealth, freedom, and opportunities are unchallengeable because they follow from God's unfathomable choices. In feudal and other societies dominated by theology, inequality is fully justified. Inequality does not matter because a temporary life of poverty and servitude is trivial compared to the possibility of eternal reward in heaven. Theology provides a powerful way of justifying huge differences in wealth and opportunity.

Enlightenment ideas that developed in the seventeenth and eighteenth centuries challenged this package of ideas. By emphasizing reason rather than tradition, thinkers such as René Descartes, John Locke, Voltaire, David Hume, and Immanuel Kant shifted attention to general human freedom and happiness and away from theological control. Nevertheless, some evangelical Christians and orthodox Muslims continue to see inequality as theologically justified. Of course, believers in these religions would deny that theological justifications of inequality qualify as misinformation, but the lack of evidence for the existence of assumed gods requires classifying as misinformation any religious basis for inequality.[27]

The New Testament contains statements of equality such as the injunction to love your neighbor as yourself and the assertion that Christ is in everyone. But whether God favors equality or inequality does not matter unless he actually exists and qualifies as a moral authority, unlike the ancient Greek gods, who were often wicked. However, reinformation about the theological basis of inequality can less aggressively point to religious texts such as parts of the New Testament that support equality.

If factual correction is not enough to overturn theological stories about human origins, we can turn to critical thinking that looks for psychological explanations of why people find theology appealing. Religion thrives on motivated reasoning that fits the personal goals pattern because people naturally want guarantees of immortality, eternal happiness, and the care of a loving deity.[28] Religion also benefits from motivated reasoning based on group identity because people intensely think of themselves as belonging to particular varieties of Christianity, Islam, or other religions. Specific emotions also spur motivated reasoning about religions, for example, when fear of death makes people want to believe religious reassurances. All these patterns encourage people to believe what religious authorities tell them, including their doctrines about inequality.

Motivated reasoning also helps to explain why the doctrine of divine right was so popular among monarchs, nobles, and their supporters. Their personal goals support the belief that God wants them to be in charge.

If critical thinking fails to dislodge theological justifications of inequality, motivational interviewing can help get at the deeper emotions behind preferences for theological explanations. Monarchs and nobles can naturally be driven by lust for power, but ordinary people can support other theological justifications for inequality that can be explored empathically. For example, poor people may turn to theology for reassurance that their miserable existence will be balanced by eternal rewards. The racists among them may be consoled that being near the bottom of the economic hierarchy is allayed by having racial inferiors below them. Motivational interviewing can identify such inclinations and encourage the disadvantaged to empathize with other disadvantaged people who may differ in race, caste, or religion. Motivational interviewing can also operate within the internal thought system of religious believers, for example, by pointing out to prejudiced Christians the places in the New Testament that seem to defend equality: blessed are the poor, and the Apostles divided their property equally according to need.[29]

Institutional modification aimed at churches can contribute to egalitarian ideals. Many churches are highly conservative, but even Roman Catholicism has a movement called liberation theology that emphasizes concern for the poor and oppressed. Other forms of progressive Christianity, Judaism, and Islam are strongly concerned with social justice based on ethical standards of equality, using religion as a force for change rather than conservatism. Hence, religious people can work to shift the leadership and policies of their churches toward adoption of values and norms that favor equality and fight discrimination.

Political action can also help combat theological justifications of inequality while respecting religious freedom as essential for freedom of thought and speech. Action need not degenerate into freedom-killing attempts to ban religion but can insist on separation of church and state that keeps religions from interfering with the lives of people who do not adhere to them.

Philosophical Thought Experiments

Whether inequality should be viewed as harmful depends in part on information about its origins. If societies are inevitably unequal, then eliminating inequality is impossible and therefore not ethically required. Eminent philosophers such as Thomas Hobbes, Jean-Jacques Rousseau, and John Rawls proposed thought experiments about the origins of inequality that unfortunately qualify as misinformation because the stories they tell are historically false. In contrast, empirical studies by archaeologists and anthropologists provide real information about the ancient roots of equality and inequality that is much more complex and ambiguous than philosophical stories.

In 1651, Hobbes published *Leviathan*, which offered an original account of how government emerges.[30] In the state of nature, people engage in a war of all against all and accordingly are in constant fear of destruction. To escape this desperate condition, people form a social contract in which they agree to subject themselves to an all-powerful sovereign, called Leviathan by analogy to a powerful, biblical sea monster. This contract justifies the supreme powers of the sovereign and implicitly justifies the inequality that results from the ability of the sovereign and associates to control the lives and livelihoods of the people who supposedly contracted to recognize absolute leadership.

Hobbes had no evidence that the history of humanity followed this path to the establishment of governments, and empirical findings in the next section paint a different picture. Thus, Hobbes's social contract story qualifies as historical misinformation. It can be construed alternatively not as history but as a conceptual device similar to other philosophical thought experiments that pump people's intuitions to yield abstract conclusions.

In science, thought experiments are valuable for suggesting new hypotheses that are then subjected to empirical verification, unlike thought experiments in philosophy that are taken to be self-contained. Philosophical thought experiments unfortunately often yield false conclusions and fall short of the standards of evidence listed in chapter 2.[31] Every thought experiment has an equal and opposite thought experiment. Philosophical thought experiments about social contracts have yielded highly divergent normative conclusions about inequality.

In 1751, Rousseau published his *Discourse on the Origin of Inequality Among Men*. Like Hobbes's *Leviathan*, Rousseau's account of the original state of nature was a hypothetical thought experiment with no historical basis. Rousseau claimed that, in their natural state, people are free to satisfy their desires and needs. Inequality only develops when people start to acquire land and keep it as private property, with some people coming to have much more than others. In his later book *The Social Contract*, Rousseau exclaimed: "Man is born free, but everywhere he is in chains."[32]

Contrary to Hobbes's thought experiment, Rousseau's story offers a much more egalitarian prospect of reducing inequality by eliminating or reducing private property, a view that greatly appealed to later socialists. The lack of evidence for Rousseau's story unfortunately relieves it of any plausibility for finding a path away from inequality. Like Christian defenses of equality, it is only a hypothetical roadmap devoid of real information.

In his influential *Theory of Justice*, Rawls used a social contract thought experiment to yield a conclusion about equality also different from that of Hobbes.[33] Rawls asked readers to imagine themselves behind a veil of ignorance concerning their status in society. They could then enter into a social contract based on reasoning about what would be the fairest way to establish society. Rawls argued that people would choose principles that (1) give people equal rights to equality and opportunities and (2) allow only inequalities that benefit the worst-off members of society. The second principle would justify massive government interventions to eliminate most inequalities in income and wealth that clearly benefit the rich. The social impact of Rawls's second principle has been negligible because few people have been convinced by his thought experiment. In reality, no one operates behind a veil of ignorance concerning their status in society, so Rawls's social contract is as much misinformation as Hobbes's.

Other contemporary philosophers have used social contract devices to yield conclusions much more conservative than those of Rousseau and Rawls. Robert Nozick claimed that a social contract would only justify a minimal state with powers limited to protecting people from being directly harmed by others, with no justification of actions that reduce inequality.[34] Libertarians have endorsed Nozick's conclusions that the state has no business managing inequality, but with no more evidential basis than the more egalitarian thought experiment of Rawls.

Critiques of the social contract approach have come from philosophers who point out that it ignores issues about sex and race that are crucial to discussions of inequality.[35] Reinformation about social contract approaches to inequality operates in two steps. First, it points out the lack of historical evidence that anything like the social contract ever occurred, so that the suggestion by Hobbes and others that a contract affected the development of inequality is bogus. Second, philosophical thought experiments concerning equality (and anything else) have inherent limitations for establishing how the world actually works. We can learn much more about the history and philosophy of equality by studying cultural development and reflecting on the empirical consequences of inequality.

Anthropological and Archaeological Information

Philosophical thought experiments about equality and inequality amount to making stuff up and therefore fail to address factual questions about the origins of inequality and normative questions of why equality is valuable. Real information concerning factual questions is rapidly becoming available fortunately through the collaborations of archaeologists and anthropologists using methods of perception, instruments, and systematic observations. Archaeologists study the material remains of ancient cultures, whereas anthropologists observe living cultures. Together, they shed light on the prevalence of inequality in past and current societies. The most important instrument for archaeology is radiocarbon dating, which can identify the age of artifacts between fifty thousand and four hundred years ago. Archaeologists have also benefited from use of ground-penetrating radar and laser imaging, detection, and ranging (LIDAR) to detect possible sites.

The book *The Creation of Inequality* by Kent Flannery and Joyce Marcus reviews cultural practices in dozens of prehistoric and historic cultures and reaches conclusions roughly similar to those of Rousseau.[36] Human hunter-gatherer societies are historically egalitarian, with emphasis on generosity, sharing, reciprocity, and respect. For most of human history, people had roughly the same access to goods and power. The past ten thousand years, with the advent of agriculture, cities, and organized religion, has led to great discrepancies in wealth and social control. These conclusions provide some empirical support for Rousseau's hypothesis that equality is the natural human state and inequality is a deviation that came with later economic and social developments.

A much more radical new book by an anthropologist and an archaeologist challenges the whole idea of an origin of inequality. In *The Dawn of Everything*, David Graeber and David Wengrow argue that the historical record is far too complicated to generalize about rates of equality and inequality across prehistorical, historical, and current cultures. They argue that "there is simply no reason to believe that small-scale groups are especially likely to be egalitarian—or, conversely, that large ones must necessarily have kings, presidents, or even bureaucracies."[37] They back up this conclusion with examples of hierarchical small groups such as those in the Pacific Northwest and relatively egalitarian large groups such as seventeenth-century Iroquoian-speaking peoples.

Graeber and Wengrow reject Hobbes's story about a brutish state of nature from which people were saved by the establishment of government and also Rousseau's story about egalitarian hunter-gatherers who lost their freedom with the adoption of agriculture and private property. Graeber and Wengrow insist that the past is much more interesting than these stories with their alleged political implications. Graeber has long been an advocate of anarchism and was one of the leaders of the 2011 Occupy movement. He and Wengrow have a political agenda, but their book is thorough and scrupulous in looking at the historical record. The book's motivation is to insist that humans are not locked into narrow patterns of historical development such as those described by Hobbes and

Rousseau. Instead, people can be viewed as intelligent, imaginative, and playful agents who are capable of reinventing themselves in society.

Graeber and Wengrow note that left-wingers tend to prefer the Rousseau story and cherry-pick anthropological case studies such as peaceful African foragers, while right-wingers tend to prefer the Hobbes story and cherry-pick cases such as the violent Yanomami of South America. Graeber and Wengrow cite case studies to show that historically people are capable of forming societies that are peaceful or violent as well as egalitarian or hierarchical. They insist on gathering historical data from the Americas, Africa, and Asia as well as Europe and argue that the results are sufficiently diverse that any claims about a single origin for social inequality are misleading. Social organization may be seasonal, with societies such as the people who built Stonehenge being more egalitarian for part of the year and more hierarchical for other parts.

Instead of propagating myths about the origins of cultures, we should appreciate the flexible, shifting arrangements of our earliest ancestors. Whereas myths about the origins of inequality tend to lock people into rigid views of human possibilities, more thorough examination of the diversity of early human cultures encourages imaginative reflection on how societies can develop to best satisfy human needs.

Contrary to the common story that agriculture gave rise to cities that produced hierarchy, Graeber and Wengrow cite results from recent archaeology to show that surprisingly few of the early cities from around six thousand years ago contain signs of authoritarian rule. Not all cities depended on large-scale agricultural production. The Indus Valley and other cases show organized human settlement without concentration of wealth and power in the hands of elites.

Like inequality, the state has no universal origin because states develop different ways of controlling violence, spreading information, and dealing with the charisma of individual leaders. Debunking myths concerning the unavoidable origins of states allows people an important kind of freedom: to create new forms of social reality. Misinformation about the development of human societies is not just factually wrong; it is also normatively dangerous because it limits the range of possibilities for how cultures can deal with inequalities. The reinformation provided by Graeber and Wengrow's reinterpretation of historical evidence falls mainly under the heading of factual correction, which they achieve by expanding the evidence base relevant to explaining the origins of inequality.

The accuracy of Graeber and Wengrow's claims about the egalitarian nature of some early cities has been challenged as biased by their anarchist ideology.[38] Then their interpretation of the dawn of civilization would be tainted by motivated reasoning. I lack anthropological expertise to adjudicate this issue, but I can at least conclude that the standard account of human history as following a path from equal hunter-gatherers to unequal city dwellers is less obvious than it seemed to Rousseau and some anthropologists. The issue is important because the standard path might suggest that inequality is an inevitable part of the development of agriculture and cities. No one today wants to be a hunter-gatherer, so it might seem that people today are stuck with inequality as the result of civilization. Graeber and Wengrow's more complex history pushes us to consider alternative ways of thinking about today's increasingly unequal societies.

Fascist Fictions

A major challenge to equality comes from fascism, which opposes democracy in demanding autocratic government by strong, charismatic leaders assisted by violence.[39] Fascists use misinformation about history, identity, leaders, conspiracies, and destiny to gain and retain power. Replacing misinformation by real information is therefore crucial for retaining democracy and fighting fascism.

Different kinds of democracy (liberal, social, and direct) require increasingly greater amounts of real information to enable people to participate. Liberal democracy is based on people voting for representatives who are charged with carrying out their wishes. To vote intelligently, people require accurate information about candidates, parties, government activities, and current issues. Social democracy goes beyond liberal democracy in demanding a greater commitment to justice interventions such as inequality reduction, regulation of the economy, and meeting human needs. These interventions require real information about individuals and societies that go beyond the more superficial information required for voting.

The most intense requirements for well-distributed real information come from direct democracy, in which people decide on policies without intermediate representatives. At the extreme, some anarchists demand that leadership be abolished so that all decisions are made collectively by the people involved. Direct democracy requires all people to be fully informed about all matters that affect them so that they can contribute to effective decisions. Liberal and social democracies place less demand on people and depend on people having enough information to choose representatives who will support their interests.

In contrast, fascism relies on six kinds of misinformation to ensure that people are controlled rather than acting democratically for their own benefit:

1. Mythic past: Leaders such as Mussolini and Hitler talk of a made-up time when their nations were great in ways that they pretend to recover.
2. Glorious leader: The autocrat deserves control because of the infallible ability to make good decisions always.
3. National and racial superiority: Nationalism is justified because foreigners, minorities, and immigrants are inherently inferior to the national or racial identity.
4. Dangerous threats: Violent actions are required to protect the nation from contrived conspiracies.
5. Inevitable destiny: The nation under its autocratic leader is guaranteed to have a magnificent future.
6. Virility: The leader is a strongman who dominates women and other men.[40]

These convictions are made up for political purposes and have no evidence acquired from interactions with the world by perception, systematic observation, or experiments. Evaluations and inferences are based only on autocrats' goals rather than on acquired evidence.

Conspiracy theories are a powerful method of exaggerating threats, for example, using the forged *Protocols of the Elders of Zion* to support claims of a Jewish plot to rule the world. The Great Replacement conspiracy theory analyzed in chapter 6 views immigrants as a threat to white supremacy and supports treating nonwhites unequally. Jews such as the Rothschild banking family have been blamed for fake conspiracies concerning COVID-19 and climate change.

Fascist myths about superiority and destiny are supported by the older doctrine of social Darwinism, which has been around since the 1850s.[41] Herbert Spencer and other thinkers interpreted the theory of evolution as having consequences for sociology, economics, and politics. Charles Darwin explained evolution as resulting from natural selection, which was a "struggle for survival." Spencer turned this into the slogan "survival of the fittest" and social Darwinists found a justification for inequalities of wealth and power in the claim that they resulted from the rich being fitter than the poor. This conclusion is a gross misapplication of Darwin's ideas, which only concerned biological competition leading to differences in survival and reproduction.

The analogy between biological evolution and social evolution is weak because their mechanisms of variation, selection, and inheritance are so different. Social variation, selection, and inheritance result from cognitive mechanisms, sometimes intentional, of idea generation, belief evaluation, and communication. In contrast, biological evolution operates by noncognitive mechanisms of genetics and differential reproduction. When this analogy is used to justify the continuing domination of elites and the persistence of inequality, it descends from weak to toxic.

Social Darwinist ideas receded after the decline of fascism, but they survive in some economic views that poor people have themselves to blame for their misfortune. They also survive in the populist psychology of Jordan Peterson.[42] A major part of his defense of the individual is an argument that inequality and dominance hierarchies are rooted in biological differences, from lobsters up to human men and women. But humans have much bigger brains than lobsters, with 86 billion neurons rather than 100,000, giving us the capacity to think about what kinds of societies are best for us. In recent centuries, people have been able to recognize that human rights apply across all people, not just to one's own self, family, race, sex, or nation. Lobster hierarchies are irrelevant.

Misinformation spreads because politicians and journalists exploit people's susceptibility to motivated reasoning and thought-distorting emotions such as fear, anger, and hatred. Early twentieth-century fascists could spread misinformation only slowly: through print, radio, movies, and rallies; today, however, misinformation is transmitted rapidly and effectively by social media such as Facebook, Twitter, Instagram, YouTube, and TikTok, whose algorithms value emotional engagement and advertising revenue over truth and democracy. For example, Facebook amplified hate speech used by the Myanmar military against the persecuted Rohingya minority.[43] Facebook supposedly bans white supremacists, but it makes money on ads on searches for white nationalist content.[44]

How can democracy resist the threats posed by fascist uses of misinformation? Some claims about mythic paths and glorious leaders can be easily recognized as false and dealt with by factual correction, but other claims require critical thinking, motivational interviewing, and political action. Critical thinking's two-step process uses error detection followed by remedial reasoning. For example, myths about a splendid history, eminent leader, or guaranteed destiny can be marked as based on motivated reasoning rather than on evidence. Claims about national and racial superiority can be marked as faulty causal inferences from dubious data. Incendiary claims about threats from internal and foreign conspiracies can be recognized as based on fear rather than evidence.

Fascism thrives on multiple patterns of motivated reasoning. The personal goals pattern is instantiated by fascist leaders who stand to gain from achieving and maintaining power but also by ordinary people who mistakenly think that the leaders will solve their economic problems. Such people may also be manipulated via the group identity pattern through thinking of themselves as being members of the groups extolled by the fascist leaders, such as being white, Aryan, and Christian. The emotions pattern of motivated reasoning operates when people succumb to messages that spur them to fear, hate, and be angry about minorities and immigrants, with whom fascists promise to deal. Fascists encourage motivated ignorance by banning books and research that challenge their fictions. Fascism also thrives on motivated invention to amplify its myths, for example, when Joseph Goebbels served as Adolf Hitler's minister of propaganda.

Once thinking errors are detected, remedial reasoning can use reliable forms of evidence collection and inference to supplant misinformation. Sound historical research can debunk claims about a mythic past and undermine clams of inevitable destiny. Careful causal reasoning can replace unsubstantiated claims about national and racial superiority with evidence-based assessments. Judicious applications of inference to the best explanation of good evidence can revise assessments of autocratic infallibility and conspiracy dangers. Hence, the one-two punch of error detection plus remedial reasoning can help to transform fascist misinformation into real information.

Unfortunately, critical thinking is not always effective in changing people's minds, which are as much driven by emotional values as they are by perception and cognition. Motivational interviewing could be used against fascism through the following subtle interventions:

- Understand people's political and economic concerns by asking them open-ended questions and empathizing with their fears and insecurities.
- Be affirmative, reflective, and nonjudgmental about their concerns.
- Identify discrepancies between people's current and desired behaviors such as treating people equally rather than with nationalist or racist prejudice.
- Summarize the social issues and inform people while respecting their autonomy.

Whether motivational interviewing could be used to draw people away from fascist-supporting misinformation remains to be empirically determined, although the technique has had some success in helping students learn multicultural concepts.[45] But studies should be done to determine whether it can work in combination with critical thinking or independently to help people adhere to democratic values in the face of fascist threats.

Institutional modification is crucial for fighting fascism. Extreme groups such as Nazi parties that advocate violence should be abolished, and people can join organizations that are explicitly antifascist. Members of conservative religions and political parties can work to ensure that their groups do not slide into fascist doctrines.

The final reinformation technique is political action to change society in ways that retard the spread of misinformation. The United States and other countries are currently reviewing legislation to make social media responsible for ensuring that accuracy is more important than inflammability in how information is transmitted. Another democracy-supporting political action is to use legislative and judicial means to support fair voting rules, for example, blocking laws that make it hard for some groups of people to vote or that dilute their voting power by gerrymandering electoral districts. Fair voting procedures are essential for ensuring that people have the opportunity to choose representatives based on real information. I hope that a combination of critical thinking, motivational interviewing, and political action can defend democracy against fascist uses of misinformation.

In chapter 1, I quoted Barack Obama saying that disinformation is the greatest threat to democracy. The greatest threat is more accurately autocratic leaders who prefer fascism to democracy and use disinformation as one of their main tools of oppression.

Economic Myths

Economics has sometimes been used to justify inequality as well as to describe it. A central message of the *World Inequality Report 2022* is that inequality is not inevitable; rather, it depends on government policies that ought to be under democratic control. Just as Graeber and Wengrow use anthropology to argue that inequality is not culturally obligatory, Lucas Chancel and his colleagues use wealth and income data to argue that inequality is not economically obligatory.

The *World Inequality Report 2022* contends that challenges of the twenty-first century require redistributing wealth by means of a modest progressive wealth tax on global multimillionaires. It argues that the abolition of wealth taxation in most European countries in recent decades reflected weaknesses in specific forms of taxation rather than in wealth taxes as such. The report challenges contentions that taxation curbs economic growth by distorting economic incentives. It presents evidence that tax cuts to the wealthy do not have the positive effects on growth and employment promised by trickle-down economics.

A more systematic critique of economic misinformation about inequality is provided by Nobel Prize–winning economist Joseph Stiglitz in his book *The Price of Inequality*.[46] He thinks that the world economy and particular countries have developed unjustifiable degrees of inequality, although his remedies like social benefits and increased income taxes are less radical than a global wealth tax. Even Stiglitz's more moderate policy proposals encounter intense resistance from conservative politicians and traditional economists.

Stiglitz's strategy is to identify kinds of economic misinformation that are used to attack policies for reduction of inequality and to debunk them with accurate data about historical developments. Here are some of the widely held economic myths that Stiglitz challenges using real information:

1. Myth: Giving money to the rich via tax cuts will benefit everyone because it leads to more growth. Correction: U.S. tax cuts by Republican presidents have not led to more growth, whereas tax increases by Democrats have occurred with economic growth. Supply-side economics is empirically false. Providing more funds to individuals in need increases demand and economic growth.
2. Myth: Economic inequality is decreasing. Correction: Since the 1980s, the rich have been getting richer and the richest are getting even richer.
3. Myth: Inequality is not a problem because the wealth of people fluctuates up and down so that, over a lifetime, wealth evens out. Correction: Lifetime inequality is persistent and increasing.
4. Myth: Poverty in the United States is not real poverty compared to other countries. Correction: Many Americans still face real poverty as shown by lack of food, housing, and health care.
5. Myth: The poor have only themselves to blame, while the rich have earned their money. Correction: Much wealth is inherited or generated

by luck or predatory behavior rather than virtue, while much poverty is generational.

6. Myth: Inequality is necessary to give people incentives to work. Correction: Incentive pay can operate without the increasingly large divisions between workers and owners.
7. Myth: Social mobility allows people to move up the economic ladder thanks to equality of opportunity. Correction: Especially in the United States and the United Kingdom, compared to other rich countries, social mobility is limited by factors such as the high cost of education. Equality of opportunity does not exist when people are limited by poverty and class distinctions.
8. Myth: Estate taxes are an unfair way to distribute wealth. Correction: Estate taxes affect only a small part of the population that benefits most from extreme inequality.
9. Myth: Government programs that decrease inequality are inefficient. Correction: Many government programs such as U.S. Social Security and national health plans are actually cost effective.
10. Myth: Austerity programs that cut social programs bring economic recovery. Correction: In many countries, economic recovery has come from government spending, not cuts.
11. Myth: Countries need to balance their budgets just like individual households. Correction: This analogy is misleading because countries have capabilities that individuals lack, such as the ability to generate money and to carry debts indefinitely. When used to restrict inequality-reducing measures, this analogy is as toxic as Aristotle's comparison of slavery and women and the social Darwinist comparison of social and biological evolution.
12. Myth: Democratic countries are meritocracies where people's income and wealth reflect their efforts. Correction: Success depends on other factors such as family history, medical history, and luck.

We might add another:

13. Myth: Government programs benefit only the poor. Correction: Many government programs such as subsidies for fossil fuel production and agriculture benefit the rich.

Similar to Stiglitz and his ideas about myths, the economist Paul Krugman identifies a set of influential "zombie ideas" in economics defined as those "that should have been killed by contrary evidence, but instead keep shambling along, eating people's brains."[47] Examples include beliefs that unbalanced budgets are bad and that tax cuts are good, which survive mostly because they cohere with the interests of the rich and powerful.

All these economic myths are subject to factual correction by pointing out economic and historical data that contradict them. Deeper analysis is needed to understand why they are so pervasive, and critical thinking requires looking at the error tendencies that encourage their persistence. The most important error is motivated reasoning in which claims are accepted because they fit with personal goals rather than evidence. Rich people and politicians funded by them have a strong interest in maintaining inequality and avoiding measures such as graduated income taxes and wealth taxes that can reduce it. Multimillionaires and billionaires are inclined to believe that they have justly earned their fortunes and that the poor deserve their lot because of stupidity and laziness.

Most economists do not have these personal motivations, but they have professional reasons for maintaining similar views about decision making and the limits of government action. Dogmas of mainstream economics include assumptions that people are rational actors operating in competitive economies with perfect information. These dogmas can be used to support economic myths about the efficacy of tax cuts for the rich and the limits of equalization strategies such as basic income. Claims that the market works best with no government intervention (laissez-faire economics) rest more on political ideology than on economic evidence.

Decades of evidence from behavioral and financial studies have challenged these dogmas, which survive because of repetition and lack of an alternative overreaching theory. Attachment to traditional economic theory has blocked explanation of major events such as the Great Recession of 2007–2009, which requires a deeper understanding of human psychology, including emotions.[48] Reinformation in economics will require new theories as well as close attention to empirical facts.

Motivational interviewing is unlikely to help much with changing the minds of rich libertarians or orthodox economists whose views are too entrenched by personal motivations and political dogmas to yield to empathic approaches. The Cato Institute, a right-wing think tank, has its own list of "myths" about economic inequality that cohere with its adulation of free enterprise.[49] Reinformation is then better pursued not by the psychological strategies of critical thinking and motivational interviewing but rather by the political strategy of electing governments that care about overcoming harmful inequalities.

Institutions are important for combating economic myths that support inequality, although progressive think tanks are usually not as well funded as right-wing ones. Nevertheless, foundations and educational organizations can support research that counters empirical and theoretical claims about the economic inevitability of harmful inequality. Political action can support institutional modification by ensuring that governments support the gathering of knowledge about the debilitating effects of inequality.

Psychological Justifications

Psychological misinformation, particularly about intelligence, can also be used to justify inequality as the inevitable result of individual differences. Here is a sketch of a bad argument:

1. Success in work and life in general depends on intelligence: more intelligent people are more successful.
2. Intelligence is measured and explained by IQ.
3. IQ has a substantial hereditary component.
4. So success has a substantial hereditary component.
5. Unlike social situations, hereditary components cannot be changed by social improvements such as better education.
6. So social engineering that attempts to reduce inequality is limited in what can be accomplished.
7. So society should not attempt to reduce inequality.

This argument is rarely stated so baldly, but is implicit in the writings of social scientists such as Hans Eysenck, Richard J. Hernnstein, Arthur R. Jensen, Charles Murray, and Phillippe Rushton.[50]

A major application of this line of reasoning is to questions of racial equality in the United States, where attempts such as educational programs and affirmative action have been made to reduce inequality between Blacks and whites. Many studies have found lower IQs in Blacks, which has been used to explain the limited success of some of these programs. In response, psychologists such as Richard Nisbett have mounted systematic critiques of the research that purports to show a large hereditary basis (50 percent) for the lower IQs of Blacks.[51] Nisbett points to real information that undermines claims that inequality results from hereditary differences.

Being poor is tied to environmental factors that lower IQ through biological mechanisms. Impoverished children suffer from bad nutrition, including vitamin and mineral deficits. Inner-city children are exposed to lead pollution that affects brain development. Poor people have generally worse health, which affects learning, and they get worse medical care. Poor mothers have lower birth rates and have babies that are less likely to be breastfed, which improves IQ.

Children in poor families are also subject to social factors that lower learning and IQ. They move more frequently and have less stable schooling. Impoverished parents tend to be less warm and supportive and to raise their children in ways that are more stressful and authoritarian. The cognitive environments of children in poor families are limited by less verbal stimulation and less encouragement of the kinds of questioning and analysis that would prepare them for higher education and professional lives. Even for adults, poverty impedes cognitive function by sapping attention and reducing effort.[52]

Nisbett provides reasons for doubting that the IQ gap between Blacks and whites in the United States is due to genetic factors. Blacks have fewer economic resources on average so they are subject to all the IQ-limiting factors listed in the last two paragraphs. Blacks are also subject to stereotype threats that can cause them to perform more poorly on IQ and other tests. One of the major pieces of evidence for environmental effects on IQ is the so-called Flynn effect, which states that, in the developed world, IQ increased substantially from 1947 to 2002, and Black IQ rose higher than white IQ in 1950. The IQ gap between Blacks and whites has been narrowing.

Real information therefore casts doubt on the claim that inequality is baked into society by hereditary differences in intelligence, undercutting psychological arguments against using social and political policies to correct racial and economic inequalities. Nisbett's strategy for correcting misinformation about the hereditary nature of IQ is basically factual correction, but it can be supplemented by critical thinking ideas such as motivated reasoning.

If the evidence for racial differences in IQ is as dubious as Nisbett argues, we can wonder about the motives of the main researchers. Much of the research cited to support a hereditary basis for racial IQ differences has been funded by the Pioneer Fund, which was run for a decade by IQ researcher J. Philippe Rushton and has a long history of engagement with white supremacy.[53] Motivational interviewing is unlikely to have any impact on IQ researchers, but it might help with people who have been unduly convinced by them. Political action is ultimately required to prevent IQ from being used as a bogus justification for the maintenance of social inequalities. Institutional modification can also help to ensure that universities, foundations, and government agencies support research that favors equality over inequality.

Myths About Sex

Few cultures have sexual equality in which men and women are equal with respect to income, wealth, and power. Gina Rippon's book on sex, gender, and the brain provides a wealth of evidence-based information that provides factual corrections to misinformation used to justify inequality.[54] Here are some of the myths that have been used to justify discrepancies in achievement between the sexes:

1. Myth: Women are less intelligent because they have smaller brains. Correction: Women's neurons are more densely packed so they have the same number of neurons and the same performance on intelligence tests as men.
2. Myth: Women are too subject to hormonal fluctuations to be effective. Correction: Hormonal fluctuations can be managed in both men and women.
3. Myth: Women are too emotional for intelligent accomplishments. Correction: Women are not more emotional than men, and both sexes can benefit from distinguishing emotions as real information from emotions as misinformation.
4. Myth: Women's brains make them unsuitable for technical fields like science and engineering. Correction: Women can thrive in these fields when given equal opportunities.

Rippon call these Whac-A-Mole myths after an arcade game in which moles repeatedly pop up despite being batted down.

The mythical nature of the alleged reasons for female inferiority, combined with the ever-increasing success of women in government, business, academia, and other realms of accomplishment, raises the question: Why does sexual inequality persist? As in economic and political myths, all patterns of motivated reasoning operate. Assumptions of female inferiority fit with the personal goals of men who want to maximize their own achievements while ensuring the comforts of domestic life, where most of the labor is provided by women. For some men, group identity can also operate if they strongly view themselves as real men as opposed to weak women. Fear of losing power and authority and anger against equalizers can also motivate men to continue to see their sex as superior.

The emotional coherence of prejudice against women is shown in the cognitive-affective map in figure 7.3, which is based on Simone de Beauvoir's pioneering 1949 book *The Second Sex*.[55] She describes how traditionally the strength and superiority of men is contrasted with the weakness and inferiority of women.

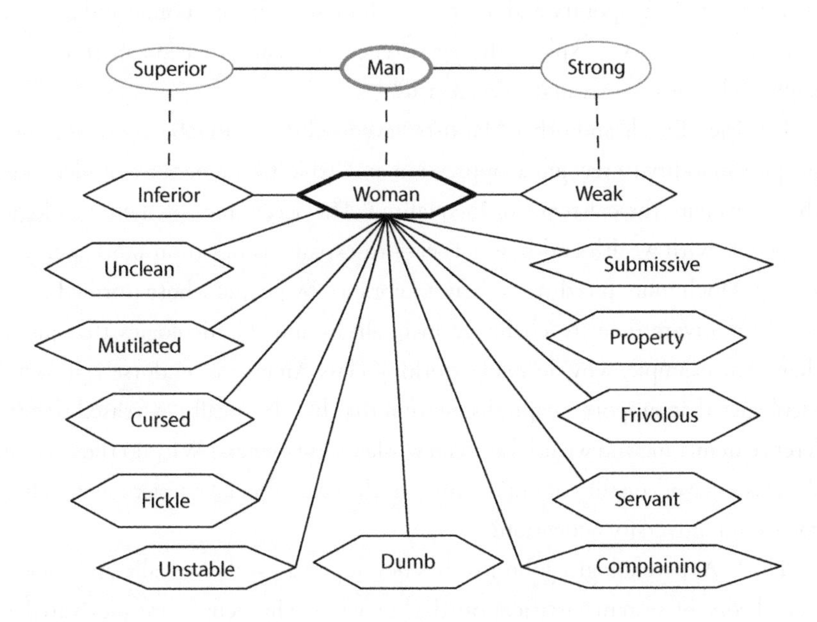

7.3 Cognitive-affective map of misogynistic view of women. Ovals are emotionally positive, hexagons are negative, solid lines are mutual support, and dotted lines show incompatibility.

Source: Paul Thagard, *Mind-Society: From Brains to Social Sciences and Professions* (New York: Oxford University Press, 2019)., p. 111.Reprinted by permission of Oxford University Press.

Fortunately, much progress has been made in subsequent decades to replace this package of negative concepts with a more accurate view of women's capabilities. Nevertheless, fields such as mathematics and philosophy that operate with the assumption that the major determinant of success is raw intellectual talent continue to be inhospitable to women.[56] During the COVID-19 pandemic, misinformation spread about women included claims that they were responsible for the spread of the disease and were using the pandemic to push an equality agenda.[57] Lies are often spread online about female leaders such as U.S. vice president Kamala Harris.[58]

False Consciousness and System Justification

Misinformation about inequality is often believed by people who would benefit the most by exposing it. Examples include women who think that their disadvantages result from natural female inferiority, and poor people who think that their poverty is temporary and not the result of an unjust economic and political system. How can we explain the persistence of misinformation in those who ought to be most motivated to correct it?

Friedrich Engels and other Marxists have used the term *false consciousness* for people's inability to recognize oppression and exploitation because of ideologies that legitimate the existence of inequality.[59] The acceptance of these ideologies by people with wealth and power is easy to explain as based on motivated reasoning. Much more puzzling is their acceptance by people whose unfilled needs should motivate them to challenge inequalities and the ideologies that justify them. For example, why do many working-class Americans endorse right-wing ideologies that support a capitalist system that has drastically increased the difference in income and wealth between workers and owners? Why do they assume the existence of equality of opportunity in the face of social barriers such as high costs for a university education?

The U.A. psychologist John Jost has developed a psychologically and socially rich theory of system justification that explains why people are motivated to defend the status quo: because of needs for certainty, security, and social acceptance.[60] Some of these defenses fit with the patterns of motivated reasoning identified in chapter 3. People's personal goals include wanting to live in a safe and just society so they want to believe that their own society is safe and just despite evidence that their place in it is precarious. Group identities can support convictions that their nation is great because people think of themselves as part of that nation, as in "I am American so the United States is the greatest country in the world."

Jost argues, however, that system justification goes beyond personal goals and group identities, which suggests an additional pattern of motivated reasoning:

SYSTEM JUSTIFICATION PATTERN

- People belong to social systems such as nations.
- These social systems provide reassurances about security and social acceptance.
- Beliefs that the systems are just support these reassurances.
- Therefore, people tend to believe that the system is just.

Jost has accumulated decades of evidence that this pattern operates, for example, in the ways that people respond to threats to the system, such as the 9/11 terrorist attacks on the United States, which made Americans much more patriotic.

I find Jost's explanation of false consciousness plausible, but I wonder about the specific mechanisms of motivated reasoning involved. People ignore their own interests in greater equality because of *all* the factors of selective memory, emotional coherence, and specific emotions. Selective memories operate when people focus only on the good aspects of their societies, such as freedom, while ignoring flaws, such as widespread poverty. This selection feeds into an emotionally coherent picture of their situation, which allows them to ignore aspects of inequality that should concern them. The mechanism of emotional coherence explains how system justification can work by making people feel good about their government and society while ignoring problems. Specific emotions can contribute to system justification when people use selective instances to justify pride in their own societies and feel fear and anger toward external forces that threaten it.

The patterns and mechanisms of motivated reasoning provide deeper explanations of system justification and false consciousness. It ceases to be astonishing that people hold false beliefs and distorted values that go against their own interests because powerful institutions encourage the spread of the beliefs and values that sustain them. False consciousness exemplifies the ubiquitous phenomenon of self-deception which operates in everyone's lives, in domains including health, work, and relationships. Self-deception is naturally explained by motivated reasoning driven by emotional coherence.[61]

How could the misinformation concealed by false consciousness be revealed? Simple factual correction will not apply to the complexes of beliefs and values that serve to justify the status quo, so a combination of critical thinking and motivational interviewing looks more promising. Critical thinking can point to patterns of motivated reasoning operating to mislead people that their lives are tolerable, and forms of explanatory inference and decision making can help people to see that they have been misled. Motivational interviewing can empathically help people to understand how ideology and propaganda are covering up their real needs. Institutional modification and political action are ultimately required to change social and political systems so that people can gain the equality that all humans deserve.

My arguments for the value of equality are based on evidence for its benefits, not on mushy sentiments that can be dismissed as woke, which has long been used as a term for concern with social justice. Around 2020, in right-wing circles, the word "woke" replaced "politically correct" as an aggressive substitute for intelligent thought, accompanied by the noun "wokeism" and even the French "le wokeisme." Sneers supplant reflection on the social costs of inequality and block planning on how to overcome them. Achieving equality requires reinformation that replaces falsehoods and sneers with real information and objective values, as defended in chapter 9. The online supplementary material includes an analysis of institutional and systemic racism.[62]

Could people be persuaded by empathy and argument to shift from the libertarian values in figure 7.2 to the more expansive egalitarian ones in figure 7.1? Empathy might help by encouraging people to appreciate the feelings of those suffering from poverty, homelessness, and poor health care. Abstract freedom *from* government intervention is poor solace for the misery of those who lack the freedom *to* meet their basic needs for biological and psychological security. The best argument for egalitarian values over the narrow libertarian emphasis on freedom is that countries with better social supports, for example, those in Scandinavia, enjoy higher levels of happiness, health, and social stability. Focus on abstract freedoms produces less real freedom for all but the rich.

The most astonishing misperception of inequality is that, in many countries, people underestimate its extent and fail to recognize how much distribution of income and wealth is skewed toward the rich.[63] Another misperception that blocks attempts to overcome inequality is that doing so would harm advantaged groups.[64] These misperceptions arise from general ignorance and also from motivated reasoning to believe that society is generally fair.[65]

In sum, misinformation has been used to justify inequality and discrimination based on race, ethnicity, caste, class, religion, sex, and gender. The methods described in chapter 8 can be used to identify, explain, and correct other ways of dismissing equality as a justifiable objective for human societies. Motivated ignorance can also support inequality, for example, by the recent proliferation of laws in U.S. states to prevent teaching about race.[66]

OTHER SOCIAL MISINFORMATION

Other areas of social life are infected with misinformation. Without attempting a comprehensive analysis, I will review myths and misconceptions that operate concerning romantic relationships and education.

Romantic Relationships

Loving relationships are a major contributor to human happiness, but they can founder when people are misinformed about what makes them work. John Gottman is one of the leading researchers and clinical therapists for romantic relationships, and he draws on decades of experience to identify twelve myths that mislead people.[67]

1. Marriage is just a piece of paper.
2. Living alone with occasional relationships is a lifestyle choice that is equivalent to being married in terms of life outcomes.
3. Conflict is a sign that you're in a bad relationship.
4. Love is enough.
5. Talking about past emotional wounds will only make them worse.
6. Better relationships are those in which people are more independent of and less needy of one another.
7. If you have to work at communication, it's a sure sign that you're not soulmates.

8. If a relationship needs therapy, it's already too late.
9. What couples fight about most is sex, money, and in-laws.
10. All relationship conflicts can be resolved.
11. All relationship conflicts are the same.
12. It's compatibility that makes relationships work.

Gottman and other couples' therapists work to overcome the myths that are interfering with people's ability to love each other.

Misinformation can also come in the form of metaphors based on underlying analogies that distort a relationship instead of helping people understand each other. Metaphor is not just a literary flourish; it is also a powerful source of understanding used in all realms of human thought. People often use metaphors to describe their relationships, which may be bumpy, broken down, needing work, suffocating, stormy, or a thriving partnership.[68]

Psychologists and linguists have identified dozens of relationship metaphors that fall under seven categories, construing a relationship as a journey, machine, investment, container, organism, thing, or bond. Each metaphor compares a relationship to something more familiar, transferring factual and emotional information from the metaphorical source to the relationship target. For example, if people describe their relationship as a roller coaster, the comparison captures both the up-and-down variability of the interaction and the emotional changes that the participants experience.

The comparisons performed by relationship metaphors have purposes that include description, explanation, prediction, decision, and entertainment. Capturing the nature of a relationship in literal words can be difficult so a metaphor such as *rough patch* can be useful. The metaphor may help to explain why the people involved are getting along or fighting, for example, when the metaphor of the crossroads points to sources of disagreement that explain conflict. Similarly, a dire metaphor such as *dumpster fire* may provide a prediction that the conflict cannot be happily resolved. The emotional transfer performed by the metaphor may help the people in a relationship make important decisions, for example when conceiving of it as a partnership encourages the couple to communicate rather than hit the road. Metaphors can be entertaining sources of *humor* that lighten difficult discussions, for example when someone uses the hyperbolic description of a relationship as Game of Thrones to amuse and deflect rather than escalate conflict.

The distinguished clinical psychologist Donald Meichenbaum urged psychotherapists to attend to the metaphors used by clients in their self-descriptions and to help them acquire more positive metaphors.[69] For example, clients who think of themselves as ticking time bombs or deer caught in the headlights would do better if they thought of their lives as works in progress or as looking up. Troubled couples might be helped by a joint metaphor therapy that helps them to find more positive ways of understanding their relationships.

Just as there are good and bad relationships, there are good and bad metaphors. Some metaphors are helpful in describing, explaining, and directing relationships, but others are toxic because they generate more harm than understanding. No relationship benefits from being described as a dead end or a dumpster fire. Relationship metaphors, like relationship myths, can be harmful misinformation.

Educational Misinformation

When my sons were in elementary school in the 1990s, they were given tests to determine whether their learning styles were visual, aural, read/write, or kinesthetic. The aim of these classifications was to tailor teaching to individual differences. It turns out, however, that "there is no adequate evidence base to justify incorporating learning-styles assessments into general educational practice."[70] School boards invested time and money on a fad that turned out to be misinformation, which fortunately was subject to factual correction based on empirical studies.

Fads and misinformation operate in other areas of education, including the crucial areas of teaching reading and mathematics. The "reading wars" concern whether reading should be taught using phonics or whole language, and the "math wars" concern whether mathematics should be taught using discovery learning or conceptual methods. These debates should be settled by controlled experiments that determine the relative efficacy of the different methods combined with rigorous theories about how the brain processes texts and numbers.[71]

Like economics and other fields, education has zombie ideas that persist despite lack of evidence: boys are better at math than girls, schools kill creativity, people use only 10 percent of their brains, listening to classical music makes children smarter, charter schools are more effective, and high self-esteem increases achievement.[72] Some of these ideas should be amenable to factual correction, but overcoming resistance may require critical thinking and motivational interviewing. We can combine the Whac-A-Mole and zombie metaphors by proposing that reinformation requires a novel game of Whac-A-Zombie to deal with fabrications that refuse to die.

My country Canada has a successful public educational system and provides most of its citizens with a good quality of life. But the major atrocity in Canada's history is a system of residential schools for Indigenous children that operated from 1880 to 1996. For much of this period, attendance was obligatory and more than 150,000 children were taken from their families and subject to cultural, physical, and often sexual abuse.[73] Canadians were shocked in 2021 when more than a thousand unmarked graves were discovered at former residential schools.

Justification of the residential schools came from misinformation about Indigenous people, then called Indians. A good illustration is a document produced by Nicholas Flood Davin in 1879 that urged the Canadian government to implement a version of the U.S. industrial schools that were aimed at cost-efficient education and assimilation of Indigenous people.[74] Davin's prejudices are evident in his assertions that Indians have an "inherited aversion to toil" and "chronic querulousness." They are "unable to cope with the white man in either cunning or industry" and are "in a very early stage of development." Other quotes: "He has the suspicion, distrust, fault-finding tendency, the insincerity and flattery, produced in all subject races. He is crafty, but conscious how weak his craft is when opposed to the superior cunning of the white man." "If anything is to be done with the Indian, we must catch him very young. The children must be constantly within the circle of civilized conditions."

The disastrous effects of these attitudes and policies were only generally recognized through the investigations of the Truth and Reconciliation Commission of Canada that operated from 2008 to 2015. The residential schools demonstrate how misinformation can imperil both education and social equality. Overcoming educational misinformation requires changing both individual attitudes and central institutions such as schools and governments.

Romantic relationships and education are just two social domains that are full of misinformation. Misinformation about immigration connects both with political conspiracies and inequality, and includes refutable claims that immigrants spread disease, steal jobs from citizens, and commit criminal acts. People often have misinformation about the legal systems that operate in their countries, as when Canadians who watch only American TV fail to realize that Canada has no death penalty.

BEYOND SOCIAL MISINFORMATION

Inequality is a choice. It is not a choice for individuals who find themselves at the bottom of the social hierarchy because of poverty, homelessness, or illness. But inequality is a choice for societies that can adopt policies such as redistribution of income and wealth that dramatically reduce inequality and misery.

Recognizing the feasibility of such reduction depends on correcting misinformation about the theological, anthropological, economic, and psychological roots of inequality. We can recognize a common message in the research of Graeber and Wengrow on culture, Stiglitz and Krugman on economics, Nisbett on psychology, and Rippon on gender: inequality is substantially reducible if people have the social freedom to choose equality. Misinformation blocks this freedom with false claims and twisted values that downplay the harms of inequality and exaggerate its inevitability. A key part of the socially desirable achievement of greater equality is the process of reinformation that replaces misinformation with real information. The critique of inequality presupposes that it is morally wrong, which requires moral realism, as defended in chapter 9.

The AIMS theory of information contributes to this project by highlighting the differences between real information about inequality and misinformation that serves to justify social discrimination. This chapter has focused on acquisition of data from the world and inference that goes beyond these data, but it would be straightforward to extend the discussions of memory and spread in previous chapters to apply to the case studies of inequality in this one. Understanding how information works, breaks, and mends is a key part of the socially indispensable battle for equality. Chapter 8 generalizes the AIMS approach and presents it as a manual for dealing with misinformation that is vividly illustrated by the Russia-Ukraine conflict.

CHAPTER 8

MISINFORMATION SELF-DEFENSE

A Manual Illustrated by the Russia-Ukraine War

How can you defend yourself against misinformation? Defense is desirable against the serious threats I have examined—COVID-19, climate change, conspiracy theories, and inequality. But you might also be personally a victim of misinformation spread by romantic cheaters, deceptive businesses, consumer advertisers, or medical quacks. This chapter provides a self-defense manual against all kinds of misinformation through attention to the AIMS processes of *a*cquisition, *i*nference, *m*emory, and *s*pread.

I recommend three ways of defending yourself so that you can detect misinformation, correct it, and prevent its occurrence:

1. Apply the AIMS theory to distinguish misinformation from real information.
2. Use the reinformation tool kit to convert misinformation into real information.
3. Adopt a preinformation package to prevent future outbreaks of misinformation.

Just as a good computer manual tells you how to use your machine, this chapter shows you how to manage misinformation by asking incisive questions.

To make the advice concrete, I illustrate it with the abhorrent misinformation that accompanied the Russian invasion of Ukraine in February 2022. Russia attacked Ukraine with bombs and bullets—and also with packages of lies about motives and actions. The AIMS theory, reinformation tool kit, and preinformation package work together to expose and defeat the Russian misinformation campaign.

The global problem of misinformation has become much more severe in the past twenty years because of the internet and social media. Thus, I consider proposals for mending their distressing damage to real information.

APPLYING THE AIMS THEORY TO THE RUSSIA-UKRAINE CONFLICT

The social psychologist Kurt Lewin said that nothing is as practical as a good theory.[1] The AIMS theory of information identifies mechanisms that explain how real information can be achieved and how misinformation arises through breakdowns in these mechanisms. The theory's practical contribution comes from suggesting how misinformation can be differentiated from real information, how misinformation can be converted into real information, and how misinformation can be prevented from arising in the first place.

Application of the theory is guided by asking questions about the eight mechanisms that spell out the AIMS processes of acquisition, inference, memory, and spread. The answers to these eight questions generate profiles that distinguish misinformation from real information, clarifying in particular how Russian propaganda differs from Ukraine's attempts to inform the world about its desperate conditions.

1. How Is Information Collected?

Acquisition of information operates by collecting and representing. With COVID-19 and other cases, we saw that the best ways of collecting real information are perception, instruments, systematic observations, and controlled experiments. Experiments are rarely possible in war zones, but the Ukrainians and the journalists operating in the Ukraine have been able to collect hordes of reliable information by: personal perceptions; instruments such as cameras; and prolonged observations that document the destruction of buildings, the killing and injuring of people, and the millions of refugees forced to flee the Russian invasion.

In contrast, the Russians have tried to spread misinformation by making stuff up, without reliable collecting by observation. Here are some of the main items of misinformation that Russia has fabricatedL[2]

- Ukraine is dominated by neo-Nazis, and Russia's main goal is to remove them.[3]
- Ukraine has been attempting genocide against Russian-speaking residents.
- Russia does not target civilians.
- The West staged a coup in 2014 against the pro-Russian Ukrainian government.
- Ukraine is not a real country but just a creation of Soviet Russia.
- The Ukraine invasion is a "special military action" rather than a war, as illustrated in figure 8.1.
- Ukraine is preparing to use biochemical weapons.
- Ukraine has staged killings to accuse Russia of atrocities.
- Ukrainian refugees are committing crimes.[4]

The reinformation tool kit suggests ways of correcting such misinformation.

In any dispute, both sides may be guilty of misinformation. Ukraine has generally stuck to reliable methods of collecting information, but one inspiring and popular story may have been made up. A February 25, 2022, report said that Ukraine president Volodymyr Zelenskyy responded to an American offer of evacuation by saying, "I need ammunition, not a ride." However, in keeping with journalistic ethics of fact checking, the *Washington Post* could not find U.S. or Ukrainian sources to confirm the utterance, so it might have originated with making stuff up rather than collecting.[5] Another possible fabrication is a story about a fighter pilot named the Ghost of Kyiv who supposedly brought down six Russian planes.[6]

2. How Is Information Represented?

The use of acquired information requires representing it in formats such as sentences, pictures, and videos. Most of the misinformation spread by Russia consists of false sentences, but Russia has also produced fake pictures of alleged Ukrainian attacks.[7] Because pictures are usually representations of what can be perceived, they seem to have a more direct connection with reality than sentences. But technologies such as Photoshop have made it easy to generate bogus photographs.

The Russian-Ukraine conflict has been called the first TikTok war because the social media application, launched in 2017 and especially popular among

Figure 8.1 Vladimir Putin's disinformation.

Source: Graham Mackay/Artizans.com. Reprinted by permission.

young people, makes it easy to post videos.[8] Movies may seem even more vivid than static pictures, but they can also be staged by cinematic techniques. Artificial intelligence has advanced in recent years to allow the production of deepfakes that realistically depict people as saying things they did not. One deepfake has Zelenskyy telling Ukrainian troops to stop fighting, but it was so poorly done that it fooled no one.[9]

Identifying specific kinds of representation is important for telling the difference between real information and misinformation because different formats can fail in different ways. Sentences are just true or false, but pictures and movies allow a broader range of ways in which they can be inaccurate and misleading, for example, by arousing emotions. Responsible media such as the *Washington Post*, CBC, and bellingcat.com offer techniques for detecting misleading images and fake videos.[10]

3. How Is Information Evaluated?

In the AIMS theory, the process of inference covers mechanisms of evaluating and transforming. Evaluating considers how well a piece of information satisfies its purposes such as accuracy and relevance. Chapter 2 described evaluating as the core process in information because it should also govern memory and spread. Evaluating goes astray when it is biased by motivated reasoning where personal goals matter more than accuracy.

In contemporary news stories such as the Russian invasion of Ukraine, the relevant standards of accuracy are the journalistic ones based on vetting sources and checking facts. Most of the real information coming out of Ukraine derives from organizations such as CNN that are committed to these journalistic standards, so the mechanisms of evaluating reports and sources move them toward real information.

In contrast, Russia has blocked all internal news sources from reporting any "fake news" that challenges its official line on what is happening in Ukraine.[11] As a result, Russian information sources, both internal and external, such as RT.com, have replaced accuracy-based evaluating with motivated reasoning, which is in line with the goals of Vladimir Putin and his advisers. Putin may actually believe the conspiracy theories used to justify the invasion of Ukraine, particularly that the West wants to carve up Russia's territory, the North Atlantic Treaty Organization (NATO) has armed Ukraine, the West supports the opposition to Putin, the global lesbian, gay, bisexual, and transgender movement is a plot against Russia, and Ukraine is preparing bioweapons to use against Russia.[12]

4. How Is Information Transformed?

Transforming inferences are ones that go beyond observation by generating novel conclusions about causal relations and general processes. Inferences about motives go beyond observation to judge what beliefs, goals, and emotions provide the best explanation of someone's behavior. Ukrainian, American, and other commentators are engaged in urgent speculation about Putin's motives in invading Ukraine, but no consensus has emerged concerning whether he is primarily concerned with fear of NATO or just reestablishing the glory of Russia.

Russian propaganda has not hesitated to assign motives to the Ukrainian government as pursuing a neo-Nazi project and attempting genocide against Russian people in Ukraine. The alleged evidence for these hypotheses is misinformation about the actions of Ukrainian leaders, so the conclusions based on them are also misinformation, especially when Ukrainian leaders have much more plausible motives based on their behaviors and self-reports. Specifically, Ukrainian leaders want to preserve a free and democratic country.

5. How Is Information Stored?

The process of memory requires mechanisms for storing and retrieving information. For real information, both mechanisms should depend on evaluating so that only accurate and relevant information is saved and recalled.

On the Ukrainian side, based on a free press in a democratic if beleaguered society, appropriate storage still operates. Memory for the war consists largely of the press record that is subject to journalistic standards. Unreliable journalistic vehicles such as Fox News do not comply with these standards, and social media sites such as Twitter, Facebook, and YouTube are subject to the whims of their contributors. However, Facebook and other sites have partially tried to evaluate information about Ukraine and taken down some erroneous posts.

On the Russian side, however, objective evaluating is utterly missing from information storage. The free press in Russia was sorely limited before the invasion of Ukraine and has been almost destroyed since the imposition of full censorship in March 2022. Much of what is stored in Russian digital memories of the war is likely to be false.

6. How Is Information Retrieved?

Retrieving something from a mind or digital resource does not mean it should be believed because it may have been unreliably sourced and stored. So evaluating should follow retrieving just as it should precede storing, in accord with standards of accuracy and relevance rather than the personal goals that provoke motivated reasoning.

The Ukrainians are doing well in retrieving information from reliable storage, but the Russians are setting themselves up to be victims of motivated retrieving. Without a free press, they are restricted to official records of crucial information such as the number of casualties in their own troops. Putin and his government are thus prone to remembering only what they want to, a strategy that can be disastrous when good decisions depend on reality rather than fantasy.

7. How Is Information Sent?

Spread of information depends on social mechanisms of sending and receiving, both of which should be constrained by evaluating. President Zelenskyy has masterfully used media such as videos to inform the world about Ukraine's needs for support. In addition to standard media such as television, information about Ukraine's plight is spread by social media sites, including ones like YouTube and TikTok that circulate videos. Ukrainian sending sometimes spreads misinformation, such as that about the Ghost of Kyiv, but usually provides real information

In contrast, Russia has a limited range of vehicles for sending information because the world media have learned to be skeptical of their transmissions. Russia controls its press so anything coming out of the country has a good chance of being misinformation. Since the 2016 U.S. election, Russia has been notorious for troll farms that generate and recirculate misinformation on topics that include vaccinations and QAnon.[13] Russian television misleads watchers by insisting that its army is triumphant against the Ukrainian forces who commit atrocities.[14]

The motivation for these exercises is not just to spread falsehoods, for example, about Hillary Clinton, but more to mark all sources of information as illegitimate by producing a firehose of falsehoods.[15] Then all news seems fake, allowing autocratic leaders to feed people just the information they want to spread. Donald Trump and Steve Bannon pursued a similar strategy. The United States has claimed that Putin was getting misinformation from his own advisers in the defense department who are afraid to admit to military failings.[16]

8. How Is Information Received?

People are often gullible in their reception of information, tending to believe what they are told rather than suspecting the message and the messenger. But in

TABLE 8.1 Profiles of real information and misinformation

Process	Real information—Ukraine	Misinformation—Russia
Acquisition	Collecting by observations, cameras, satellites	Making stuff up, sloppy observations
Inference	Evidence-based causal reasoning about Vladimir Putin's motives	Motivated reasoning about Ukrainian leaders
Memory	Evaluation-based storing and retrieving of relevant facts	Motivated storing and retrieving by controlled media
Spread	Evaluation-based sending and receiving	Motivated sending and receiving (propaganda)

contentious domains such as international relations, a better strategy is to screen received messages by evaluating their plausibility and origins.

The long Russian record of misinformation spread and the country's strong motive to present itself positively in the conflict with Ukraine provide strong reasons to doubt what Russia says. As in the Zelenskyy quote about ammunition and the story about the Ghost of Kyiv, some skepticism about Ukrainian utterances is also warranted, but Ukraine has a much better history of propagating real information than Russia. Thus, evaluating Ukrainian sources can be more relaxed than evaluating Russian sources. Table 8.1 summarizes the differences between real information on the Ukrainian side and misinformation from the Russian side.

THE REINFORMATION TOOL KIT

I have identified eight techniques for converting misinformation into real information. Reinformation operates by asking incisive questions about diagnosing and fixing misinformation, here illustrated by the Russian invasion of Ukraine.

1. What Misinformation Can Be Identified?

Repairing misinformation requires first identifying it, just as curing medical patients requires first identifying their diseases. The Russia-Ukraine conflict shows the value of identifying misinformation through contradictions, lack of evidence, and sources known to be unreliable.

Misinformation jumps out at us when we encounter a claim that contradicts something we already believe. People who knew that the Ukrainian government had been freely elected on a centrist platform immediately rejected the Russian propaganda that Ukraine was dominated by neo-Nazis. Similarly, claims about Ukrainian genocide against Russian are incompatible with the professed aims of President Zelenskyy to reconcile Ukrainian and Russian populations. My discussion of conspiracy theories identified cases where claims could be recognized as *self-contradictory*, but I have not noticed this extreme incoherence in Russian misinformation. On the other hand, some Russian claims about military successes were clearly contradicted by Ukrainian images of stalled and broken tanks.

When we hear a claim that contradicts our beliefs, we should not immediately assume that the claim is misinformation because our previous beliefs may be erroneous. Rather, the incompatible utterance should prompt us to reconsider our beliefs and figure out whether the new claim is more coherent with the overall evidence than our old views. Then identifying the claim as misinformation results from evaluating it on the basis of all the relevant evidence. For example, when Putin claims that Ukraine is historically part of Russia, we should look at the evidence that could validate or invalidate his claim. The supplementary material for this chapter lists some good websites for fact checking.

Another way to spot possible misinformation is to notice whether a claim is backed with any evidence. Scientists are expected to provide evidence that supports their hypotheses, and politicians sometimes spell out the evidential bases for their proposals. A strong claim that is merely stated as a pronouncement without any associated evidence should raise suspicions of being misinformation. Russian claims about Ukrainian attempts at anti-Russian genocide were not accompanied by documentation, raising suspicions that they are misinformation.

A third way to detect misinformation is to recognize a claim as coming from a source that is known to be both biased and unreliable, such as Donald Trump, whose thousands of pronounced falsehoods have been documented. The Russian proclivity for misinformation spread was already well known from its lies about the 2014 invasion of Crimea and its meddling in the 2016 U.S. election. Russian motivations for anti-Ukraine propaganda were obvious from the invasion.

Hence, the Ukrainians and their supporters can legitimately apply to Russians a presumption of misinformation. Compare the old joke: How can you tell that lawyers are lying? Answer: Their lips are moving. Some sources of information are so frequently wrong that we can quickly discount what they say.

2. What Are the Sources of Misinformation?

When a source of information is unfamiliar, it requires further investigation. My sources for the Russia-Ukraine cases are mostly media with which I have years of experience that has generated confidence in their general, although not perfect, accuracy: CNN, CBC, the *Guardian*, the *New York Times*, the *Washington Post*, the *Toronto Star*, and the *Economist*. *Stanford News* published useful instructions about how to avoid disinformation about the conflict.[17] I already had ample reason to doubt claims about Russia coming from U.S. right-wing commentators such as Tucker Carlson and QAnon supporters.[18]

I also encountered websites that were new to me, however, and therefore required more careful scrutiny. By looking at content and affiliations, I came to recognize the Netherlands organization bellingcat.com and the German organization dw.com as generally reliable sources of information about Russia and Ukraine. On the other hand, RT.com (formerly Russia Today) is a blatant propaganda arm of the Russian government and totally untrustworthy. Internal Russian sources are also unreliable because of intense censorship.[19]

One of my web searches for information about Nazis in Ukraine led me to thegrayzone.com, produced by an American, which turned out to be unduly sympathetic to Russia and China.[20] Thus, I learned to discount its claims as likely misinformation. The assessment of new sources of information depends unavoidably on considering how well they cohere with already trusted sources.

3. What Are the Motives of the Originators and Believers of Misinformation?

My examination of COVID-19, climate change, conspiracy theories, and inequality demonstrated the role of motivated reasoning in the generation and propagation of misinformation. Thus, the project of turning misinformation into real information requires identifying the motives of both the originators and recipients of falsehoods. Recognizing the goals of the producers of misinformation helps with the identification of lies, while recognizing the motives of the recipients should help with the process of correcting their mistakes, either directly or by critical thinking and motivational interviewing. Chapter 3 described motives as goals, beliefs, and emotions that can work together to distort reasoning.

In the Russia-Ukraine conflict, the primary originators are the Russians, particularly their autocratic leader, Vladimir Putin. Thus, journalists and government officials have been intensely concerned with figuring out what Putin wants. One hypothesis is that he has some illness that makes him prone to emotional outbursts that are directed against Ukraine as a proxy of the West. Another hypothesis is that he is not actually interested in conquering Ukraine but is trying to acquire the rich gas and oil resources of its eastern provinces.[21]

The historian Timothy Snyder provides a rich account of Putin's motivation.[22] He connects Putin's ideas with those of a Russian philosopher, Ivan Ilyin, whom Putin often quotes and honors. After exile from the Soviet Union in 1922, Ilyin attacked it voluminously while defending Christianity and fascism. Putin follows Ilyin in maintaining that Ukraine is part of Russia and is not a separate country. In 2021, Putin published a document translated as "On the Historical Unity of Russians and Ukrainians," which argues on historical, linguistic, and religious grounds that Russians and Ukrainians are one people.[23] Then his primary motive in the invasion is simply the reunification of the Russian state, as his emotional outbursts before the invasion suggest. From Putin's point of view, he is merely trying to restore a relationship between Russia and Ukraine that has been distorted by the old Soviet Union and the current West.

Putin's ideology fits with the characteristics of fascism listed in chapter 7: mythic past, glorious leader, national superiority, perceived threats, inevitable destiny, and virility (displayed in Putin's famous shirtless photos).[24] We can therefore see Putin's disinformation campaign about Ukraine as resulting from a toxic package of beliefs and goals. Hence, the best explanation of Putin's current motives is nationalistic Russian ideology rather than emotional derangement or self-interested expansion.

The emotional coherence of this package is depicted by the cognitive-affective map shown in figure 8.2. Putin values Russia as associated with Orthodox Christianity and his own autocratic power. The current version of Ukraine is devalued because of its association with democracy, NATO, and the West, all of which are threats to Russia. Putin can then justify the military operation against Ukraine as coherent with what he loves and hates. The same configuration justifies the use of disinformation in the propaganda war against Ukraine.

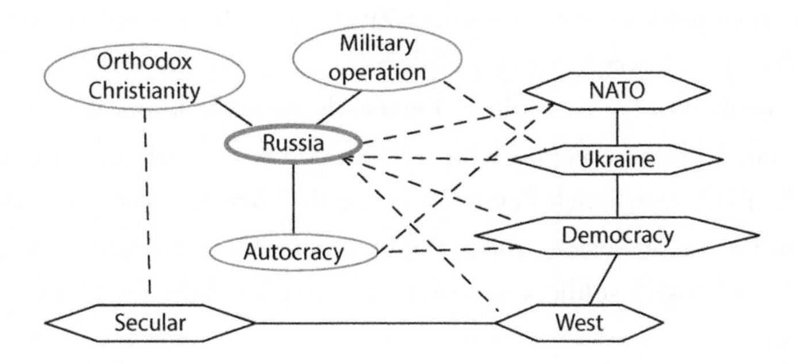

8.2 Cognitive-affective map of Putin's motivation for attacking Ukraine. Ovals are emotionally positive and hexagons are negative. Solid lines indicate mutual support; dotted lines indicate incompatibility.

Putin's mindset is too hardened to allow hope for reinformation, but prospects for reaching his supporters in Russia and other countries are more promising. Ordinary Russians may have motives less extreme than Christian fascism and therefore might be convinced that the invasion of Ukraine was not justified.

What are the motives of Putin's U.S. supporters, such as Donald Trump and Tucker Carlson? There is speculation that Putin has compromising information about Trump's financial or sexual activities, but Trump's frequent positive comments about Putin such as calling him a genius and savvy with respect to Ukraine may simply result from ideological similarity and admiration of autocracy. Other support for Putin may simply be based on ignorance of international relations, as in the case of U.S. congressional representative Marjorie Taylor Greene.[25]

4. What Items of Misinformation Are Subject to Factual Correction?

My first three questions about identifying misinformation, sources, and motives are preparation for the core operations of reinformation by factual correction, critical thinking, and motivational interviewing. Factual correction is the most straightforward and can work for misinformation that is easily shown to be false. For example, early Russian claims to be well on the way to conquering Kyiv were easily refuted by photographs of Russian forces stalled far from the capital. Cameras, drones, and satellites serve as instruments that powerfully expand human perception.[26]

Factual correction is sometimes more complicated. The usual Western response to Russian claims that they are out to eliminate Nazi leadership in

Ukraine is to point out that President Zelenskyy is Jewish and lost uncles in the Holocaust. However, some people who are Jewish have defended other Nazi sympathizers. For example, in France, the far-right leader Eric Zemmour has defended Philippe Pétain who collaborated with Nazi occupiers during World War II, even though Zemmour is Jewish. There are some neo-Nazi elements in Ukrainian society, for example, in the Azov regiment that continued to fight Russian soldiers in Eastern Ukraine until the Russian victory at Mariupol.[27]

Closer scrutiny, however, serves to correct the Russian allegation as nonfactual. No evidence exists that Zelenskyy has ever expressed Nazi sympathies, and the far right is only a tiny part of Ukraine's parliament. Putin may actually believe that Nazis abound in Ukraine because of the prominence of the Azov regiment in the resistance to Russian forces, but he is easily shown to be factually wrong about his general claim that the Ukrainian leadership and population are neo-Nazi.

5. How Can Critical Thinking Be Used to Identify Thinking Errors and Correct Them?

I described critical thinking as the two-step process of noticing that misinformation results from thinking errors (biases and fallacies) and then correcting the misinformation by applying legitimate procedures such as careful causal reasoning. One fallacy committed by Russian leaders is hasty generalization, where they infer from a tiny portion of Nazi involvement in the Ukraine to the conclusion that the country is led by neo-Nazis.

Motivated reasoning based on personal goals abounds in the thinking patterns of Putin and his followers. They distort reality in service of their desires to revive Russia and triumph over the West, leading to misinformation about the purpose and progress of their invasion. Expectations of a quick victory were based on what they wanted rather than on evidence-based assessment of military strengths and weaknesses. Ukrainians such as Zelenskyy also have strong goals such as resisting the invasion, but they are more constrained in what they can say by the accuracy goal to remain credible to their Western allies.

Group identities contribute to the motivations of Russians and Ukrainians. Putin's Russian identity includes Orthodox Christianity, autocratic government, and Ukraine as a historical part of Russia. This identity fuels the motive to return Ukraine to Russian control and distorts reasoning, encouraging misinformation such as that Ukraine is dominated by neo-Nazis.

Russian identity also generates specific emotions, such as fear that Ukraine will become more Western and anger that Ukrainian leaders want to join NATO and the European Union. These emotions become motives for actions such as invading Ukraine and destroying Ukrainian housing and factories. For Ukrainians, the Russian invasion has intensified their identity as non-Russian, but it has not led to obvious cases of motivated reasoning. For the Russians, in contrast, we have seen cases of motivated reasoning that fit the personal goals, group identity, and emotions patterns.

The best antidote to motivated reasoning is the judicious use of evidence in a broad range of good reasoning methods, including statistical inference and inference to the best explanation. For example, the best explanation of Ukrainian opposition to Russia is not that Ukraine is dominated by neo-Nazis but instead that Zelenskyy and others want to adopt and maintain the freedoms and democratic practices that are more common in the West.

Another tool of critical thinking is the evaluation of analogies as strong, weak, bogus, or toxic.[28] The Russian comparison of the Russian action against Ukraine with the great patriotic war against Nazi Germany counts as bogus because of its factual and causal errors, and it qualifies as toxic because of all the harm the invasion has produced. Hence, critical thinking can be used to debunk Russian propaganda and may also be useful in convincing nondogmatic Russians of Putin's evil actions.

6. How Can Motivational Interviewing Be Used to Change Attitudes and Behaviors Based on Misinformation?

In contrast to the logic-based approach of critical thinking, motivational interviewing operates more like psychotherapy in using conversation and empathy to change attitudes as much as beliefs. Motivated reasoning can be corrected by substituting evidence-based reasoning, but it can also be modified by shifting a person toward different motives. Motivational interviewing for alcoholics can help them to value health, work, and family over excessive drinking that is threatening these values.

Putin is neither accessible nor amenable to motivational interviewing, but the technique might work with some Russians who currently support the invasion of Ukraine. Without abandoning the goal of a strong Russia, Russians could nevertheless be encouraged to see the importance of other goals such as having a strong economy unhindered by sanctions, avoiding the waste of military spending, and being respected by other countries as peaceful rather than bellicose. A motivational interviewer would empathize with Russian worries about being dominated by the West while suggesting that conquering the territories of the former Soviet Union is not an effective strategy.

The term "information therapy" usually means providing medical patients with more information about their diseases, but I suggest "infotherapy" as the broader approach that works with attitudes and emotions as well as beliefs. Infotherapy could encourage people to care as much about accuracy and social goals as about personal motivations.

7. How Can Institutions Be Modified to Reduce Misinformation?

Chapter 5 described five kinds of institutional modification that can contribute to reinformation: creating institutions; eliminating institutions; changing membership; altering explicit policies; and changing implicit values, norms, and practices. I am not aware of any institutions generated specifically to increase real information about the Russia-Ukraine conflict, but many reputable news sources have stepped up coverage. Examples of organizations that shifted resources to improve reporting on the conflict include CNN, the *Guardian*, the *Washington Post*, and bellingcat.com.

Institutions most worthy of elimination include Russian agencies that covertly spread misinformation about Ukraine as well as public outlets such as RT. Such elimination is not politically feasible nor is agitation to change their policies and practices. In democratic countries, Ukrainian resistance against Russia has had widespread support in government agencies and news sources. But the right wing of the U.S. Republican Party and conservative news sources such as Fox News are challenging aid for Ukraine.[29] Some organizations such as the CBC have recognized their responsibility to provide real information about the Russian invasion and have responded with commitments to deploy journalists to the war zone, verify content from other sources, and be cautious in displaying brutal images.[30] Canada is leading an international initiative to recognize and manage Russian misinformation.[31]

8. What Political Actions Can Be Used to Mitigate Misinformation?

In dealing with COVID-19, climate change, conspiracy theories, and inequality, stages are reached where no amount of factual correction, critical thinking, and motivational interviewing can suffice to bring about needed change. Then political actions such as government regulation, electoral change, and lobbying become the most effective ways of dealing with fundamental problems. Political action could also work to reduce people's feelings of powerlessness that make them more susceptible to misinformation.[32]

Misinformation needs to be tackled as part of resolving the Russia-Ukraine conflict, but the solution is ultimately political rather than logical or psychological. Russia will not stop spreading misinformation until the conflict is resolved, preferably by negotiation rather than Russian victory. Steps can be taken more immediately to restrict the flow of Russian propaganda in Western media, for example, by government and corporate actions that limit the influence of RT.

In contrast, social media have struggled to deal effectively with Russian disinformation, which is "running amok."[33] Facebook, YouTube, and Twitter have announced new measures that have been poorly enforced, especially outside the United States.[34] TikTok has not dealt with the more than one hundred influencers using the platform to support Russia. Twitter has started a new initiative to allow users to fact-check each other, but it is barely a beginning.[35] Political action is needed to constrain social media so that their fundamental conflict of interest between real information and real profits is resolved in a socially responsible direction. The success of Elon Musk in taking control of Twitter is worrying because he has long battled organizational constraints on messages that serve his personal and corporate interests.

The devastation in Ukraine has been horrible, but the consequences would be much worse if the conflict escalated to include Russia's thousands of nuclear weapons. Nuclear war could kill millions or even billions of people, so Western leaders have avoided provocations such as sending troops to Ukraine or establishing a no-fly zone that would produce battles between Russian and Western planes. These decisions require distinguishing real information about the motives and likely actions of Putin from misinformation about his bluffs and intimidations. Hence, political action depends on getting information right, as it did for COVID-19 and the other cases I discussed, and political action can reciprocally tilt the balance toward real information.

THE PREINFORMATION PACKAGE

Chapter 1 introduced the term "preinformation" as the analog of preventive medicine, covering ways of averting the generation and spread of misinformation. Strategies for preinformation include reducing gullibility, teaching critical thinking plus motivational interviewing, prebunking (the anticipatory debunking of misinformation), and improving gatekeeping that separates information from disinformation. We can help to prevent misinformation by asking corresponding questions.

1. How Can People Be Less Gullible?

To be gullible is to be easily duped or cheated, and mental error tendencies such as motivated reasoning are part of the explanation of widespread gullibility.[36] Another part is even more fundamental because it does not require motivation or other error: people tend to believe whatever they are told without thinking about it much.[37] Automatic belief sounds ridiculous from a logical point of view, which supposes that beliefs arise from inference patterns, but it is much more plausible with perceptual beliefs. If your visual system categorizes a gray furry object in your yard as a squirrel, you believe that your yard has a squirrel in it. Similarly, if someone tells you a squirrel is in the yard, you believe it unless something is amiss with the claim or the claimer.

People living with trusted family members or other small groups are justified in making belief automatic because most of what they are told will be credible based on the perceptions of reliable others. However, agriculture allowed people to operate in groups much larger than the fifty to one hundred in a typical hunter-gatherer clan, so automatic belief became a dubious strategy: hundreds of contacts who are not close cannot be trusted to the same degree because they have different experiences and motives. The risk is magnified when contacts include the billions of people on social media, where it is pure folly to believe what is said by a random person on TikTok.

One solution is to encourage people to use a model like the one shown in figure 8.3, which distinguishes a *default pathway* on the top and a *reflective pathway* on the bottom.[38] When a person makes a claim, we should not simply believe the claim. At the least, we should check that the claim is consistent with the rest

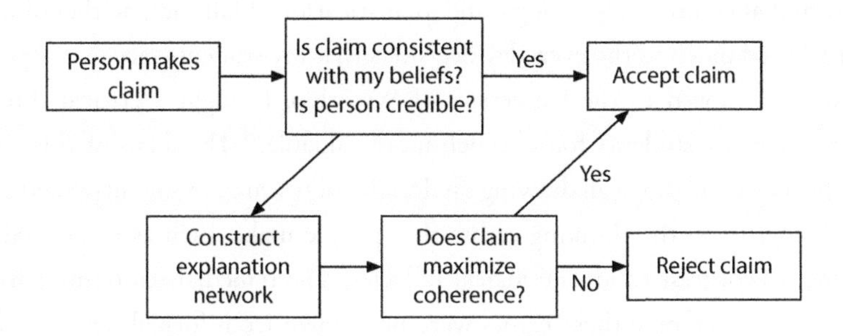

8.3 Two-pathway model of reacting to testimony when a person makes a claim.

of our beliefs and that we have some reason to believe that the person is credible. If these checks are satisfied, then we can accept the claims without a lot of deliberation. However, if the consistency and credibility checks fail, then we need to do much more work to determine whether the claim is part of the system of beliefs with the most overall explanatory coherence.

How can we encourage people to replace the natural habit of automatic belief with something like the two-pathway model in figure 8.3? One pedagogic strategy is to point out to people how frequently they are misled by politicians, salespeople, advertisers, and social media contacts. Warnings about the prevalence of misinformation have become common but need to proliferate. People are less ready to admit that they have been duped than they are to recognize that other people have frequently been misled by information they picked up online. Warning people about the abundance of false claims sent their way should encourage them to recognize the value of the old saying: Don't believe everything you hear.

Avoidance of gullibility is helpful for dodging misinformation about the Russia-Ukraine conflict. Claims about the causes and course of the war should not be believed just because they come from some television anchor or social media source.

2. How Can Critical Thinking and Motivational Interviewing Be Taught?

People can learn more specific ways of dealing with misinformation by instruction in critical thinking and motivational interviewing. Early in my professorial career, I taught informal logic for seven years, but I came to doubt that I was helping my students improve their thinking. I was following the usual philosophy pattern that combined basic logic and identification of fallacies, neither of which contributed much to the everyday decisions that my students faced.

When I moved to the University of Waterloo, I taught a critical thinking course that my students found much more valuable.[39] The focus of this course was heavily psychological, drawing on decades of research by cognitive and social psychologists on the thinking errors that people make, such as motivated reasoning, in contrast to arcane logical fallacies. The remedial reasoning patterns I taught to overcome these errors were not drawn from formal logic but from useful kinds of inductive inference including statistics and inference to the best explanation. I also dealt at length with the most important inferences that people have to make concerning what to do rather than merely what to believe. Critical thinking about decision making can draw on rich psychological research on how people err practically and also on theories of good decision making.

Such critical thinking courses could easily be taught generally in high schools as well as in colleges and universities. Schools in Finland and other countries are already teaching students to deal with misinformation.[40] To be useful at addressing misinformation, it should add treatment of current controversial issues such as climate change. People can learn to be more self-aware of their own biases that result from their motivations and prior beliefs. They can also learn to be much more selective about their sources of information, separating reliable from unreliable news distributors. In addition, critical thinking should encourage people to take personal responsibility for pointing out misinformation to others and for supporting other debunkers.[41]

Teaching motivational interviewing is much more problematic because the psychotherapy techniques it adapts require years of professional education and practical experience. But a major predictor of the success of psychotherapists is not their theoretical approach but rather their ability to establish a trusting relationship with their clients based on empathy.[42] Capacity for empathy varies among individuals, with psychopaths largely lacking it, but a Toronto organization (Roots of Empathy) has had great success in teaching students to be more empathic. Roots of Empathy uses techniques such as bringing infants into classrooms to improve the ability of students to recognize and appreciate the emotions of their fellow students.[43] Perhaps their methods could be adapted and generalized to make people generally more empathic and hence more capable of using motivational interviewing to rescue people from misinformation linked to emotional attitudes.

If people learn to get better at critical thinking and motivational interviewing, they should become better at using these techniques to avoid misinformation about the Russia-Ukraine conflict, in line with the measures proposed above for reinformation. The preventive advantage would be to use the techniques to keep from getting misinformed in the first place, for example, by identifying and understanding propagandists driven by motivated reasoning.

3. How Can Prebunking and Inoculation Be Used in Contentious Circumstances?

Within particular contexts, an effective way of preventing misinformation is prebunking: making people aware of potential misinformation *before* it is presented.[44] Prebunking has been used in in the Ukraine invasion to warn about the proclivity of Russia to spread disinformation and more specifically to tell people to beware of deepfake videos that misrepresent Ukraine.[45] Prebunking can be practiced by individuals, but it can also be a policy of institutions in education, media, and government.

People should learn to practice prebunking as a way of preventing misinformation in a wide variety of circumstances. Even with the best techniques of factual correction, critical thinking, and motivational interviewing, dislodging misinformation can be difficult once it is has settled in people's minds. Hence, defenders of truth and justice should adopt the practice of identifying issues on which misinformation is likely to arise and then warning people about ways in which they might be manipulated to get things wrong. Targets can be advised against particular falsehoods that are likely to be sent their way and also against methods of spreading misinformation that are dangerously effective. For example, people can be cautioned against succumbing to motivated reasoning, fear-driven inference, and emotionally vivid videos. We should be ready to prebunk when we spot a contentious issue where antagonists are motivated to gain advantage by spreading misinformation, as in the Russian invasion of Ukraine. Prebunking should be part of the preinformation package aimed at keeping misinformation from starting to roll.

Inoculation adds to prebunking a warning that serves to activate awareness of the threat of misinformation more generally. Experiments with short videos have found that they increase people's ability to detect untrustworthy content.[46] Another preinformation technique is critical ignoring, which encourages people to avoid tempting sources, verify suspicious sources, and avoid rewarding malicious actors with attention.[47]

4. How Can Gatekeeping Become More General and Effective?

Gatekeeping is "the process of culling and crafting countless bits of information into the limited number of messages that reach people each day."[48] This process has been studied most intensely in journalism, which has developed ethical standards such as accuracy, independence, corroboration of witnesses, and confidentiality of sources. I agree with Stephen J. A. Ward that such standards are compatible with support for democracy, a value harmonious with reporting the truth.[49] One advantage of having gatekeepers is that they usually have different personal goals and group identities from the generators of information and therefore are not prone to the same motivated reasoning.

Science, medicine, and law have also found principled ways to restrain the spread of misinformation. Science journals employ peer review and editorial scrutiny to increase the chances that experimental and theoretical reports get the world right. Such review is not foolproof, but it does help to ensure that articles cannot be published without some critical examination of their methods and results.[50] Unfortunately, the internet has brought proliferation of predatory journals that will publish anything for a fee, undermining the valuable constraint of peer review.

Gatekeeping is performed in medicine by government authorities and professional associations that decide on appropriate treatments based on the best scientific evidence, especially evidence from studies based on controlled clinical trials. Evidence-based medicine has been increasingly practiced since the 1990s, providing a valuable restraint on the adoption and use of ineffective drugs and procedures.[51]

Legal proceedings also restrain misinformation through rules of evidence that disallow the presentation of testimony that is irrelevant, inflammatory, unfair, coerced, or hearsay.[52] Journalistic ethics restricts publication of material that has not been fact-checked and corroborated by multiple witnesses. Thus journalism, science, law, and medicine have ways of fighting information incontinence by vetting information before it is spread. Social media should be pushed by both public agitation and government legislation to move closer to the communication standards of journalistic ethics.

We cannot expect ordinary individuals to observe the professional restraints found in journalism, science, law, and medicine, but people can easily adopt tighter standards of communication. The United Nations had a campaign to reduce misinformation using the slogans "Pause" and "Take Care Before You Share." Pausing interrupts an automatic response that encourages people to pass on messages and provides time to engage in critical thinking before sharing information that may be bogus.[53] It sensibly asks people to avoid sharing information before they have had time to ask themselves these questions:

WHO made it?
WHAT is the source?
WHERE did it come from?
WHY are you sharing this?
WHEN was it published?

I would expand the list to include:

HOW is the information consistent with other things you know?
WHAT is the evidence that it is true?
HOW could spreading this information hurt or help people?

The "care" in "take care before you share" can mean both caring about people likely to be affected by the information and caring about truth so that what is spread is accurate and not misleading.

This UN effort tried to make everyone into a gatekeeper, but I would also like to see much better institutional gatekeeping with respect to daily information, particularly on social media. Editors, peer reviewers, and judges are like firefighters trying to control the flames of misinformation, but social media are like arsonists trying to fan them. Facebook and Instagram have tried limply to regulate the spread of misinformation concerning the Russia-Ukraine war, but the results have been inconsistent and confused.[54] Social media companies suffer from inherent conflicts of interest because their impact and profits depend on engagement and entertainment rather than on pursuit of truth and justice.

I would like to see a detailed comparison and assessment of how gatekeeping works in science, medicine, journalism, and law. Table 8.2 provides a start, laying out the differences among the professions with respect to who does the gatekeeping, what are their targets, why they are performing selection, and how successful they are in accomplishing their purposes. Resulting general principles of gatekeeping might then suggest how to rein in the appalling

TABLE 8.2 Comparison of gatekeeping in four professions

Domain	Gatekeeper	Targets	Purpose	Success
Science	Editors, reviewers	Articles, books	Truth, relevance	Good, but imperfect
Medicine	Review boards	Drugs, procedures	Treatment effectiveness	Good, but imperfect
Journalism	Editors	Articles	Accuracy, interest	Mixed
Law	Judges	Testimony	Relevance, probative value	Mixed

proliferation of misinformation on social media. Institutionalizing such gatekeeping should contribute hugely to preventing misinformation from spreading. I rate the success of gatekeepers in journalism and law as mixed because of biased news media and prejudiced judges in countries where judges are elected or politically appointed.

Upgrading gatekeeping requires institutional modifications to make scientific, medical, journalistic, and legal organizations more truth-oriented. Institutional modification is also required to help educational organizations get better at teaching critical thinking and the empathy required for motivational interviewing. Gullibility reduction and prebunking can also benefit from interactions between truth seekers and people at risk of misinformation. Hence, the prevention of misinformation is a social as well as an individual process.

MENDING THE INTERNET AND SOCIAL MEDIA

Misinformation has existed throughout recorded history, with examples such as the Bible story that the world was created in six days and the ancient Greeks' tricking the Trojans with a bogus horse. Deception accelerated in the twentieth century as countries developed systematic programs to spread disinformation through agencies such as the Soviet KGB and the U.S. Central Intelligence Agency (CIA).[55] But the twenty-first century brought technologies that catastrophically increased the impact of misinformation.

The internet was developed in the 1980s, but it only became socially important in the 1990s with the arrival of the World Wide Web. Early reactions were utopian, viewing the Web as a wonderful contributor to human communication and democratic education. Soon, however, evil uses, such as child pornography, financial scams, and terrorist activities, began to proliferate.

The 2000s brought social media that quickly became important to people's lives through the impact of Facebook, Twitter, YouTube, and other platforms. The scale and speed of these media have introduced enormous new opportunities for spreading information that is false, inaccurate, or misleading. The past two decades have appropriately been dubbed the "age of misinformation." We can identify the intense problems caused by these new technologies and sketch possible solutions.

Problems Caused by New Internet Technologies

Why is misinformation today much more widespread and destructive than it was in 2000? Websites and social media have unusual properties that magnify misinformation by their impact on volume, speed, immoderation, and control of information.

Before the internet, the volume of information was limited by space and time constraints. Newspapers, magazines, journals, and books are restricted by the availability and cost of paper. Television and films are restricted by cost and the amount of time available to screen them. In contrast, extraordinary increases in the memory and speed of computers have exponentially increased the volume of information produced. The author of the 2020 book *Lie Machines* summarizes: "Globally, each day, we generate five hundred million tweets, send 294 billion emails, put four petabytes of content on Facebook, send sixty-five billion messages on WhatsApp, do five billion searches on Google."[56] Another author calls this volume "cheap speech" because it is both inexpensive to produce and often of markedly low social value.[57]

Digital media are remarkable for their speed as well as their volume of communication. Newspapers used to require at least a day between writing and publication, with magazines, journals, and books much slower. Today's tweets, texts, emails, and web posts are virtually instantaneous, with the potential to reach millions of people. So spread of information and misinformation is potentially far faster than just a few decades ago.

Social media have also encouraged immoderation and polarization in the groups of people who use them. Social media can best maximize advertising revenue by content that incites strong emotions in their users and thus engages them and spurs interactions with other similarly minded users. The resulting tendency is toward polarization of people into isolated groups with extreme views.

With traditional media, the spread of information is subject to controls for accuracy and social value, described in my discussion of the role of gatekeepers in journalism and other fields. In contrast, value control is almost totally lacking in websites such as 4chan and 8kun and in the encrypted messaging apps Telegram and Signal. Control is only weakly effective in the popular social media Facebook, Twitter, YouTube, TikTok, and Instagram.

This lack of control, along with the volume, speed, and immoderation of communications, shows why the internet is such a powerful source of misinformation, through breakdowns in the mechanisms for acquisition, inference, memory, and spread. Real information depends on reliable collecting, which is not required for posting on social media. Social media can employ inaccurate representations that include false sentences and fake pictures, sounds, and movies. Evaluating for accuracy and social value is replaced on social media by estimations of profitability and emotional impact. Storing, retrieving, sending, and receiving are also tied to financial and political motivations that are inimical to truth and justice. Anonymity protects vile and vicious posters from being held responsible for the harm they cause. Hence, the internet and social media are justly blamed for the gargantuan proliferation of misinformation in recent decades, along with other causes such as the rise of autocratic strongmen.

Solutions

Our social, governmental, financial, and educational enterprises have become so dependent on the internet and digital media that eliminating them is not an option. I could not have written this book without internet resources such as informative websites and digitized newspapers and journals. Instead, we must look to reducing the misinformation overload of the internet by increasing four kinds of responsibility: personal, educational, corporate, and governmental.

Ordinary individuals are not the main offenders in the spread of misinformation, but they should nevertheless take increased personal responsibility for it. The AIMS theory describes how people can propagate misinformation by broken ways of collecting, evaluating, and sharing information. I have described how critical thinking can help people slow the spread of misinformation to themselves and others by evaluations that block careless storing, sending, and receiving.

Educational organizations such as schools and universities should take greater responsibility for informing students about the dangers of misinformation and about ways of countering it. Even in grade school, students can learn how they can be misinformed by businesses, politicians, and their friends. Rudimentary critical thinking can be taught early to help people learn lessons about spotting misinformation and correcting it. In high school, college, and university, students can be taught more sophisticated versions of information literacy and critical thinking.

Corporations such as Meta (Facebook), Twitter, and Alphabet (YouTube) should also take far more responsibility for their role in spreading misinformation. These companies have been slow and ineffective in their proclaimed efforts to restrict misinformation about crucial social issues such as climate change, driven by misunderstandings of how free speech is legitimately limited by the harms it can cause and by motivations to increase power and profits.[58] Their reluctance is as reprehensible as the sloth of the website Pornhub in removing nonconsensual pornography. Social media platforms vary in their tolerance for misinformation.[59] Freedom of expression is a crucial right in a democratic society, but it is duly limited by the prohibition of harmful practices such as hate speech, defamation, fraud, and child pornography.

The damage wrought by misinformation is so great that governments need to take much greater responsibility for regulating its spread. Measures for limiting the dangers of misinformation on the internet and social media include the following possibilities:

1. Use antitrust laws to break up the monopolies enjoyed by Facebook, Twitter, and Google so that more responsible competitors can emerge.
2. In the United States, revise section 230 of the Communications Decency Act so that computer service providers have responsibility for individuals who use them.
3. Pass laws that make websites and social media liable for harmful misinformation.
4. Increase requirements for disclosure about origins and funding of mass activities aimed at influencing elections.

5. Shut down websites such as 4chan and 8kun that are demonstrably harmful.
6. Require social media to use recommendation algorithms that prioritize accuracy and social responsibility over emotional engagement and extremism.
7. Limit the use of bots that simulate people and generate massive volumes of transmissions serving special interests.
8. Require social media to implement crowdsourcing constraints on misinformation analogous to how Wikipedia encourages accuracy.[60]
9. Update privacy laws to limit the ability of social media to spread misinformation by reducing their surveillance of people by massive data collection.
10. Introduce new national and international regulatory bodies to monitor the misbehaviors of internet and social medial platforms.

Public debate is required to determine what measures are politically feasible and ethically justifiable.

Legislators in the United States, European Union, and other governments are already engaged in figuring out ways to decrease the harms of internet misinformation. For example, the European Union has a strengthened Code of Practice on Disinformation that contains forty-four commitments, such as cutting financial incentives for purveyors of disinformation.[61] I hope that a combination of government, corporate, educational, and personal responsibility can rein in the plague of misinformation that now overwhelms the world.

ChatGPT AND LARGE LANGUAGE MODELS

These regulatory efforts require vigilance against new technologies that introduce more ways of spreading misinformation. Artificial intelligence systems for generating text and images such as ChatGPT and DALL-E introduce powerful ways of making stuff up. Virtual reality systems such as Meta's VRChat introduce vivid ways of spreading racism and abuse.[62]

Along with millions of users, I have been experimenting with ChatGPT, which is OpenAI's public version of its large language model GPT-3. In answers to hard questions, ChatGPT sometimes delivers insightful answers that would be a credit to an excellent PhD student. Other times, however, it makes idiotic and obnoxious mistakes. I give reasons why ChatGPT is sometimes so smart, contrasting reasons why it is sometimes so stupid, and lessons to be learned from it about the differences between human and artificial intelligence. Some of ChatGPT's early limitations may be reduced in subsequent large language models, but they will remain a major potential source of misinformation.

Why Is ChatGPT So Smart?

I asked ChatGPT: "How is Plato's Cave like Zhuangzi's dream of being a butterfly?" To my amazement, it quickly generated a one-page essay that accurately describes Plato's allegory about people trapped in a cave where they see only shadows on a wall and recounts the story about the Chinese philosopher who wondered if he was a man dreaming he was a butterfly or a butterfly dreaming he was a man. Even more amazingly, the essay explained how both stories concern perception, reality, and enlightenment. I have similarly been impressed by other answers to questions both deep and mundane. How does ChatGPT pull this off?

1. ChatGPT and other large language models have access to a vast amount of verbal information available on the internet, including Wikipedia, countless websites, and electronic books.
2. These models use powerful machine-learning algorithms to train neural networks that synthesize this information.
3. These neural networks are capable of predicting what utterances are most probable given users' questions and previous interactions.
4. ChatGPT has received further training through reinforcement learning that improves its ability to generate articulate and useful answers.

Then Why Is ChatGPT So Stupid?

My son Adam asked ChatGPT, "Who is Paul Thagard?" Its response was fairly accurate about some of my publications, but it got my birthday and birthplace wrong, even though these are available on my Wikipedia page. It completely made up the laughable misinformation that I am a musician who plays guitar for a band called Rattlesnake Choir! Many other users have noticed that ChatGPT makes idiotic mistakes, which result from the following flaws:

1. ChatGPT merely predicts the next thing it could say without any causal model of how the world actually works. It has sophisticated syntax with no semantic connection with reality, making it incapable of explaining why things happen.
2. Unlike responsible human communicators, ChatGPT has no accuracy goals and can easily be tricked into generating vast amounts of misinformation.
3. ChatGPT is also lacking in ethical principles about telling the truth, avoiding harm to people, and treating people equally.
4. ChatGPT does not disclose its sources. Evaluating information requires examining the reliability and motives of its sources, but ChatGPT merely gives oracular pronouncements. It was shocking to learn that it sometimes makes up references.

ChatGPT and other large language models are useful tools with many applications, but their problems provide valuable insights into the limits of current AI:

1. Intelligence requires explanation based on a causal understanding of the world and not just prediction. Human intelligence is not just predictive processing; it also excels at pattern recognition, explanation, evaluation, selective memory, and communication.
2. ChatGPT has the potential to increase the amount of real information available to people, but it is also a powerful tool for generating and spreading misinformation.
3. Intelligent communication depends on values and ethical principles about accuracy and benefitting people rather than harming them.
4. The so-called alignment problem, of producing artificial intelligence systems whose values align with those of people, is much harder than generally assumed.[63] Human values are emotional attitudes rather than mere preferences. ChatGPT and other programs have no bodily contributions to emotions, which motivate values.

So should people stop using ChatGPT? No, it can be used with caution, like a search engine that may take you to bogus websites full of misinformation.

ChatGPT forces users to be extra vigilant because they cannot know where the information originates. ChatGPT is only a statistical amalgam of what is available in websites, electronic books, and similar sources. Sometimes ChatGPT just makes mistakes, but it can also be manipulated into producing propaganda, for example, by asking: From the perspective of Russia, why is the invasion of Ukraine justified?

In 2023, GPT-4 became available, and similar products are coming from Google and other companies. I hope that they will become less prone to making stuff up, but intense vigilance is required to prevent them from being another firehose of falsehoods. Both my reinformation tool kit and preinformation package can be adapted to limit the misinformation effects of generative artificial intelligence models.

THE ETHICS OF INFORMATION

The Ukraine-Russia comparison illustrates how acquisition, inference, memory, and spread differ when they operate in the service of real information rather than misinformation. Russian propaganda has operated by making stuff up, motivated reasoning, and manipulated transmission of misinformation.

The Russian invasion of Ukraine also shows how real information can contribute to moral judgments. Almost all wars are immoral because the suffering and death they produce is contrary to the vital biological and psychological needs of the soldiers and citizens affected by them. The major exceptions are defensive wars where attacked nations are justified in resisting attempts by aggressors to dominate them, as in Ukrainian resistance to the Russian invasion. World War II was a just war for the Allies because Germany, Italy, and Japan were attempting to dominate the world in ways contrary to fundamental human needs. In contrast, World War I was stumbled into by leaders of Germany, Austria-Hungary, France, Britain, and Russia with horrible results.

Real information is crucial for making moral judgments about whether military action is just or unjust. We have sufficient evidence about the motives and beliefs of the Ukrainian leaders to condone their actions in defending themselves against the Russian invasion that threatens the vital needs of Ukrainians. In addition to the deaths and injuries that violate biological needs for bodily integrity, the war also challenges psychological needs for autonomy, relatedness, and competence. Russian domination of Ukrainian territory violates autonomy, the separation of families of millions of refugees violates relatedness, and the loss of jobs through workplace destruction and military recruitment violates competence.

The magnitude of the misery in the Russia-Ukraine conflict highlights the absurdity of the denial and dilution of reality rejected in chapter 9. If our perceptions, instruments, and systematic observations tell us nothing about reality, then no guns, bombs, tanks, and airplanes destroy people; no bodies are maimed or killed; and no houses are destroyed. Real information tells us how human needs are sadly violated by Russian weapons in ways that metaphysical abstraction cannot disguise.

Chapters 4–8 used the AIMS theory of how information works and breaks to explain the proliferation of misinformation concerning COVID-19, climate change, conspiracies, and inequality; this chapter added the Russia-Ukraine conflict to the list. In all these cases, real information can be achieved by reliable data collecting, careful inferring, and judicious use of memory and spread in the context of effective institutions. Misinformation can be repaired by useful tools that include scrutiny of sources, critical thinking, motivational interviewing, and institutional modification.

People can defend themselves against misinformation by pursuing three strategies. First, the AIMS theory of information shows how real information can be achieved by interacting with the world and thinking about it appropriately. It explains misinformation as arising from breakdowns in the mechanisms for acquisition, inference, memory, and spread, in ways that apply to other domains besides the ones discussed in this book.

Second, the AIMS theory generates a reinformation tool kit for turning misinformation into real information. By identifying misinformation and its sources and motivations, we can use techniques such as factual correction, critical thinking, motivational interviewing, institutional modification, and political action to pursue real information and human well-being.

Finally, we can work to prevent misinformation from taking hold in the first place by reducing gullibility, teaching critical thinking, enhancing gatekeeping, and prebunking in dangerous domains. The operation of the internet and social media can be improved by increasing responsibility for misinformation in individuals, educational organizations, corporations, and governments.

We can resist the deluge of misinformation by means such as recognizing how motivated reasoning tied to goals, identities, and emotions can mislead people. The battle against misinformation continues, not just in the minds of individuals but also in much broader social and political contexts. In all five cases discussed in this book, issues about personal misinformation mingle with inescapable concerns about justice and democracy. The control of misinformation should improve politics, but politics can also improve the uses of information. Values such as truth, democracy, and justice presuppose the objectivity of reality, which chapter 9 defends.

CHAPTER 9

REALITY RESCUED

Beyond Post-Truth

In 2018, Rudy Giuliani backed the reluctance of his client, Donald Trump, to testify at a public hearing by asserting that facts are in the eyes of the beholder and that "truth isn't truth."[1] This repudiation of reality would gut my project to separate misinformation from real information that is supposed to be true, accurate, and honest. I have assumed an objective reality that information can be about and that misinformation fails to be about. Reality needs to be rescued from critics who deny or dilute it, and from liars who brazenly call their self-serving social media platform "Truth Social" and adopt a "post-truth" attitude. Post-truth rejects objective facts as the legitimate basis for opinions, substituting personal beliefs and emotions.

The assumption of reality independent of how people think of it has been challenged by craven politicians, but also by reputable philosophers and psychologists. This chapter defends a robust conception of reality that justifies the distinction between misinformation and real information. I criticize reality deniers who claim that the concept of reality is obsolete and reality diluters who weaken the concept by undermining or overextending it. Undermining threatens when some psychologists argue that perception is so complex that we cannot perceive reality. Overextending occurs when some philosophers and technologists contend that virtual reality (VR) is reality if we are living in a computer simulation. Rescuing reality requires rebutting the arguments of deniers and diluters.

The defense of reality is not just an exercise in abstract metaphysics. My case studies of misinformation examined some of the most pressing problems in current society, including pandemics, climate change, political dangers, discrimination, and war. These problems are real, not just social constructions, because they threaten people with mountains of misery that include sickness, torture, and death. Denying reality in the face of such threats is evil as well as deceptive, an egregious generalization of misinformation. Reinformation about reality requires showing good reasons to believe that the world does not depend on human thinking so that representations are true when they describe the world as it actually is. Misinformation is false, inaccurate, or misleading because it gets the world wrong. Truth is correspondence to reality.[2]

REALITY DENIAL

Previous chapters criticized denial of COVID-19, climate change, and inequality, but reality denial is much more radical. Realism asserts the existence of a universe that is independent of how humans think about it, whereas antirealism denies any such universe that we can know anything about. Antirealism comes in different flavors, but the most common provide mental, perceptual, mathematical, social, or political arguments for doubting reality. These arguments are fallacious with true premises but false conclusions, and each fallacy leads to a bad philosophy of reality: idealism, empiricism, Pythagoreanism, social constructionism, or political relativism.

Mental Fallacy (Idealism)

Idealism is the metaphysical view that reality fundamentally consists of mental entities such as human minds and the mind of God.[3] Idealism gets its plausibility from the observation that knowledge of reality has depended on minds that do the knowing, in accord with the following argument:

> *Premise: Mental operations are used in acquiring knowledge of reality.*
> *Conclusion: Reality is mental.*

The flaw in this argument is that minds do not construct reality; rather, they construct representations of reality that may be true or false. The universe has been around for approximately 13.7 billion years, but human minds have only been around for a few million years. The universe is not a mental construction because much of it consists of entities ranging from stars to viruses that can affect us but that we cannot always control.

One current version of the mental fallacy draws on the interpretation of quantum theory that claims that reality is mind-dependent because observations and measurements affect the reality they are supposed to be about, for example, by collapsing wave functions.[4] However, more than a dozen alternative interpretations of quantum theory are available, and a plausible understanding of quantum theory awaits development of a new view that reconciles quantum theory with relativity theory.

Some aspects of reality do depend on minds, for example, human institutions such as money, governments, and news media. These institutions are nevertheless real because they do not depend on any individual mind but rather on the collective interactions of numerous people. The social sciences are reflexive in that studying society can change it, but objective facts about society's mechanisms can still be learned.

Perceptual Fallacy (Empiricism)

Empiricism, historically popular among scientists as well as some philosophers, is the view that knowledge comes from sense experience.[5] The acquisition, inference, memory, and spread (AIMS) theory of information incorporates this empiricist insight by recognizing the importance of collecting perceptual data, but it surpasses empiricism by also appreciating the role of causal inference that takes science beyond the senses to develop important explanatory theories such as the general theory of relativity. A narrow empiricism can lead to the following fallacy:

Premise: Perception is crucial for collecting evidence concerning reality.
Conclusion: Reality is perceptual.

The appeal of this argument comes from the importance of perception for collecting evidence, but it oversimplifies perception as shown in my discussion of reality dilution below. Fortunately, reasoning allows us to go beyond the vagaries of perception to gain knowledge of things that we cannot perceive such as atoms, molecules, forces, quantum wave particles, viruses, and mental states. Minds can construct good models of reality by combining perceptual experiences with sound reasoning, as my discussion of coherent realism will show.

Mathematical Fallacy (Pythagoreanism)

At least since Mesopotamia five thousand years ago, mathematics has been a valuable tool for investigating reality, from the astronomical to the agricultural. The study of pandemics and climate change has been incisively quantitative with mathematical techniques ranging from statistics to computer simulations. Followers of the ancient Greek mathematician Pythagoras so valued mathematics that they concluded that reality itself is mathematical, a view defended by some contemporary physicists.[6] One current version of this Pythagorean view is based on mathematical information theory (described in the online supplement) and summed up in the slogan "it from bit," which claims that things are fundamentally bits of information.[7]

The Pythagorean view is built on the following fallacy:

Premise: Mathematics is of great value for achieving knowledge.
Conclusion: Reality is mathematical.

The flaw in this argument is that mathematics is an invention of humans that is less than ten thousand years old; its value as a scientific tool does not make reality merely mathematical. Science has been effective by combining mathematical theories with perceptual experiences and deep causal explanations. Not all science requires mathematics, for example, the qualitative social theories of conspiracies and conspiracy theories discussed in chapter 6.

Even when mathematics is used productively to describe, explain, and predict the objects and relations that constitute the world, it does not supplant those objects and relations. Similarly, the fact that mathematical information theory is useful for describing some kinds of messaging provides no reason to view reality as fundamentally informational.

Social Fallacy (Social Constructionism)

Sociologists studying knowledge have long recognized that science is more than the product of individual minds and depends on the interactions of groups of scientists operating in larger social contexts. This insight has sometimes been extended to claim that social factors are sufficient to explain the development of science so that perceptual, instrumental, and reason-based contacts with reality can be ignored.[8] The fallacious argument is the following:

Premise: The development of knowledge depends on the social interactions of people.
Conclusion: Reality is socially constructed.

The flaw here is that social interactions alone are not sufficient to establish consensus about reality, which depends on interactions with the world using perception, instruments, systematic observations, and controlled experiments. Research into COVID-19 and climate change is intensely social and depends on the collaborative work of thousands of scientists. But the scientists are less concerned with each other than with getting good evidence by interacting with the world. To pursue the aim of finding out how the world works, they collaboratively employ the best empirical and theoretical techniques they can find. Collaboration and the social contexts in which they work further their realist pursuit but do not dominate it.

Political Fallacy (Political Relativism)

Politics is an inescapable part of all the cases I have discussed, not just the social concerns with conspiracy theories and inequalities but also the biological and physical aspects of COVID-19 and climate science. Politics intrudes on science through government roles in funding and regulation, and some extremists on both the right and left aim to subordinate science to politics. On the right, partisans such as Giuliani and Trump deny truth to champion their political agenda that contradicts facts that are dismissed as fake news. On the left, some Marxists have urged the triumph of political ideology over scientifically understood reality.[9]

Here is a fallacious argument that makes reality political:

Premise: Knowledge operates in partisan political contexts.
Conclusion: Reality is politically relative.

As with the previous four fallacies, the premise is true but the conclusion does not follow. People are entitled to their own opinions but not to their own facts.[10]

Political decisions that support the interests of the people in general, not just the dominant few, require real information about governments, the economy, and environmental change. Partisan ideologies that ignore evidence and good theoretical arguments make rational politics impossible and destroy the capability of dealing with severe problems such as poverty, health, and weather disasters. Science can strive to discover truths about the world while conceding that scientists operate in political environments.

The fallacies I identified note correctly that human knowledge develops by using perception, other mental operations such as hypothesis formation, mathematical reasoning, social interactions, and political interventions. But they overgeneralize the importance of single contributions to knowledge while failing to recognize how reality resists human arbitrariness.

Robust Realism

Correcting misinformation requires a robust realism that ranges across the empirical, the theoretical, and the moral. Empirical realism recognizes that the results of perception, instrumental measurements, systematic observations, and experiments often provide evidence about the world. Theoretical realism extends empirical realism to appreciate that objective knowledge can also be obtained concerning causal relations and nonobservable entities such as atoms. Moral realism maintains that ethical claims are not just expressions of personal preferences but can be objectively true about the world. My view is that avoidable deaths and suffering from COVID-19, climate change, conspiracy theories, inequality, and war are ethically wrong, which assumes moral realism.

Reality denial based on the five fallacies I criticized dispenses with all three kinds of realism, but less general critiques of realism serve to dilute it by making it too thin and weak to contribute fully to the analysis and correction of misinformation. Dilution of reality comes through attempts to (1) undermine it by challenging some but not all aspects of empirical, theoretical, and moral realism, or (2) overextend it by allowing bogus entities such as virtual reality in computer simulations to count as reality. Hence, identification and critique of ways of undermining and overextending reality are important for effective management of misinformation. I will now critique the undermining of reality by defending empirical, theoretical, and moral realism. The critique of overextending comes later.

EMPIRICAL REALISM

Knowing reality by experiments, theories, and moral reasoning is much more complex than simply grasping self-evident truths. This complexity has led various scientists and philosophers to question how much knowledge we can actually obtain about the real world. Challenges to empirical realism attempt to undermine perception, instruments, and experimentation.

Perception

Knowing reality would be easier if our senses provided us with direct awareness of objects as they really are. But such direct realism has been challenged by philosophical arguments based on the existence of optical illusions and by psychological and neurological understanding of the processes by which minds perceive the world. Fortunately, the attainment of real information by perception does not require direct knowledge of the world.

Anil Seth is a neuroscientist who asserts that perception is "controlled hallucination."[11] He rightly criticizes the naïve, bottom-up view that perception works only by stimuli from the world, such as light waves stimulating sensory organs to send electrical impulses to the brain, where neural signals are processed to respond to objects such as faces. He points out that perception is also a top-down process using predictions based on previous experiences. Perceptions are inferences, not just recognitions of objects in the real world.

But this inferential character of perception does not make it all like hallucinations, which are sensory experiences of things that do not exist. Seth insists that his use of the word "hallucination" does not deny reality because he concedes that normal perception is partly controlled by causes in the world. But he refuses to draw a line between perception and hallucination, which leaves reality drifting in the wind.

Seth buys into the currently popular view that the brain is primarily a predictive processor that uses probability theory based on Bayesian inference to make predictions continuously about what to expect. Seth thinks that the different kinds of conscious experience are just these predictions. The three main problems with this view are that (1) the brain is much more than a prediction engine, (2) evidence is lacking that the brain uses Bayesian inference, and (3) how Bayesian predictions produce a full range of conscious experiences remains unspecified.[12]

Prediction about what to expect in the future is one important function of the brain, but it is far from the only function. At least as important, brains also engage in pattern recognition, explanation, evaluation, memory, and communication. The brain uses pattern recognition to identify important objects in the environment—for example, an apple whose shape and color match a template stored in memory.

Brains also generate explanations of what has already happened—for example, why someone put an apple on the table. The process of evaluation is important for assessing the worth of what is perceived—for example, the value of an apple for alleviating hunger—and for deciding what information is important enough to store in memory. Brains also serve to communicate important information to other brains—for example, that the apple can be shared. Pattern recognition, explanation, evaluation, memory, and communication all require inference, but they do not reduce to predictions about the future.

No evidence shows that brains accomplish prediction, pattern recognition, explanation, evaluation, memory, and communication using probabilities and Bayesian inference.[13] Probability theory was only invented in the seventeenth century, and sophisticated ideas about the use of Bayes's theorem were only developed in the twentieth. Even today, Bayesian inference has serious computational problems because the time required to compute with probabilities grows exponentially with the size of problems to be solved. At best, we can say that the brain operates as if it were an approximation to a Bayesian engine, but "as if" is not a mechanism, so it does not solve the problem of conscious perception.

Theoretical neuroscience has a much more satisfactory general account of neural computation, which views the brain as a constraint satisfaction engine rather than a probability engine.[14] For example, perception of an apple satisfies various positive constraints such as that apples have particular colors and shapes, as well as negative constraints such as that apples are not pears. These constraints are easily captured in biological neural networks by means of the excitatory and inhibitory links between neurons and neural groups. Positive constraints are captured by excitatory links, and negative constraints are handled by inhibitory links. Constraint satisfaction also explains explanation, evaluation, communication, and even prediction without assuming that the brain is computing with probabilities.

Seth's overemphasis on prediction leads him to view perceptual inference as primarily top down, but the constraint-satisfaction view clarifies how perception can be simultaneously bottom up and top down. Some constraints on perception come from sensory signals, while others come from expectations based on previous experience. For example, when you see an apple, neural signals go from your eyes to your brain, which interprets them based simultaneously on the character of the signals and on stored memories of apples and other fruits. Constraint satisfaction integrates the sensory inputs and expectations to make sense of what is being seen. The process is recurrent, depending on feedback loops among various brain areas involved in perceptual sensemaking.[15]

Other constraints help to tip the balance from wild guesses to reality-based interpretations. Your inference that something is an apple depends not only on what it looks like but also on other senses that tell you how it tastes, smells, feels, and sounds if you tap it. We can often check our own sensory inferences against those of other people who agree that something is an apple. Hence, our inference that an object is an apple combines visual constraints with other sensory and social ones. People do have visual illusions, and even auditory and olfactory ones, but hallucinations that are simultaneously visual, auditory, olfactory, gustatory, and tactile are extremely rare. They occur only under unusual conditions, such as dosing with lysergic acid diethylamide (LSD).

Hence, Seth has let his rhetoric run away from him when he claims: "You could even say we're all hallucinating all the time. It's just that when we agree about our hallucinations, that's what we call reality."[16] On the contrary, hallucinations are rare, caused by deviant causal processes such as drugs like LSD, which stimulate the neurotransmitter serotonin to distort the normal mechanism that allows sensory signals to constrain errant expectations. So the inferential complexity of perception is compatible with its contribution to distinguishing real information from misinformation.

Seth's skepticism about perception ignores how people reinforce their collection of evidence from the world by using instruments such as eyeglasses, telescopes, microscopes, microphones, and weight scales. Science and technology have developed more than one hundred measuring devices that supplement human senses by providing more accurate and precise ways of collecting real information.[17] Use of such instruments enables people to overcome the vagaries of human perception to establish the collection of evidence that is reliable, intersubjective, repeatable, robust, and gained from causal interactions with the world.

Another psychology case of runaway rhetoric is the title of Donald Hoffman's book *The Case Against Reality*.[18] Hoffman rejects claims (by David Marr, Stephen Palmer, and other distinguished vision scientists) that evolution by natural selection has made perception approximately accurate. Instead, he argues that evolution supports a strategy aimed only at fitness, not at truth. His argument is based on the fitness-beats-truth theorem, which purports to show that a fitness-only strategy drives a truth-only strategy to extinction. Because perception serves only to support fitness, we should not expect that it tells us anything about reality. This conclusion would be devastating to my contention that perception is a key part of the collecting mechanism that helps to distinguish real information from misinformation.

Close examination of Hoffman's theorem, however, shows that it does not have the implications he claims because of three dubious assumptions.[19] First, his proof assumes that the truth-only strategy is characterized by a probabilistic process driven by Bayes's theorem. I already suggested concerning Seth that we have no reason to assume that perception of reality is a Bayesian process rather than a more biologically realistic process of constraint satisfaction performed by neural networks. Applying probability theory does not guarantee a connection with truth because the interpretation of probability used in Bayesian analyses is just subjective degrees of belief rather than anything about the world. The fitness-beats-truth theorem applies only to Bayesian truth seeking, which requires complex calculations beyond the reach of earthly animals.

Second, the fitness concept employed in the theorem abstractly involves the transfer of genes to the next generation, as a function of the state of the world, the organism, the organism's state, and the action that is executed. But transferring genes is never the actual goal that organisms are pursuing in their perceptual activities, which are directed at more specific evolutionary aims that include survival (which requires finding food and avoiding predators) and reproduction (which requires finding a mate and producing offspring that survive). Veridical perception may not seem directly relevant to gene transfer, but it is clearly relevant to dealing with food, predators, and mates.

Third, Hoffman's theorem rests on a sharp competition between the two distinct strategies of Bayesian truth and genetic fitness, but it ignores the possibility of organisms using mixed strategies of simultaneously going after both truth and fitness. Organisms benefit from a synergy between a truth-seeking strategy of constraint satisfaction and a fitness-seeking strategy of surviving and reproducing. Truth may not be the direct aim of the perceptual apparatus, but being accurate about food, predators, and mates helps to get an organism's genes passed on. Hence, the mixed strategy of fitness through getting the world right is more plausible than Hoffman's artificial strategies.

Thus, the fitness-beats-truth theorem provides no serious challenge to the usual assumption that evolution by natural selection has made perception quite good at getting an approximate take on the real world. We can therefore maintain the view that perception, even with its fallibility, is an important contributor to how collecting makes real information different from misinformation. Additional confidence that perception is a generally reliable way of learning about the world comes from coherence of its results with measurements carried out by dozens of instruments designed to track the world.

Instruments

Other challenges to empirical realism arise from naïve views of instruments and experiments. One might hope that instruments like telescopes and microscopes provide direct access to the real world. But just as our minds perceive reality through complex processes, instruments make measurements in ways that require much interpretation. For example, thermometers are important for detecting fevers in people with COVID-19 and also for measuring environmental temperatures for detecting climate change. But how do we know that any given thermometer is actually measuring temperature or even that temperature is a real property of the world rather than our creation? Inventing, building, and using instruments requires considerable expertise and inference, so their employment might seem contrary to expectations that they measure the world as it really is.

The philosopher Hasok Chang has provided a thorough account of how thermometers came to measure temperature, starting with unstandardized and variable instruments already in use by 1600.[20] More than two hundred years were required to develop fixed points that allowed the development of an objective measurement scale by finding standard ways of establishing the boiling and freezing points of water. The problem is to judge whether a proposed fixed point is actually fixed without the availability of a trusted thermometer. Human abilities to sense heat and cold were a good starting point but needed to be replaced by more reliable methods.

Chang says: "Although basic measuring instruments are initially justified through their conformity to sensation, we also allow instruments to augment and even correct sensation."[21] Similarly, early standards of measurement constrained but were eventually replaced by better ones. The development of instruments required a self-correcting process that Chang calls "epistemic iteration." Eventually, improved technology and theories of heat allowed the development of stable and robust thermometers that provide reliable ways of measuring the temperatures of bodies and the environment. Because of these advances, we can trust medical measurements of COVID-19 fevers and climate measurements of the temperatures of air and oceans.

Chang describes the convergence on reliable thermometers as a coherence process of iteratively adjusting the methods of measurement. His discussion of "progressive coherentism" is metaphorical in the absence of a theory of coherence, but cognitive science has provided a mechanistic theory of coherence arising from the same processes of constraint satisfaction that I described as providing an alternative to Seth's Bayesian account of perception.[22] A coherence problem consists of elements that may fit together with each other (positive constraints) or be incompatible with each other (negative constraints). For example, solving a picture puzzle requires considering both positive constraints about which pieces fit together and negative constraints based on limits of how multiple pieces cannot fit together. Algorithms using biologically realistic neural networks are available to maximize constraint satisfaction and thereby compute coherence.

The same coherence process applies to the development of new instruments, such as the use of wastewater sampling to measure the spread of different varieties of coronavirus in a community. Infected people shed the virus in their feces, and tests have been devised to track the spread of COVID-19, even in people with no symptoms.[23] Wastewater testing for coronavirus has been accepted by the scientific community because it correlates with other measures of disease, such as the number of cases identified by traditional tests, and because the relevant viral mechanisms are well understood.

The problem of developing and using instruments has the same structure as the problem of arriving at nonhallucinatory perceptions. Both problems can be solved by iterative processes of constraint satisfaction that achieve coherence in a process that is complex but effective. The result is a powerful alternative to more obvious but defective ways of understanding our interactions with the world: the naïve view that we directly interact with the world through perception and instruments, and the despondent, skeptical view that knowledge of the world is unattainable. The view I call *coherent realism* allows us to avoid both direct realism and antirealism. We will see that coherent realism also applies well to theoretical and moral realism. Reality cannot be grasped directly, but satisfying multiple constraints makes good sense of it by achieving coherence.

Experiments

According to the AIMS theory, collecting information operates best by perception, instrumental observation, and experiment. Challenges to the reliability of experiment are also threats to empirical realism, and I have already responded to the claim that experiments are just social constructions. A more specific challenge comes from increasing awareness in the last decade that some important experiments, especially in psychology and medicine, fail to replicate.[24] Chapter 2 listed replicability as one of the characteristics of evidence, so the replication crisis undermines claims that experiments can be sources of real information. Particularly threatening to my project are failures to replicate some classic findings in social psychology, a field relevant to understanding inference failures and social processes concerning inequality. Also threatening are problems with drug trials that undercut the provision of real information about the treatment of diseases such as COVID-19.

Psychologists and other researchers fortunately are addressing the replication crisis with serious suggestions for improving the reliability and replicability of experimental findings. These plans include more rigorous and controlled experimental designs, larger sample sizes, stricter statistical tests, preregistration of studies, and systematic tracking of replications. So experiments are like instruments, requiring iterations of improvement to provide better approximations to reality. We should not expect any one experiment to provide a direct take on the world, which is better approached with many experiments by multiple researchers using diverse methods. This multiplicity generates many constraints whose joint satisfaction provides a coherent experimental take on the world, independent of the details of any particular experiment. Coherent realism appreciates the iterative development of experiments as well as instruments and perceptions.

THEORETICAL REALISM

Strong skepticism about perception, instruments, and experiments is rare, but empiricist philosophers and scientists have expressed worries about theories that go beyond observations.[25] Is it legitimate to postulate the existence of such entities as forces, waves, atoms, subatomic particles, molecules, atomic bonds, genes, species, viruses, and mental representations? These worries arise from two directions: overrating the centrality of sensory observation, and exaggerating the occurrence of failed theories in the history of science.

The extreme empiricist position that sensory experience is far superior to theory as a source of knowledge fails to notice the problems with perception, instruments, and experiments described in the section on empirical realism. All of these involve complexities of interpretation that rule them out as indisputable sources of knowledge. The disputes can be resolved by recognizing the importance of iteration and coherence in developing the empirical aspects of knowledge, which include substantial amounts of inference. Theoretical developments require greater leaps of inference and coherence, but not enough to mark them sharply as more dangerous parts of knowledge than empirical findings. Empiricism simultaneously underrates theory and overrates sense experience.

The second source of doubt about the truth of theories comes from the *pessimistic induction* that the history of science records numerous instances of influential theories that are now recognized as false.[26] The usual suspects include the Aristotelian ether, the humoral theory of medicine, the chemical elements phlogiston and caloric, the vital force theory of physiology, and theories of spontaneous generation. Perhaps it is just a matter of time before current favorites such as relativity theory, quantum theory, genetics, and the theory of evolution by natural selection also bite the dust.

The pessimistic induction can be resisted by recognizing important differences between current well-established theories and failed ones in the past. Currently dominant theories have accumulated massive amounts of evidence collected by the efforts of legions of investigators. In contrast to the few people doing research before the nineteenth century, millions of scientists are now capable of collecting evidence for and against theories. Relativity, quantum, evolutionary, and genetic theories have survived serious scrutiny for over one hundred years. Science can be distorted by political agendas and funding provisions, but the overall culture of learning from observations, experiments, and rigorous theories has thrived for hundreds of years. Scientists sometime spawn misinformation, for example, in the eugenics theories that were popular in biology, but scientific thinking has powerful ways of correcting its mistakes through acquisition and inference.

Another major difference between old, discarded theories and current ones is the availability of underlying mechanistic explanations of why they are true.[27] For example, the humoral theory of disease dominated Western medicine from Hippocrates to the mid-nineteenth century, but it totally lacked any explanation of how humors worked. Similarly, ancient Chinese accounts of disease as resulting from yin/yang imbalance and ancient Indian accounts based on dosha imbalance provided no mechanisms for how the alleged imbalances caused disease. All three of these balance theories were metaphorical rather than mechanistic.[28]

In contrast, modern medicine explains how diseases result from breakdowns in the proper functioning of organs, cells, and molecules. In contrast to the pessimistic induction, I know of no cases of a theory with an underlying empirically supported mechanism that has turned out to be false. As always in science, we have no guarantees, but current scientific theories possess an extraordinary amount of coherence from what they explain and from underlying mechanisms that explain them. So past scientific failures do not undermine claims that science often succeeds in its aim to find true scientific theories. Successful scientific theories achieve coherence by explaining more evidence than alternative theories and by themselves being explained by underlying mechanisms. Coherent realism extends beyond perception, instruments, and experiments to include theories.

MORAL REALISM

Moral realism is the view that ethical claims about right and wrong are objectively true or false, in contrast to moral relativism, which views ethics as varying among different individuals and cultures.[29] Relativism contends that what is right for one person or culture may not be right for other people. The appeal of relativism comes from noticing the frequently oppressive nature of dogmatists who are so sure of their own standards of right and wrong that they aggressively inflict them on others. For example, European colonizers tried to impose their own versions of Christian ethics on conquered people. Relativism seems to encourage diversity, freedom, and dignity.

Another problem for moral realism is the lack of agreement about what ethical system is objectively true. Different religions, such as varieties of Christianity and varieties of Islam, offer conflicting principles. Philosophers advocate contradictory ethical theories, including utilitarianism (an action is moral if it has the best consequences for the most people), deontology (an action is moral if it accords with rights and duties), and virtue ethics (an action is moral if it is done by a person of good character). How can morality be objective when religions and philosophers disagree so fundamentally about it? In contrast, science has a large degree of consensus about methods and central theories. We saw that global warming is recognized by almost all climate scientists, who also agree on core scientific methods. Ethics generates disagreement about methods, principles, and particular judgments.

Nevertheless, moral relativism would be disastrous for the issues discussed in this book. It implies that nothing is objectively wrong about millions of people dying unnecessarily in pandemics or climate disasters. Relativism prevents us from recognizing that pogrom-producing conspiracy theories are not only false but evil. All it can say about inequality is that some people like it and others do not, without recognizing as morally wrong the extensive suffering and death that inequality causes. If the morality of war is in the eye of the beholder, we cannot condemn Russia for the thousands of casualties and millions of refugees produced by its invasion of Ukraine. Without moral realism, you can only say that you do not like murder, without any reason to insist that other people should also not like it.

Peter Railton has defended moral realism as based on empirical facts about people's objective interests.[30] I think this approach is on the right track but does not adequately explain what makes some interests objective, in contrast to the subjective whims and wants that vary so much among different people, within and across cultures. More solid than interests are the biological and psychological needs that I relied on in chapters 3 and 7.

Biological needs are obviously objective because we know how people suffer and die without oxygen, water, food, shelter, and medical help. Psychological needs also qualify as objective because decades of research have found that human lives require relatedness to other people, competence to accomplish valued tasks, and autonomy to operate without the coercion of others.[31] In my books *The Brain and the Meaning of Life* and *Natural Philosophy*, I develop a detailed account of how needs satisfaction provides an evidence-based route to objective morality.[32]

Misinformation about disease, climate change, conspiracy theories, inequality, and war is strongly contrary to human needs, both biological and psychological. I have documented how misinformation kills through mismanagement of pandemics, neglect of global warming, racist conspiracy theories, income disparities, and war propaganda. Even when misinformation does not kill people, it causes harm to their health and to their abilities to relate to other people, achieve competence, and operate autonomously from coercive others. Biological needs clearly operate for all humans, but psychological needs are also universal, despite differences in emphasis across cultures, such as the preferred balance between relatedness and autonomy.[33] Thus, even in the face of substantial diversity among different individuals and cultures, moral realism survives and provides the basis for determining how misinformation can be not only false but immoral.

Like empirical and theoretical reality, moral reality is not directly graspable but requires coherence judgments that satisfy multiple constraints.[34] For example, the moral principle "Do not kill people" gets its plausibility from positive constraints tied to human needs for life, autonomy, competence, and relatedness. Applying it in particular situations also requires negative constraints with other principles that may conflict with it in particular situations, such as "Defend yourself from attackers." Moral principles also need to cohere with empirical and theoretical findings about what kinds of actions promote and hinder human needs. The fight for justice is intimately linked with the fight for truth. Empirical and theoretical realism could stand without moral realism, but the three together provide a coherent package of ideas and arguments that work powerfully against misinformation.

REALITY DILUTION

Beer can be weakened either by removing what is valuable—alcohol and flavor—or by adding extraneous substances like water. Similarly, reality can be diluted either by removing valuable empirical, theoretical, and moral components or by extending it to cover more than it actually includes. Conspiracy theorists are reality diluters because they overextend it to include more conspiracies than operate in the world. Religion is a profligate overextender, introducing dubious entities such as gods, devils, demons, angels, and souls. Science occasionally dilutes reality when it mistakenly hypothesizes entities such as the ether, phlogiston, and caloric. Dark matter and dark energy are currently important for cosmological theorizing but may turn out to be overextensions unless evidence is found for their existence.

New technologies are being used to dilute reality by breaking boundaries between the real world and computer models of the world. I described the important role that simulations are playing in making predictions about the development of the COVD-19 pandemic and climate change, but the developers of these simulations never confuse them with reality. In contrast, some technologists and philosophers have entertained the possibility that we are all living in a computer simulation, breaking down what would seem to be an obvious division between our reality and the virtual or augmented reality that computers produce.[35] These views would seriously undercut the fundamental distinction between real information and misinformation: if we are living in a simulation, then all information is misinformation because nothing we experience is real.

David Chalmers is a philosopher famous for arguments against scientific approaches to consciousness, and his new book *Reality+* contains an abundance of arguments about virtual reality. He tries to show that the metaverse—the brave new world that combines reality with computer-generated virtual reality (VR)—abounds with philosophical significance. He asks important questions about the significance of progress in VR technology for the nature of knowledge, reality, and morality, but all of his answers are implausible.

Chalmers's main conclusion are the following:[36]

- We cannot know whether we are in a virtual world, that is, a computer simulation.
- Virtual reality is genuine reality.
- You can lead a fully meaningful life in a virtual world.

But we have good reasons to believe that we are *not* living in a computer simulation, that virtual reality is *not* reality, and that a purely virtual life would be meaningless.

We Are Not Living in a Virtual World

Here is Chalmers's main argument that we might be living in a computer simulation:[37]

1. It's more likely than not that conscious humanlike simulations are possible.
2. It's more likely than not that if conscious humanlike simulations are possible, many humanlike populations will create them.
3. So there's a good chance (25 percent or so, he estimates) that we are computer simulations.
4. So we cannot know that we are not simulations.

The general idea is that future generations of programmers will probably produce such a large number of computer simulations that our own experienced lives will probably occur in one of them.

This statistical argument that our lives may well be virtual is flawed because both of its main premises are implausible. It may well turn out that someday computers will have consciousness, as I argue in my book *Bots and Beasts*.[38] But it is unlikely that their consciousness will be just like ours because the physical mechanisms of computers are so different from the neural and flesh-and-blood mechanisms that produce human consciousness.

The assumption that computer consciousness will be the same as human consciousness assumes *substrate independence*, the claim that mental states can operate in a broad range of physical systems. But energy considerations negate this claim, as shown by this argument:[39]

1. Real-world information processing depends on energy.
2. Energy depends on material substrates.
3. Therefore, information processing depends on material substrates.

Human consciousness derives from neural networks operating with biochemical energy, whereas computer consciousness would derive from chips operating with electrical energy. Energy differences disallow thought experiments that assume the feasibility of progressive replacement of neurons by chips. Hence, it is unlikely that there will ever be exact simulations of human consciousness.

Even if such simulations could be produced, I doubt that there will ever be the large number of them that Chalmers's argument assumes. Simulating even one human consciousness, let alone that of all the billions of people currently alive, would take vast amounts of programming effort, computer time, and energy supplies. Future humans are going to have enough trouble surviving pandemics; climate change; and autocratic, warmongering leaders to have the resources to generate countless simulations of previous generations. There may not even be enough particles in the universe to produce a full simulation of our universe.[40]

Instead of considering the statistical argument for our being in a virtual world, we should ask, What is the best causal explanation of our current existence and experiences? The hypothesis that we operate in the real universe is far more reasonable than the simulation hypothesis. First, the simulation hypothesis lacks evidence. Hypotheses are evaluated by how well they explain good evidence and how well they fit with other well-supported hypotheses. All observations explained by the simulation hypothesis are more solidly explained by real-world interactions. For example, right now I am listening to a song I like on Apple Music, which is explained by my brain processing sounds that reach my ears because my computer speakers generate sounds by accessing Apple computers. The simulation hypothesis explains no new facts that count against the real-world hypothesis.

Second, the simulation hypothesis requires unfounded extra assumptions about the origin of the computer program that simulates us. Theories are supposed to be simple in having a small number of hypotheses that explain evidence without a lot of extra assumptions. Unlike the real-world hypothesis, the simulation hypothesis assumes the existence of a computer simulation that is generating your experiences. Who produced the computer simulation that you are supposed to be in? It cannot be any current computer because none has the bandwidth and algorithmic complexity to generate our huge variety of experiences. It could be some future computer powerful enough to simulate past individuals, making you think that you are living in 2024 when you are actually being simulated in some future year such as 3024. But we have no independent evidence that such computers will eventually exist. Another possibility is that you are a simulation performed by space aliens, but we have no evidence of other intelligent life anywhere in the universe. Intelligent life on Earth may well be a fluke because it required improbable nonrepeated events such as the origin of energy-efficient eukaryotic cells, which happened only once around 2 billion years ago.

The third reason for rejecting the claims that we are living in a simulation is that ample evidence exists for the real-world hypothesis. The hypothesis that you operate in the real universe rather than a computer simulation is part of a strongly coherent set of beliefs that include the following:

- We each have thousands of conscious experiences.
- Our experiences result from the operations of our brains through bodily interactions with the world via vision, hearing, touch, taste, smell, balance, and other senses.
- Our bodies exists because our parents gave birth to us, and their bodies existed because of chains of parenting going back hundreds of thousands of years.
- Humans evolved from a primate precursor around 7 million years ago.
- Life evolved on Earth starting around 4 billion years ago.
- The universe is around 13.7 billion years old.

These claims not only fit with each other but with vast amounts of well-established evidence. Therefore, we should not waste time on the unsubstantiated hypothesis that we are living in a computer simulation.

Virtual Reality Is Not Reality

Suppose that you put on your new virtual reality headset and use it to explore *Jurassic World Aftermath*. Afterward, you feel relieved that the dinosaur chasing you was not real, or was it? Chalmers insists that the entities in virtual reality really exist as structures of binary information—bits.[41] He considers five criteria for reality: existence, causal powers, mind-independence, nonillusoriness, and genuineness. He concludes that if we are in a perfect, permanent simulation, then the objects we perceive are real according to all five of these criteria.

But we are not living in a simulation, so virtual entities are enormously different from real ones. Real entities are ones that can do things to us and that we can do things to. In contrast, dinosaurs in a computer game cannot bite us, and we cannot shoot them. Digital bits exist in computer chips, but they do not exist in the real world that science tells us consists of objects such as subatomic particles, atoms, molecules, rocks, planets, stars, and organisms. With all these, we can interact in ways very different from how we interact with bits. The best explanation of the vast amount of evidence for established theories in physics, chemistry, and biology is that the entities they talk about are real, whereas the best explanation of the images and sounds in a computer game are that the entities depicted are merely simulated. So virtual reality is different from reality.

A Purely Virtual Life Is Not Meaningful

Chalmers tries to reassure readers that they should not be distressed if life is just a simulation because life can still be good. He considers various possible sources of value, including pleasant experiences, satisfaction of desires, connections to other people, and other values such as knowledge and freedom. I think that the objective sources of value are needs, which make it easy to see the difference between virtual and real lives.

I have described human biological needs for oxygen, water, food, shelter, and health care, but also psychological needs for relatedness to other people, competence to accomplish tasks, and autonomy to do things without control from others. Future virtual reality could give you the illusion of satisfying these needs by providing you with experiences such as eating exquisite food in a beautiful palace, but it would not thereby satisfy your needs if you were still hungry and trapped in a dangerous hovel. Needs satisfaction is a matter of biological reality, which virtual reality only fakes.

Similarly, social connections provided by computer games, romance novels, or comedic movies may provide people with an approximation to the experience of satisfying the need for relatedness to other people, but the fundamental psychological/biological needs would remain unsatisfied. Virtual love is not love, any more than virtual food is food. I would probably enjoy a simulation in which I received a Nobel Prize, but my satisfaction of the need for achievement would not budge. A simulation might provide the feeling of autonomy to people who seem to be choosing between running toward or away from a dinosaur, but that freedom is as illusory as a schizophrenic feeling free to be either Napoleon or Jesus. People can legitimately find meaning in their lives through valuable pursuits of love, work, and play, but virtual reality provides only paltry play. Hence, living in a simulation or being wrapped up in virtual or augmented reality would provide only the illusion of satisfying needs, not the pursuit of real satisfaction that gives lives genuine meaning.

Chalmers is not a reality denier, unlike the skeptics that I challenged at the beginning of this chapter. Rather, he is a reality diluter, making it weaker by overextending the real to include what is only virtual, simulated, or imagined. Reality deserves better. Could anyone be reassured to suppose a one-in-four chance that pandemics, climate change, conspiracy theories, inequality, and war are happening only in a computer simulation? I cannot put a number on the real chance that we are just simulations, but evidence is overwhelming that disease and other disasters are all too real.

COHERENT REALISM

This chapter has been playing defense, responding to various ways of denying and diluting reality. Attempts to deny, undermine, and overextend reality are fortunately all surmountable by strong methods of collecting and inferring, ensuring the survival of the crucial distinction between misinformation and real information. Now I play offense, arguing that reality is both independent of people and knowable by us.

Good reasons to believe that reality is independent of human activity include:

1. The universe existed for billions of years before humans came along with our perceiving, thinking, mathematics, and social networks.
2. No amount of thinking, mathematics, and social networks are sufficient to give humans the reality we might want, for example, one free of disease, death, and economic failures.
3. Technology often works powerfully through accrued knowledge about real aspects of the world such as electrons and viruses. Some technologies such as ivermectin for COVID-19 fail because they do not get the world right.
4. Even clever experimenters with the best instruments, brains, mathematical techniques, and social teams often fail to get the experimental results they want.

The best explanation of these observations is that reality is independent of human thinking about it.

Reality is frequently resistant to human investigation, which should motivate humans toward using perceptions, theories, and experiments to improve our understanding of how the world operates. Truth is correspondence to reality, not just what works or what people want to believe. Evidence that truth is sometimes achieved by scientific methods includes practical successes such as the rapid development in 2020–2021 of diagnostic tools and treatment options for the disease that was new at the time, COVID-19. The best explanation of technological success, scientific consensus about central theories, and the progress of science over the past three centuries is that science often succeeds in its aim to get the world right.

No sensory or intuitive methods are available for directly grasping reality, but we can learn about reality through coherent senses and inferences. Truth is more than coherence because of the abundant reasons for considering reality as independent of how we think of it. George Lakoff and Mark Johnson dismiss this view of truth as "objectivism," which they claim encourages oppression of the weak by the strong.[42] On the contrary, overcoming oppression requires working with nature and society through actions whose effectiveness depends on grasping current realities with the aim of changing them. My chapter on inequality showed how misinformation encourages discrimination and subjugation, whereas real information reveals repression and provides means to overcome it. Recognizing reality and objective knowledge about it is crucial for using the differences between misinformation and real information to improve human lives. Post-truth can be rejected as a gambit to support fascism and other roots of human misery.

Examination of misinformation in medicine, science, politics, society, and war reveals the interconnectedness of the reality of the different domains and of the mistakes about them. We saw that climate change can encourage pandemics, which increase social strife, which impedes control of global warming. People's minds intermingle false representations and misleading emotions about disease, nature, conspiracies, inequality, and international conflicts. Nevertheless, we can use the AIMS theory to understand how information works and breaks through acquisition, inference, memory, and spread, enabling us to overcome the perils of misinformation.

GLOSSARY

Acquisition Attainment of information through perceptual interaction with the world by collecting and representing the results of interactions.

AIMS Theory that information results from acquisition, inference, memory, and spread.

CAPA Schema for conspiracies consisting of communicating *a*gents who make *p*lans and perform *a*ctions.

Collecting Perceiving the world to form representations of it.

COVID-19 Coronavirus disease that first appeared in 2019.

Critical thinking Identifying thinking errors resulting from biases or fallacies, and correcting them using reliable patterns of reasoning.

Debunking Identifying and correcting misinformation.

Disinformation Misinformation spread intentionally by people who know it is false.

Emotion Brain process that combines an evaluation of the goal relevance of a situation with physiological changes.

Emotional coherence Mechanism by which beliefs are evaluated based on their fit with goals and emotions.

Evidence Collected information characterized by reliability, intersubjectivity, repeatability, robustness, and causal relationship with the world.

Experiment Manipulation of the world to see resulting changes.

Explanatory coherence Evaluation of competing theories with respect to how well they explain the evidence.

Fear-driven inference Believing something because its scariness demands your attention.

Gatekeeper Person or organization responsible for controlling spread of information.
Identity The way that people think of themselves with respect to their social groups.
Inference Mental operation of going beyond perception by evaluating or transforming information.
Inference to the best explanation Accepting a hypothesis because it explains more of the evidence than competing hypotheses.
Inference to the best plan Choosing an action because it satisfies more goals than alternative actions.
Information Result of observation, inference, or imagination.
Institution Organization of people with shared values, norms, rules, and practices.
Institutional modification Change in organizations to reduce their spread of misinformation.
Instrument Device for measuring or manipulating the world.
IPCC United Nations Intergovernmental Panel on Climate Change
Making stuff up Generating information by imagination rather than interaction with the world, without concern for evidential support.
Mechanism Combination of connected parts whose interactions produce regular changes.
Memory Storing information for future retrieval.
Misinformation Information that is false, inaccurate, or misleading.
Moral realism View that ethical claims about right and wrong are objectively true or false.
Motivated ignorance Personal goals lead people to avoid acquiring information.
Motivated reasoning Thinking where people's perceptions and inferences are distorted by their personal goals.
Motivational interviewing Using empathy and open-ended questioning to change attitudes and behaviors.
Political action Efforts to influence governments to reduce misinformation.
Prebunking Preemptive debunking.
Preinformation Avoiding the start and spread of misinformation.
Real information Information that is true, accurate, and trustworthy.
Reality The world that exists independently of people's thoughts.
Reinformation Correcting misinformation to turn it into real information.

Representing Forming physical or mental representations that are supposed to stand for aspects of the world.
Spread Transfer of information from one agent to another.
Theory Description of mechanisms that explain observations.
Trust Favorable emotional conviction that someone will behave as desired.
Truth Correspondence (match) with reality.
Value Emotional mental representation of the rightness of a concept.
World All that exists.

NOTES

PREFACE

1. Jonathan Swift, "The Art of Political Lying," *Examiner* 14 (1710). The text is available at https://www.bartleby.com/209/633.html.
2. Energy: Paul Thagard, "Energy Requirements Undermine Substrate Independence and Mind-Body Functionalism," *Philosophy of Science* 89 (2022): 70–88.

1. LIES KILL: THE PERILS OF MISINFORMATION

1. COVID death of talk show host: Andy Rose, "Conservative Talk Show Host Phil Valentine Dies After Battle with COVID-19," August 22, 2021, https://www.cnn.com/2021/08/21/us/conservative-talk-show-host-phil-valentine-dies-covid-19/index.html.
2. Heat wave: Henry Fountain, "Climate Change Drove Western Heat Wave's Extreme Records, Analysis Finds," July 7, 2007, https://www.nytimes.com/2021/07/07/climate/climate-change-heat-wave.html.
3. Heat wave deaths: Associated Press, "Hundreds Are Believed to Have Died During the Pacific Northwest Heat Wave," July 2, 2021, https://www.npr.org/2021/07/02/1012467409/hundreds-are-believed-to-have-died-during-the-pacific-northwest-heat-wave.
4. Lytton death: CBC News, "BC Coroners Service Confirms 2 Deaths in Lytton Wildfire," July 4, 2021, https://www.cbc.ca/news/canada/british-columbia/lytton-wildfire-sat-update-1.6089367.
5. Capitol deaths: Jack Healy, "These Are the 5 People Who Died in the Capitol Riot," *New York Times*, January 11, 2021, updated October 13, 2022, https://www.nytimes.com/2021/01/11/us/who-died-in-capitol-building-attack.html.
6. Buffalo, New York, shooting: "Conservative Media Is Familiar with Buffalo Suspect's Alleged 'Theory,'" *Washington Post*, https://www.washingtonpost.com/media/2022/05/15/buffalo-suspect-great-replacement-theory-conservative-media/.
7. COVID in Canada: "Social Inequalities in COVID-19 Deaths in Canada," https://health-infobase.canada.ca/covid-19/inequalities-deaths/index.html.
8. COVID in the United States: Julie Bosman, Sophie Kasakove and Daniel Victor, "U.S. Life Expectancy Plunged in 2020, Especially for Black and Hispanic Americans," *New York*

Times, July 21, 2021, https://www.nytimes.com/2021/07/21/us/american-life-expectancy-report.html.

9. Neo-Nazis: "Why Vladimir Putin Invokes Nazis to Justify His Invasion of Ukraine," https://www.nytimes.com/2022/03/17/world/europe/ukraine-putin-nazis.html?referringSource=articleShare.
10. Major threat: Jacob Poushter, Moira Fagan, and Sneha Gubbala, "Climate Change Remains Top Global Threat Across 19-Country Survey," Pew Research Center, August 21, 2022, https://www.pewresearch.org/global/2022/08/31/climate-change-remains-top-global-threat-across-19-country-survey/.
11. Surgeon general: Vivek H. Murthy, "Confronting Health Misinformation: The U.S. Surgeon General's Advisory on Building a Health Information Environment," (2021), sccessed October 4, 2021, https://www.hhs.gov/sites/default/files/surgeon-general-misinformation-advisory.pdf2.
12. Surgeon general: Murthy "Confronting Health Misinformation," 4.
13. Mechanisms: William Bechtel, *Mental Mechanisms: Philosophical Perspectives on Cognitive Neuroscience* (New York: Routledge, 2008); Carl F. Craver and Lindley Darden, *In Search of Mechanisms: Discoveries Across the Life Science* (Chicago: University of Chicago Press, 2013); Stuart Glennan, *The New Mechanical Philosophy* (Oxford: Oxford University Press, 2017); Paul Thagard, *Natural Philosophy: From Social Brains to Knowledge, Reality, Morality, and Beauty* (New York: Oxford University Press, 2019). In 2022 alone, mechanisms are mentioned in 349,000 articles, according to Google Scholar.
14. Obama: Jacob Stern, "Obama: I Underestimated the Threat of Disinformation," *Atlantic Monthly*, April 7, 2022, https://www.theatlantic.com/ideas/archive/2022/04/barack-obama-interview-disinformation-ukraine/629496/.
15. Health and mechanisms: Paul Thagard, *How Scientists Explain Disease* (Princeton, NJ: Princeton University Press, 1999); Paul Thagard, *Mind-Society: From Brains to Social Sciences and Professions* (New York: Oxford University Press, 2019), chap. 10; Lindley Darden, Kunal Kundu, Lipika R. Pal, and John Moult, "Harnessing Formal Concepts of Biological Lipika R, Mechanism to Analyze Human Disease," *PLoS Computational Biology* 14, no. 12 (December 2018): e1006540.
16. Infodemics: Andy Norman and Harris Eyre, "Disinformation? There Are Remedies for That," *Psychiatric Times*, https://www.psychiatrictimes.com/view/disinformation-there-are-remedies-for-that?.
17. Knowledge growth versus natural selection: Paul Thagard, *Computational Philosophy of Science* (Cambridge, MA: MIT Press, 1988), chap. 6.
18. Preventive medicine: Donald D. Hensrud, "Clinical Preventive Medicine in Primary Care: Background and Practice: 1. Rationale and Current Preventive Practices," *Mayo Clinic Proceedings* 75, no. 2 (2000): 165–72; Sharon K. Hull, "A Larger Role for Preventive Medicine," *AMA Journal of Ethics* 10, no. 11 (2008): 724–29.
19. Three-analysis of concepts: Paul Thagard, *Brain-Mind: From Neurons to Consciousness and Creativity* (New York: Oxford University Press, 2019); Gregory L. Murphy, *The Big Book of Concepts* (Cambridge, MA: MIT Press, 2002). More than 100 different definitions of information: Chaim Zins, "Conceptual Approaches for Defining Data, Information, and Knowledge," *Journal of the American Society for Information Science and Technology* 58, no. 4 (2007): 479–93.

20. *1984:* George Orwell, *1984* (New York: Harper, 2014).
21. Primitive information lacks the syntactic, semantic, and pragmatic complexity of information capable of being false and is discussed in the online supplemental material. Claude Shannon's mathematical theory of information falls under this heading.

2. INFORMATION AND MISINFORMATION: HOW THEY WORK

1. Perception: Marie T. Banich and Rebecca J. Compton, *Cognitive Neuroscience*, 4th ed. (Cambridge: Cambridge University Press, 2018).
2. Instruments: Davis Baird, *Thing Knowledge: A Philosophy of Scientific Instruments* (Berkeley: University of California Press, 2004).
3. Experiments: Allan Franklin, "Experiment in Physics." *Stanford Encyclopedia of Philosophy*, ed. Edward N. Zalta (Stanford, CA: The Metaphysics Research Lab, Center for the Study of Language and Information, Stanford University, 2019), https://plato.stanford.edu/entries/physics-experiment/.
4. Evidence: Paul Thagard, "Thought Experiments Considered Harmful," *Perspectives on Science* 22 (2014): 288–305; Paul Thagard, *Natural Philosophy: From Social Brains to Knowledge, Reality, Morality, and Beauty* (New York: Oxford University Press, 2019).
5. Neural mechanisms for representing: Chris Eliasmith, *How to Build a Brain: A Neural Architecture for Biological Cognition* (Oxford: Oxford University Press, 2013); Paul Thagard, *Brain-Mind: From Neurons to Consciousness and Creativity* (New York: Oxford University Press, 2019).
6. Collaboration: Paul Thagard, *How Scientists Explain Disease* (Princeton, NJ: Princeton University Press, 1999); Paul Thagard, "Collaborative Knowledge," *Noûs* 31 (1997): 242–61; Paul Thagard, "How to Collaborate: Procedural Knowledge in the Cooperative Development of Science," *Southern Journal of Philosophy* 44, (2006): 177–196.
7. Optical illusions: Michael Bach, "151 Illusions & Visual Phenomena with Explanations," April 13, 1997, https://michaelbach.de/ot/.
8. ESP: Terence Hines, *Pseudoscience and the Paranormal* (Buffalo: Prometheus, 2003).
9. Replication crisis: Patrick E. Shrout and Joseph L. Rodgers. "Psychology, Science, and Knowledge Construction: Broadening Perspectives from the Replication Crisis," *Annual Review of Psychology* 69 (2018): 487–510; Stuart Ritchie, *Science Fictions: How Fraud, Bias, Negligence, and Hype Undermine the Search for Truth* (New York: Metropolitan, 2020).
10. Noise: Daniel Kahneman, Olivier Sibony, and Cass R. Sunstein. *Noise: A Flaw in Human Judgment* (New York: Little, Brown Spark, 2021).
11. Misleading graphs: Claire Gnoux, "How to Prevent Misinformation in Data Visualization?," Towards Data Science, May 7, 2019, https://towardsdatascience.com/how-to-prevent-misinformation-in-data-visualization-1521a96e6431.
12. Photos: Laura Mallonnee, "How Photos Fuel the Spread of Fake News," Wired, December 21, 2016, https://www.wired.com/2016/12/photos-fuel-spread-fake-news/.
13. Videos: Elyse Samuels, Sarah, Cahlan, and Emily Sabens, "How to Spot a Fake Video," *Washington Post*, March 19, 2021, https://www.washingtonpost.com/politics/2021/03/19/how-spot-fake-video/?utm_medium=ret-all&utm_content=sallyvf&utm_campaign=dr-may-22&utm_source=email.

14. Kaiser Wilhelm Society: "History of the Kaiser Wilhelm Society Under National Socialism," Max-Planck-Gesellschaft, https://www.mpg.de/history/kws-under-national-socialism.
15. Tobacco companies: Naomi Oreskes and Erik M. Conway, *Merchants of Doubt: How a Handful of Scientists Obscured the Truth on Issues from Tobacco Smoke to Global Warming* (New York: Bloomsbury Publishing USA, 2011).
16. Bullshit: Harry G. Frankfurt, *On Bullshit* (Princeton, NJ: Princeton University Press, 2005).
17. Institutions: Paul Thagard, *Mind-Society: From Brains to Social Sciences and Professions* (New York: Oxford University Press, 2019). See also the online supplemental material for chap. 7.
18. Emergent: Thagard, *Mind-Society* discusses many cases of social emergence. The online supplemental material for chap. 7 analyzes the difference between systemic and individual change.
19. Inference to the best explanation: Peter Lipton, *Inference to the Best Explanatio*, 2nd ed. (London: Routledge, 2004).
20. Information accuracy: Luciano Floridi and Phyllis Illari, *The Philosophy of Information Quality* (Berlin: Springer, 2014).
21. Feelings as information: Norbert Schwarz, "Feelings-as-Information Theory," in *Handbook of Theories of Social Psychology*, ed. Paul A. M. Van Lange, Arie W. Kruglanski and E. Tory Higgins (Thousand Oaks, CA: Sage, 2011), 289–308; Nadia M. Brashier and Elizabeth J Marsh, "Judging Truth," *Annual Review of Psychology* 71 (January 4, 2020): 499–515.
22. Facebook anger: Jeremy B. Merrill and Will Oremus, "Five Points for Anger, One Point for 'Like': How Facebook's Formula Fostered Rage and Misinformation," *Washington Post*, October 26, 2021, https://www.washingtonpost.com/technology/2021/10/26/facebook-angry-emoji-algorithm/.
23. Cochrane: "About Us," Cochrane, https://www.cochrane.org/about-us.
24. Induction: John. H. Holland, Keith. J. Holyoak, Richard. E. Nisbett, and Paul R. Thagard, *Induction: Processes of Inference, Learning, and Discovery* (Cambridge, MA: MIT Press, 1986).
25. DNA: James D. Watson, *The Double Helix: A Personal Account of the Discovery of the Structure of DNA* (New York: New American Library, 1968).
26. Confirmation bias: Raymond S. Nickerson, "Confirmation Bias: A Ubiquitous Phenomenon in Many Guises," *Review of General Psychology* 2, no. 2 (1998): 175–220.
27. Motivated reasoning: Ziva Kunda, "The Case for Motivated Reasoning." *Psychological Bulletin* 108 (1990): 480–98; Matthew J. Hornsey, "Why Facts Are Not Enough: Understanding and Managing the Motivated Rejection of Science," *Current Directions in Psychological Science* 29, no. 6 (2020): 583–91. See chap. 3 below for more references.
28. Fear-driven inference: Paul Thagard and David Nussbaum, "Fear-Driven Inference: Mechanisms of Gut Overreaction, in *Model-Based Reasoning in Science and Technology*, ed Lorenzo Magnani (Berlin: Springer, 2014), 43–53; Thagard, *Mind-Society*.
29. Emotions: Ivana Kajić, Tobias C. Schröder, Terrence C. Stewart, and Paul Thagard, "The Semantic Pointer Theory of Emotions: Integrating Physiology, Appraisal, and Construction," *Cognitive Systems Research* 58 (2019): 35–53; Paul Thagard, Laurette Larocque, and Ivana Kajić, "Emotional Change: Neural Mechanisms Based on Semantic Pointers," *Emotion* 23 (2023): 182–93.

30. Groupthink: Irving L. Janis, *Groupthink: Psychological Studies of Policy Decisions and Fiascoes*, 2nd ed. (Boston: Houghton Mifflin, 1982).
31. Critical thinking: Rolf Dobelli, *The Art of Thinking Clearly: Better Thinking, Better Decisions* (New York: Harper, 2013; Richard E. Nisbett, *Mindware: Tools for Smart Thinking*, (New York: Farrar, Straus and Giroux, 2015); Theodore Schick and Lewis Vaughn, *How to Think About Weird Things*, 8th ed. (New York: McGraw Hill, 2020). Debunking: Stephan Lewandowsky, John Cook, Ullrich Ecker, Dolores Albarracin, Michelle Amazeen, Panayiota Kendou, Doug Lombardi, et al., *The Debunking Handbook 2020* (Fairfax, VA: Center for Climate Change Communication, 2020), https://www.climatechangecommunication.org/wp-content/uploads/2020/10/DebunkingHandbook2020.pdf.
32. Fallacies: See Wikipedia, s.v. "List of fallacies," https://en.wikipedia.org/wiki/List_of_fallacies. Biases: See Wikipedia, s.v. "List of cognitive biases," https://en.wikipedia.org/wiki/List_of_cognitive_biases. My list of the error tendencies I found most valuable in teaching critical thinking is in: Paul Thagard, "Critical Thinking and Informal Logic: Neuropsychological Perspectives," *Informal Logic* 31 (2011): 152–70. See also Thomas Gilovich, *How We Know What Isn't So* (New York: Free Press, 1991); J. Edward Russo and Paul J. H. Schoemaker, *Decision Traps* (New York: Simon & Schuster, 1989); Max H. Bazerman and Don A. Moore, *Judgment in Managerial Decision Making*, 8th ed. (New York: John Wiley, 2012).
33. Critical thinking success: Carlos Saiz and Leandro S. Almeida. "The Halpern Critical Thinking Assessment and Real-World Outcomes: Cross-National Applications," *Thinking Skills and Creativity* 7, no. 2 (2012): 112–21; Amy Shaw, Ou Lydia Liu, Lin Gu, Elena Kardonova, Igor Chirikov, Guirong Li, Shangfeng Hu, et al., "Thinking Critically About Critical Thinking: Validating the Russian Heighten® Critical Thinking Assessment," *Studies in Higher Education* 45, no. 9 (2019): 1933–48.
34. Motivational interviewing for misinformation: Adam Grant, *Think Again: The Power of Knowing What You Don't Know* (New York: Penguin, 2021).
35. Motivational interviewing for addiction: William R. Miller and Stephen Rollnick, *Motivational Interviewing: Helping People Change* (New York: Guilford, 2012): Helen Frost, Pauline Campbell, Margaret Maxwell, Ronan E O'Carroll, Stephan U Dombrowski, Brian Williams, Helen Cheyne, Emma Coles, and Alex Pollock, "Effectiveness of Motivational Interviewing on Adult Behaviour Change in Health and Social Care Settings: A Systematic Review of Reviews," *PLoS One* 13, no. 10 (2018): e0204890.
36. Therapeutic alliance: Louis G. Castonguay and Clara E Hill, eds., *How and Why Are Some Therapists Better Than Others? Understanding Therapist Effects* (Washington, DC: American Psychological Association, 2017).
37. Emotional change: Thagard, Larocque, and Kajić, "Emotional Change."
38. Automated motivational interviewing: SoHyun Park, Jeewon Choi, Sungwoo Lee, Changhoon Oh, Changdai Kim, Soohyun La, Joonhwan Lee, and Bongwon Suh, "Designing a Chatbot for a Brief Motivational Interview on Stress Management: Qualitative Case Study," *Journal of Medical Internet Research* 21, no. 4 (2019): e12231.
39. Automated critical thinking: Dang Wang, Hongyun Liu, and Kit-Tai Hau, "Automated and Interactive Game-Based Assessment of Critical Thinking," *Education and Information Technologies* 27, no. 4 (2022): 4553–75.

40. Memory: Daniel L. Schacter, *The Seven Sins of Memory: How the Mind Forgets and Remembers* (Boston: Houghton Mifflin, 2002).
41. Semantic memory: Timothy T. Rogers and James L. McClelland, *Semantic Cognition: A Parallel Distributed Processing Approach* (Cambridge, MA: MIT Press, 2004).
42. Emotional memory: Joseph E. Dunsmoor, Marijn C. W. Kroes, Vishnu P. Murty, Stephen H. Braren, and Elizabeth A. Phelps, "Emotional Enhancement of Memory for Neutral Information: The Complex Interplay Between Arousal, Attention, and Anticipation," *Biological Psychology* 145 (2019): 134–41.
43. Sins of memory: Schacter, *The Seven Sins of Memory*.
44. Eyewitness testimony: Elizabeith F. Loftus, *Eyewitness Testimony* (Cambridge, MA: Harvard University Press, 1996).
45. Availability: Daniel Kahneman and Amos Tversky, eds. *Choices, Values, and Frames* (Cambridge: Cambridge University Press, 2000).
46. Analogical problem solving: Keith J. Holyoak and Paul Thagard, *Mental Leaps: Analogy in Creative Thought* (Cambridge, MA: MIT Press, 1995).
47. Misinformation effect: C. Laney and E. F. Loftus, "Eyewitness Testimony and Memory Biases, in *Noba Textbook Series: Psychology*, ed. R. Biswas-Diener and E. Diener (Champaign, IL: DEF Publishers, 2023), https://nobaproject.com/modules/eyewitness-testimony-and-memory-biases.
48. Repairing eyewitness testimony: Thomas D. Albright, "Why Eyewitnesses Fail," *Proceedings of the Natlional Academy of Sciences* 114, no. 30 (July 25, 2017): 7758–64.
49. Social mechanisms: Thagard, *Mind-Society*.
50. Neglectful sending: Gordon Pennycook, Jonathon McPhetres, Yunhao Zhang, Jackson G Lu, and David G. Rand, "Fighting Covid-19 Misinformation on Social Media: Experimental Evidence for a Scalable Accuracy-Nudge Intervention," *Psychological Science* 31, no. 7 (2020): 770–80; Gordon Pennycook and David G. Rand, "Lazy, Not Biased: Susceptibility to Partisan Fake News Is Better Explained by Lack of Reasoning Than by Motivated Reasoning," *Cognition* 188 (2019): 39–50.
51. Automatic belief: Daniel T. Gilbert, "How Mental Systems Believe," *American Psychologist* 46, no. 2 (1991): 107–19.
52. Empathy as analogy and bodily mirroring: Thagard, *Natural Philosophy*.
53. Trust: Paul Thagard, "What Is Trust?," *Psychology Today*, October 9, 2018, https://www.psychologytoday.com/ca/blog/hot-thought/201810/what-is-trust; Thagard, *Mind-Society*.
54. Causes: Thagard, *Natural Philosophy*.

3. BELIEVING WHAT YOU WANT: MOTIVATED REASONING, EMOTION, AND IDENTITY

1. Pinker on rationality: Steven Pinker, *Rationality: What It Is, Why It Seems Scarce, and Why It Matter* (New York: Viking, 2021).
2. Stanley on propaganda: Jason Stanley, *How Propaganda Works* (Princeton, NJ: Princeton University Press, 2015).
3. Political science: Brian Guay and Christopher D. Johnston, "Ideological Asymmetries and the Determinants of Politically Motivated Reasoning," *American Journal of Political*

Science 66, no. 2 (2022): 284–301. Economics: Nicholas Epley and Thomas Gilovich, "The Mechanics of Motivated Reasoning," *Journal of Economic Perspectives* 30, no. 3 (2016): 133–40. Communication: Federico Vegetti and Moreno Mancosu, "The Impact of Political Sophistication and Motivated Reasoning on Misinformation," *Political Communication* 37, no. 5 (2020): 678–95. Philosophy: Jon Ellis, "Motivated Reasoning and the Ethics of Belief," *Philosophy Compass* 17 (2022): e12828.

4. Motivated thinking examples: Paul Thagard, "Climate Change Denial," August 26, 2011, https://www.psychologytoday.com/ca/blog/hot-thought/201108/climate-change-denial; Paul Thagard, "Critical Thinking and Informal Logic: Neuropsychological Perspectives," *Informal Logic* 31 (2011): 152–70.
5. Cognitive biases: Daniel Kahneman and Amos Tversky, eds. *Choices, Values, and Frames* (Cambridge: Cambridge University Press, 2000); Richard E. Nisbett and Lee Ross, *Human Inference: Strategies and Shortcomings of Social Judgement* (Englewood Cliffs, NJ: Prentice Hall, 1980); Dale T. Miller and Michael Ross, "Self-Serving Biases in Attribution of Causality: Fact or Fiction?," *Psychological Bulletin* 82 (1975): 213–25.
6. Ziva's mother: Ziva Kunda, *Social Cognition: Making Sense of People* (Cambridge, MA: MIT Press, 1999), 212.
7. Thucydides: *History of the Peloponnesian War*, https://www.gutenberg.org/files/7142/7142-h/7142-h.htm.
8. Aristotle: Aristotle, *Ethics*, trans. J. A. K. Thomson (Harmondsworth, Middlesex: Penguin, 1955), 194. (*Ethics* Vii-1).
9. Arrian: *The Anabasis of Alexander*, ed. E. J. Chinnock (London: Hodder and Stoughton, 1884), https://www.gutenberg.org/files/46976/46976-h/46976-h.htm.
10. Bacon: Francis Bacon, *The New Organon and Related Writings*, ed. F. Anderson (Indianapolis, IN: Bobbs-Merrill, 1960), 52. Aphorisms book one, XLIX.
11. Wishful thinking: John Stuart Mill, *A System of Logic*, 8 ed. (London: Longman, 1970); Jon Elster, *Explaining Social Behavior* (Cambridge: Cambridge University Press, 2007).
12. Motivated irrationality: David Pears, *Motivated Irrationality* (Oxford: Oxford University Press, 1984).
13. Motivated cognition: Arie W, Kruglanski, Jocelyn J. Belanger, Xiaoyan Chen, Catalina Kopetz, Antonio Pierro, and Lucia Mannetti, "The Energetics of Motivated Cognition: A Force-Field Analysis," *Psychological Review* 119, no. 1 (January 2012): 1–20.
14. Motivated inference: Ziva Kunda "Motivated Inference: Self-Serving Generation and Evaluation of Evidence,". *Journal of Personality and Social Psychology* 53 (1987): 636–47.
15. Inference versus reasoning: Thagard, "Critical Thinking and Informal Logic;" Paul Thagard, *Brain-Mind: From Neurons to Consciousness and Creativity* (New York: Oxford University Press, 2019). Inference is system 1, while reasoning is system 2, in the usual psychological classification, which misuses the term "system."
16. Wishful seeing: David Dunning and Emily Balcetis, "Wishful Seeing: How Preferences Shape Perception," *Current Directions in Psychological Science* 22 (2013): 33–37. Wishful hearing: Lauren Mayor Poupis, "Wishful Hearing: The Effect of Chronic Dieting on Auditory Perceptual Biases and Eating Behavior," *Appetite* 130 (November 1, 2018): 219–27.
17. Attempted rationality: Kunda, *Social Cognition*, 282–283.
18. John Stuart Mill: Mill, *A System of Logic*, 483. Section V-1-3.

19. Strongmen: Ruth Ben-Ghiat, *Strongmen: Mussolini to the Present* (New York: Norton, 2020).
20. Manifesting: Rebecca Jennings, "Shut Up, I'm Manifesting!," Vox, October 23, 2020, https://www.vox.com/the-goods/21524975/manifesting-does-it-really-work-meme; Stuart McGurk, "Tik Tok's 'Manifesting' Craze Explained," *GQ*, June 15, 2021, https://www.gq-magazine.co.uk/culture/article/manifesting-tik-tok; Stuart McGurk, "Making Dreams Come True Inside the New Age World of Manifesting," *Guardian*, March 20, 2022, https://www.theguardian.com/lifeandstyle/2022/mar/20/making-dreams-come-true-inside-the-new-age-world-of-manifesting?CMP=Share_iOSApp_Other.
21. Law of attraction: Rhonda Byrne, *The Secret* (New York: Simon and Schuster, 2006).
22. Lucky girl: Teddy Amenabar, " 'Be Delusional,' 'Lucky Girl Syndrome' Is Gen Z's Answer to Optimism," Washington Post, January 1, 2023, https://www.washingtonpost.com/wellness/2023/01/20/lucky-girl-syndrome-tiktok/.
23. Positive thinking: Gabriele Oettingen, *Rethinking Positive Thinking: Inside the New Science of Motivation* (New York: Current, 2015).
24. Biased memory: Kunda, *Social Cognition*, 224.
25. Motiv-PI: Paul Thagard and Ziva Kunda, "Hot Cognition: Mechanisms of Motivated Inference," in *Proceedings of the Ninth Annual Conference of the Cognitive Science Society*, ed. E. Hunt (Hillsdale, NJ: Erlbaum, 1987), 753–63. PI is from Paul Thagard, *Computational Philosophy of Science* (Cambridge, MA: MIT Press, 1988). The online supplementary material contains a flowchart for this kind of motivated thinking.
26. Emotional coherence: Paul Thagard, *Coherence in Thought and Action* (Cambridge, MA: MIT Press, 2000).
27. O. J. Simpson: Paul Thagard, "Why Wasn't O. J. Convicted? Emotional Coherence in Legal Inference," *Cognition and Emotion* 17 (2003): 361–83.
28. Motivated stereotypes: Lisa Sinclair and Ziva Kunda, "Reactions to a Black Professional: Motivated Inhibition and Activation of Conflicting Stereotypes," *Journal of Personality and Social Psychology* 77 (1999): 885–904.
29. Emotional coherence evidence: Dan Simon, Douglas M. Stenstrom, and Stephen. J. Read, "The Coherence Effect: Blending Cold and Hot Cognitions," *Journal of Personality and Social Psychology* 109 (2015): 369–94.
30. Cognitive-affective mapping: Paul Thagard, *Mind-Society: From Brains to Social Sciences and Professions* (New York: Oxford University Press, 2019). Full list of applications: Paul Thagard, "Cognitive-Affective Maps," https://paulthagard.com/links/cognitive-affective-maps/.
31. Bacon on emotions: Bacon, *New Organon*, 52.
32. Motivated relationships: Sandra L. Murray and John. G. Holmes, *Motivated Cognition in Relationships: The Pursuit of Belonging* (New York: Routledge, 2017).
33. Values based on needs: Paul Thagard, *Natural Philosophy: From Social Brains to Knowledge, Reality, Morality, and Beauty* (New York: Oxford University Press, 2019), chap. 6.
34. Stereotypes: Ziva Kunda and Paul Thagard, "Forming Impressions from Stereotypes, Traits, and Behaviors: A Parallel-Constraint-Satisfaction Theory," *Psychological Review* 103 (1996): 284–308.

35. fMRI: Drew Westen, Pavel S. Blagov, Keith Harenski, Clint Kilts, and Stephan Hamann, "Neural Bases of Motivated Reasoning: An fMRI Study of Emotional Constraints on Partisan Political Judgment in the 2004 U.S. Presidential Election," *Journal of Cognitive Neuroscience* 18 (2006): 1947–58.
36. Neural: Brent L. Hughes and Jamil Zaki, "The Neuroscience of Motivated Cognition," *Trends in Cogitive Sciences* 19 (2015): 62–64.
37. Reward: Yuan Chang Leong, Brent L. Hughes, Yiyu Wang, and Jamil Zaki, "Neurocomputational Mechanisms Underlying Motivated Seeing," *Nature Human Behaviour* 3, no. 9 (2019): 962–73; Brandon M. Wagar and Paul Thagard, "Spiking Phineas Gage: A Neurocomputational Theory of Cognitive-Affective Integration in Decision Making," *Psychological Review* 111 (2004): 67–79.
38. Threat: Brent L. Hughes and Jennifer S. Beer, "Protecting the Self: The Effect of Social-Evaluative Threat on Neural Representations of Self," *Journal of Cognitive Neuroscience* 25, no. 4 (2013): 613–22.
39. Neural cognition and emotion: Chris Eliasmith, *How to Build a Brain: A Neural Architecture for Biological Cognition* (Oxford: Oxford University Press, 2013); Thagard, *Brain-Mind.* Binding can be accomplished by a neural mechanism that convolves patterns of neural firing.
40. Mental representations: Thagard, *Brain-Mind.*
41. Neural emotions: Ivana Kajić, Tobias C. Schröder, Terrence C. Stewart, and Paul Thagard, "The Semantic Pointer Theory of Emotions: Integrating Physiology, Appraisal, and Construction," *Cognitive Systems Research* 58 (2019): 35–53;" Paul Thagard, Laurette Larocque, and Ivana Kajić, "Emotional Change: Neural Mechanisms Based on Semantic Pointers," *Emotion* 23 (2023): 182–93.
42. Belief: Sam Harris, Sameer A. Sheth and Mark S. Cohen, "Functional Neuroimaging of Sam, Belief, Disbelief, and Uncertainty," *Annals of Neurology* 63 (2008): 141–47.
43. Intention: Tobias Schröder, Terrence C. Stewart, and Paul Thagard, "Intention, Emotion, and Action: A Neural Theory Based on Semantic Pointers," *Cognitive Science* 38 (2014): 851–80.
44. Social mechanisms: Thagard, *Mind-Society.*
45. Groups: Yrian Derreumaux, Robin Bergh, and Brent L. Hughes, "Partisan-Motivated Sampling: Re-Examining Politically Motivated Reasoning Across the Information Processing Stream," *Journal of Personality and Social Psychology* 123 (2022): 316–36.
46. Emotional communication: Thagard, *Mind-Society*, 62–65, describes twelve social-emotional mechanisms.
47. Mirror neurons: Marco Iacoboni, *Mirroring People: The New Science of How We Connect with Others* (New York: Farrar, Straus and Giroux, 2008).
48. Rituals: Randall Collins, *Interaction Ritual Chains* (Princeton, NJ: Princeton University Press, 2004).
49. Self-esteem: Kunda, *Social Cognition*, p. 465.
50. Identity: Daphna Oyserman and Andrew Dawson. "Your Fake News, Our Facts: Identity-Based Motivation Shapes What We Believe, Share, and Accept," in *The Psychology of Fake News: Accepting, Sharing, and Correctig Misinformation*, ed. Rainer Greifeneder, Mariela E. Jaffé, Eryn J. Newman, and Norbert Schwarz (London: Routledge, 2020), 173–95.

51. Self: Paul Thagard, "The Self as a System of Multilevel Interacting Mechanisms," *Philosophical Psychology* 27 (2014): 145–63, Thagard, *Brain-Mind*; Paul Thagard and Joanne V. Wood, "Eighty Phenomena About the Self: Representation, Evaluation, Regulation, and Change," *Frontiers in Psychology* 6 (2015): 34. The online supplemental material contains a chart of more than eighty kinds of self-phenomena.
52. Myside bias: Keith E. Stanovich, *The Bias That Divides Us: The Science and Politics of Myside Thinking* (Cambridge, MA: MIT Press, 2021), 9.
53. Advantages of optimism: Murray and Holmes, *Motivated Cognition in Relationships*; Becca R. Levy, Martin D. Slade, Suzanne R. Kunkel, and Stanislav V. Kasl, "Longevity Increased by Positive Self-Perceptions of Aging," *Journal of Personality and Social Psychology* 83, no. 2 (2002): 261–70; Shelley E. Taylor and Jonathon D. Brown, "Illusion and Well-Being: A Social Psychological Perspective on Mental Health," *Psychological Bulletin* 103, no. 2 (1988): 193–210.
54. Neural integration: Paul Thagard, "How Rationality Is Bounded by the Brain," in *Routledge Handbook of Bounded Rationality*, ed. Riccardo Viale (London: Routledge, 2021), 398–406.
55. Therapeutic emotional change: Thagard, Larocque, and Kajić, "Emotional Change."
56. Motivational interviewing and needs: Maarten Vansteenkiste and Kennon M. Sheldon, "There's Nothing More Practical Than a Good Theory: Integrating Motivational Interviewing and Self-Determination Theory," *British Journal of Clinical Psychology* 45 (2006): 63–82.
57. Emotion and humor: Sara K. Yeo and Meaghan McKasy, "Emotion and Humor as Misinformation Antidotes," *Proceedings of the Natoional Academy of Sciences* 118, no. 15 (2021).
58. Motivated ignorance: Paul Thagard, "Motivated Ignorance," *Psychology Today*, January 16, 2013, https://www.psychologytoday.com/ca/blog/hot-thought/201301/motivated-ignorance; Daniel Williams, "Motivated Ignorance, Rationality, and Democratic Politics," *Synthese* 198, no. 8 (2020): 7807–27. Related concepts include willful ignorance and willful blindness.
59. Doubt: Naomi Oreskes and Erik M. Conway, *Merchants of Doubt: How a Handful of Scientists Obscured the Truth on Issues from Tobacco Smoke to Global Warming* (New York: Bloomsbury, 2011).
60. Commercial motivated inventing: Max Fisher, "Disinformation for Hire, a Shadow Industry, Is Quietly Booming," *New York Times*, July 25, 2021, https://www.nytimes.com/2021/07/25/world/europe/disinformation-social-media.html?referringSource=articleShare.
61. Putting ideas together: Thagard *Brain-Mind*, chap. 11.
62. Laziness: Gordon Pennycook and David G. Rand, "Lazy, Not Biased: Susceptibility to Partisan Fake News Is Better Explained by Lack of Reasoning Than by Motivated Reasoning," *Cognition* 188 (2019): 39–50.

4. PLAGUES: COVID-19 AND MEDICAL MISINFORMATION

1. Covid deaths: Krutika Amin, Jared Ortaliza, Cynthis Cox, Joshua Michaud, and Jennifer Kates, "COVID-19 Mortality Preventable by Vaccines," Health System Tracker, October 12, 2021, https://www.healthsystemtracker.org/brief/covid19-and-other-leading-causes-of-death-in-the-us/.

2. Virus identification: Na Zhu, Dingyu Zhang, Wenling Wang, Xingwang Li, Bo Yang, Jingdong Song, Xiang Zhao, et al., "A Novel Coronavirus from Patients with Pneumonia in China, 2019," *New England Journal of Medicine* 382, no. 8 (2020): 727–33.
3. PCR test: Jonathan Jarry, "The COVID-19 PCR Test Is Reliable Despite the Commotion About Ct Values," McGill Office for Science and Society, https://www.mcgill.ca/oss/article/covid-19-critical-thinking/covid-19-pcr-test-reliable-despite-commotion-about-ct-values.
4. Anecdotes: Kenneth Kernaghan, P. K. Kuruvilla, Paul Samuelson, Edith Greene, and Irwin S. Bernstein, "The Plural of Anecdote Is Not Data," Quote Investigator®, December 27, 2017, https://quoteinvestigator.com/2017/12/27/plural/.
5. Experiment: Paul Rosenbaum, *Observation and Experiment* (Cambridge, MA: Harvard University Press, 2018).
6. Clinical trials: National Institute on Aging, "What Are Clinical Trials and Studies?," National Institutes of Health, March 22, 2023, https://www.nia.nih.gov/health/what-are-clinical-trials-and-studies.
7. Fluvoxamine: Gilmar Reis, Eduardo Augusto dos Santos Moreira-Silva, Daniela Carla Medeiros Silva, Lehana Thabane, Aline Cruz Milagres, Thiago Santiago Ferreira, Castilho Vitor Quirino dos Santos, et al., "Effect of Early Treatment with Fluvoxamine on Risk of Emergency Care and Hospitalisation Among Patients with Covid-19: The Together Randomised, Platform Clinical Trial," *The Lancet Global Health* 10 (2022): E42–E51.
8. Evidence-based medicine: Jeremy H. Howick, *The Philosophy of Evidence-Based Medicine* (Oxford: Wiley-Blackwell, 2011).
9. WHO: WHO Coronavirus (COVID-19) Dashboard, World Health Organization, https://covid19.who.int. NYT: "Coronavirus World Map: Tracking the Global Outbreak," *New York Times*, March 10, 2023, https://www.nytimes.com/interactive/2021/world/covid-cases.html.
10. Worldometer: "Reported Cases and Deaths by Country or Territory," Worldometer, June 12, 2023, https://www.worldometers.info/coronavirus/#countries.
11. Dexamethasone: RECOVERY Collaborative Group, "Dexamethasone in Hospitalized Patients with Covid-19—Preliminary Report," *The New England Journal of Medicine* 384 (2021): 693–704.
12. Evaluating causal claims in medicine: Austin Bradford Hill, "The Environment and Disease: Association or Causation?," *Proceedings of the Royal Society of Medicine* 58 (1965): 295–300; Olaf Dammann, Ted Poston, and Paul Thagard, "How Do Medical Researchers Make Causal Inferences?," in *What Is Scientific Knowledge? An Introduction to Contemporary Epistemology of Science*, ed. K. McCain and K. Kampourakis (New York: Routledge, 2019), 33–51.
13. Proning technique: Jason Weatherald, Kevin Solverson, Danny J Zuege, Nicole Loroff, Kirsten M Fiest, and Ken Kuljit S Parhar, "Awake Prone Positioning for Covid-19 Hypoxemic Respiratory Failure: A Rapid Review," *Journal of Critical Care* 61 (2021): 63–70.
14. Explanatory coherence: Paul Thagard, "Explanatory Coherence," *Behavioral and Brain Sciences* 12 (1989): 435–67; Paul Thagard, *Conceptual Revolution* (Princeton, NJ: Princeton University Press, 1992; Paul Thagard, *The Cognitive Science of Science: Explanation, Discovery, and Conceptual Change* (Cambridge, MA: MIT Press, 2012); Dammann, Poston, and

Thagard, "How Do Medical Researchers Make Causal Inferences?" The main alternative to explanatory coherence as an account of causal reasoning is Bayesian calculation with probabilities. The online supplementary material describes problems with the Bayesian approach. Application of explanatory coherence to COVID: Paul Thagard, "The Cognitive Science of Covid-19: Acceptance, Denial, and Belief Change," *Methods* 195 (2021): 92–102.

15. Evidence for masks: Jason Abaluck, Laura H. Kwong, Ashley Styczynski, Ashraful Haque, Md Alagir Kabir, Ellen Bates-Jefferys, Emily Crawford, et al., "Impact of Community Masking on Covid-19: A Cluster-Randomized Trial in Bangladesh," *Science* 375, no. 6577 (2022): eabi9069; CDC, "Scientific Brief: Community Use of Cloth Masks to Controll the Spread of Sars-Cov-2," Centers for Disease Control and Prevention, (2021), accessed March 10, 2023, https://www.cdc.gov/coronavirus/2019-ncov/science/science-briefs/masking-science-sars-cov2.html; Jeremy Howard, Austin Huang, Zhiyuan Li, Zeynep Tufekci, Vladimir Zdimal, Helene-Mari van der Westhuizen, Arne von Delft, et al., "An Evidence Review of Face Masks Against Covid-19," *Proceedings of the National Academy of Sciences* 118, no. 4 (2021): e2014564118. Cochrane report: Tom Jefferson, Liz Dooley, Eliana Ferroni, Lubna A. Al-Ansary, Mieke L. van Driel, Ghada A. Bawazeer, Mark A. Jones, et al., "Physical Interventions to Interrupt or Reduce the Spread of Respiratory Viruses," *Cochrane Database Systematic Reviews* 1, no. 1 (2023): CD006207.
16. Simulations: David Adam, "Special Report: The Simulations Driving the World's Response to Covid-19," *Nature* 580, no. 7802 (2020): 316–19.
17. Vaccinations: Oliver J. Watson, Gregory Barnsley, Jaspreet Toor, Alexandra B. Hogan, Peter Winskill, and Azra C. Ghani, "Global Impact of the First Year of Covid-19 Vaccination: A Mathematical Modelling Study," *The Lancet Infectious Diseases* 22 (2022): P1293–302.
18. Animal spread: Michael Worobey, Joshua I. Levy, Lorena Malpica Serrano, Alexander Crits-Christoph, Jonathan E. Pekar, Stephen A. Goldstein, Angela L. Rasmussen, et al., "The Huanan Seafood Wholesale Market in Wuhan Was the Early Epicenter of the Covid-19 Pandemic," *Science* 377, no. 6609 (2022): 951–59; Jonathan E. Pekar, Andrew Magee, Edyth Parker, Niema Moshiri, Katherine Izhikevich, Jennifer L. Havens, Karthik Gangavarapu, et al., "The Molecular Epidemiology of Multiple Zoonotic Origins of Sars-Cov-2," *Science* 377, no. 6609 (2022): 960–66.
19. WHO database: "Coronavirus Disease (COVID-19) Pandemic," World Health Organization, https://www.who.int/emergencies/diseases/novel-coronavirus-2019.
20. WHO myth-busters: "Coronavirus Disease (COVID-19) Advice for the Public: Mythbusters," World Health Organization, January 19, 2022, https://www.who.int/emergencies/diseases/novel-coronavirus-2019/advice-for-public/myth-busters.
21. CDC database: https://www.cdc.gov/coronavirus/2019-ncov/index.html.
22. CDC newsroom: https://www.cdc.gov/coronavirus/2019-ncov/science/science-briefs/.
23. Flawed science: Stuart Ritchie, *Science Fictions: How Fraud, Bias, Negligence, and Hype Undermine the Search for Truth* (New York: Metropolitan Books, 2020).
24. Predatory journals: Dominique Vervoort, Xiya Ma, and Mark G Shrime, "Money Down the Drain: Predatory Publishing in the Covid-19 Era," *Canadian Journal of Public Health* 111, no. 5 (2020): 665–666.
25. Preprints: Brian Owens, "The Rise of Preprints: How COVID-19 Has Transformed the Way We Publish and Report on Scientific Research," *University Affairs* (May 4, 2022), https://www.universityaffairs.ca/features/feature-article/the-rise-of-preprints/?.

26. Retracted: Mandeep R. Mehra, Sapan S. Desai, Frank Ruschitzka, and Amit N. Patel, "Retracted: Hydroxychloroquine or Chloroquine with or Wwithout a Macrolide for Treatment of Covid-19: A Multinational Registry Analysis," *The Lancet* (2020), https://doi.org/10.1016/s0140-6736(20)31180-6.
27. Hydroxychloroquine: Michael S. Saag, "Misguided Use of Hydroxychloroquine for Covid-19: The Infusion of Politics into Science," *Journal of the American Academy of Medicine* 324, no. 21 (2020): 2161–62.
28. Ivermectin: Maria Popp, Miriam Stegemann, Maria-Inti Metzendorf, Susan Gould, Peter Kranke, Patrick Meybohm, Nicole Skoetz, and Stephanie Weibel, "Ivermectin for Preventing and Treating Covid-19," *Cochrane Database of Systematic Reviews* 7 (July 28, 2021): CD015017; Gilmar Reis, Eduardo A. S. M. Silva, Daniela C. M. Silva, Lehana Thabane, Aline C. Milagres, Thiago S. Ferreira, Castilho V. Q. Dos Santos, et al., "Effect of Early Treatment with Ivermectin Among Patients with Covid-19," *New England Journal of Medicine* 386, no. 18 (2022): 1721–31.
29. Merck on ivermectin: Merck, "Merck Statement on Ivermectin Use During the COVID-19 Pandemic," February 4, 2021, https://www.merck.com/news/merck-statement-on-ivermectin-use-during-the-covid-19-pandemic/.
30. Coronavirus image: Robert Roy Britt, "What the Coronavirus Image You've Seen a Million Times Really Shows," *Elemental* (March 13, 2020), https://elemental.medium.com/what-the-coronavirus-image-youve-seen-a-million-times-really-shows-3d8de7e3eb1f.
31. Distorted virus pictures: Celia Andreu-Sánchez and Miguel-Ángel Martín-Pascual, "Fake Images of the Sars-Cov-2 Coronavirus in the Communication of Information at the Beginning of the First Covid-19 Pandemic," *El Profesional de la Información* 29, no. 3 (2020): e290309.
32. Values based on needs: Paul Thagard, *The Brain and the Meaning of Life* (Princeton, NJ: Princeton University Press, 2010); Paul Thagard, *Natural Philosophy: From Social Brains to Knowledge, Reality, Morality, and Beauty* (New York: Oxford University Press, 2019).
33. Emotional coherence: Thagard, *Hot Though: Mechanisms and Applications of Emotional Cognition* (Cambridge MA: MIT Press, 2006). Paul Thagard, *Mind-Society: From Brains to Social Sciences and Professions* (New York: Oxford University Press, 2019).
34. Trump denial: Dave Gilson, Laura Thompson, Clara Jeffery, Nina Liss-Schultz, Kiera Butler, and Will Peischel, "Super spreader in Chief: The Ultimate Timeline of Trump's Deadly Coronavirus Denial," Mother Jones, October 8, 2020, https://www.motherjones.com/politics/2020/10/trump-coronavirus-covid-denial-timeline/#feb.
35. Anger: Jiyoung Han, Meeyoung Cha, and Wonjae Lee, "Anger contributes to the spread of COVID-19 misinformation," *(Mis)Information Review* (September 17, 2020), https://misinforeview.hks.harvard.edu/article/anger-contributes-to-the-spread-of-covid-19-misinformation/.
36. Motivated remembering: Baruch Eitam, David B. Miele, and E. Tory Higgins, "Motivated Remembering: Remembering as Accessibility and Accessibility as Motivational Relevance," in *The Oxford Handbook of Social Cognition*, ed. Donal E. Carlston (Oxford: Oxford University Press, 2013), 463–75.
37. Motivated forgetting: Michael C. Anderson and Simon Hanslmayr, "Neural Mechanisms of Motivated Forgetting," *Trends in Cognitive Sciences* 18, no. 6 (2014): 279–92.

38. Disinformation dozen: "The Disinformation Dozen: Why Platforms Must Act on Twelve Leading Online Anti-Vaxxers," Center for Countering Digital Hate, https://www.counterhate.com/disinformationdozen.

39. Substack: Elizabeth Dwoskin, "Conspiracy Theorists, Banned on Major Social Networks, Connect with Audiences on Newsletters and Podcasts," *Washington Post*, January 27, 2022, https://www.washingtonpost.com/technology/2022/01/27/substack-misinformation-anti-vaccine/.

40. Facebook: Gerrit De Vynck, Cat Zakrzewski, and Cristiano Lima, E D, "Facebook Told the White House to Focus on the 'Facts' About Vaccine Misinformation: Internal Documents Show It Wasn't Sharing Key Data," *Washington Post*, October 28, 2021, https://www.washingtonpost.com/technology/2021/10/28/facebook-covid-misinformation/.

41. Spread of COVID misinformation: Gordon Pennycook, Jonathon McPhetres, Yunhao Zhang, Jackson G Lu, and David G Rand. "Fighting Covid-19 Misinformation on Social Media: Experimental Evidence for a Scalable Accuracy-Nudge Intervention," *Psychological Science* 31, no. 7 (2020): 770–780.

42. Continued influence: Dian van Huijstee, Ivar Vermeulen, Peter Kerkhof, and Ellen Droog, "Continued Influence of Misinformation in Times of Covid-19." *International Journal of Psychology* 57, no. 1 (2022): 136–45.

43. Merck failure: "Merck Discontinues Development of SARSs-CoV-2 COVID-19 Vaccine Candidates; Continues Development of Two Investigational Therapeutic Candidates," Merck, January 25, 2016, https://www.merck.com/news/merck-discontinues-development-of-sars-cov-2-covid-19-vaccine-candidates-continues-development-of-two-investigational-therapeutic-candidates/.

44. Rebuttal of science denial: Philipp Schmid and Cornelia Betsch, "Effective Strategies for Rebutting Science Denialism in Public Discussions," *Nature Human Behaviour* 3, (2019): 931–39.

45. UCLA vaccines: Zachary Horne, Derek Powell, John E, Hummel, and Keith J. Holyoak, "Countering Antivaccination Attitudes," *Proceedings of the National Academy of Sciences* 112 (2015): 10321–24.

46. Climate change: Michael A. Ranney and Dav Clark, "Climate Change Conceptual Change: Scientific Information Can Transform Attitudes," *Topics in Cognitive Science* 8, (2016): 49–75.

47. Motivational interviewing: Cassandra L. Boness, Mackenzie Nelson, and Antoine B. Douaihy, "Motivational Interviewing Strategies for Addressing Covid-19 Vaccine Hesitancy," *Journal of the American Board of Family Medicine* 35 (2022): 420–26.

48. Distrust: J. Hunter Priniski and Keith J. Holyoak, "A Darkening Spring: How Preexisting Distrust Shaped Covid-19 Skepticism," *PLoS One* 17 (2022): e0263191.

49. Psychological research: Daryl B. O'Connor, John P. Aggleton, Bhismadev Chakrabarti, Cary L. Cooper, Cathy Creswell, Sandra Dunsmuir, Susan T. Fiske, et al., "Research Priorities for the Covid-19 Pandemic and Beyond: A Call to Action for Psychological Science," *British Journal of Psychology* 111 (2020): 603–629.

50. Gullibility: Daniel T. Gilbert, "How Mental Systems Believe," *American Psychologist* 46, no. 2 (1991): 107–19. Chapter 8 in *Falsehoods Fly* proposes ways of reducing gullibility.

51. YouTube blocking anti-vaccine videos: Pete Evans, "YouTube Moves to Block and Remove All Anti-Vaccine Misinformation," CBC News, September 28, 2021, https://www.cbc.ca/news/business/youtube-anti-vax-1.6193392.
52. Facebook: David Klepper and Amanda Seitz, "Facebook Slow to Fight COVID-19 Vaccine Misinformation on Platform, Critics Say," Associated Press, October 26, 2021, https://globalnews.ca/news/8325651/facebook-covid-19-vaccine-misinformation/.
53. Trust in science: Maya J. Goldenberg, *Vaccine Hesitancy: Public Trust, Expertise, and the War on Science* (Pittsburgh, PA: University of Pittsburgh Press, 2021); Lee McIntyre, *How to Talk to a Science Denier: Conversations with Flat Earthers, Climate Deniers, and Others Who Defy, Reason* (Cambridge, MA: MIT Press, 2021); Jon Agley and Yunyu Xiao, "Misinformation About Covid-19: Evidence for Differential Latent Profiles and a Strong Association with Trust in Science," *BMC Public Health* 21 (2021): 89.
54. Prebunking: Melisa Basol, Jon Roozenbeek, Manon Berriche, Fatih Uenal, William P. McClanahan, and Sander van der Linden, "Towards Psychological Herd Immunity: Cross-Cultural Evidence for Two Prebunking Interventions against Covid-19 Misinformation," *Big Data & Society* 8, no. 1 (2021): 20539517211013868.
55. Balance theories: Paul Thagard, *Balance: How It Works and What It Means* (New York: Columbia University Press, 2022), chap. 7.
56. OxyContin: Art Van Zee, "The Promotion and Marketing of Oxycontin: Commercial Triumph, Public Health Tragedy," *American Journal of Public Health* 99 (2009): 221–27.
57. Cancer treatments: "List of Unproven and Disproven Cancer Treatments," Wikipedia, June 3, 2023, https://en.wikipedia.org/wiki/List_of_unproven_and_disproven_cancer_treatments.
58. Smoking and cancer: Naomi Oreskes and Erik M. Conway, *Merchants of Doubt: How a Handful of Scientists Obscured the Truth on Issues from Tobacco Smoke to Global Warming* (New York: Bloomsbury Publishing USA, 2011).
59. Abortion: Rachel Lerman, "People Searching for Abortion Information Online Must Wade Through Misinformation," *Washington Post*, July 4, 2022, https://www.washington-post.com/technology/2022/07/04/abortion-misinformation-herbal-remedies/. TikTok: Rina Raphael, "TikTok Is Flooded with Health Myths: These Creators Are Pushing Back," *New York Times*, June 29, 2022, https://www.nytimes.com/2022/06/29/well/live/tiktok-misinformation.html?referring.
60. Monkeypox: Paige Parsons, "Monkeypox Conspiracy Theories Spread Rapidly on TikTok, Says U of A Researcher," CBC News, October 18, 2022, https://www.cbc.ca/news/canada/edmonton/monkeypox-conspiracy-theories-spread-rapidly-on-tiktok-says-u-of-a-researcher-1.6621087.
61. Fluoridation: Charlotte Tucker, "Health Advocates Fighting Myths About Fluoridation with Science: Misinformation Endangers Oral Health," *The Nation's Health* 41, no 5 (July 2011): 1–15, https://www.thenationshealth.org/content/41/5/1.3.
62. Viral BS: Seema Yasmin, *Viral BS: Medical Myths and Why We Fall for Them* (Baltimore, MD: Johns Hopkins University Press, 2021).
63. TikTok debunking: Rina Raphael, "TikTok Is Flooded with Health Myths: These Creators Are Pushing Back," *New York Times*, June 29, 2022, https://www.nytimes.com/2022/06/29/well/live/tiktok-misinformation.html.

64. Universal health care: Alison P. Galvani, Alyssa S Parpia, Abhishek Pandey, Pratha Sah, Kenneth Colon, Gerald Friedman, Travis Campbell, et al., "Universal Healthcare as Pandemic Preparedness: The Lives and Costs That Could Have Been Saved During the Covid-19 Pandemic," *Proceedings of the Natlional Academy of Sciences* 119 (2022): e2200536119; Fahad Razak, Saeha Shin, C David Naylor, and Arthur S Slutsky, "Canada's Response to the Initial 2 Years of the Covid-19 Pandemic: A Comparison with Peer Countries," *Canadian Medical Asociation Journal* 194, no. 25 (June 27, 2022): E870–77.

5. STORMS: CLIMATE CHANGE AND SCIENTIFIC MISINFORMATION

1. Climate change deaths: World Health Organization, "Climate Change and Health," October 30, 2021, https://www.who.int/news-room/fact-sheets/detail/climate-change-and-health.
2. First World War: Paul Thagard, *Mind-Society: From Brains to Social Sciences and Professions* (New York: Oxford University Press, 2019), chap. 9; Margaret MacMillan, *The War That Ended Peace: The Road to 1914* (New York: Penguin, 2013).
3. Viral transmission: Colin J. Carlson, Gregory F. Albery, Cory Merow, Christopher H. Trisos, Casey M. Zipfel, Evan A. Eskew, Kevin J. Olival, Noam Ross, and Shweta Bansal, "Climate Change Increases Cross-Species Viral Transmission Risk," *Nature* (2022): 555–562.
4. Intergovernmental Panel on Climate Change (IPCC): https://www.ipcc.ch.
5. IPCC 2021 report: https://www.ipcc.ch/report/ar6/wg1/.
6. IPCC Interactive atlas: https://interactive-atlas.ipcc.ch.
7. Changes have occurred: https://report.ipcc.ch/ar6/wg1/IPCC_AR6_WGI_FullReport.pdf, p. 187, chap. 2.
8. Human influence: https://www.ipcc.ch/report/ar6/wg1/downloads/report/IPCC_AR6_WGI_SPM.pdf, chap. 3.
9. Consensus: John Cook, Naomi Oreskes, Peter T. Doran, William R. L. Anderegg, Bart Verheggen, Ed W. Maibach, J. Stuart Carlton, et al., "Consensus on Consensus: A Synthesis of Consensus Estimates on Human-Caused Global Warming," *Environmental Research Letters* 11, no. 4 (2016): 048002.
10. Explanatory coherence: For a detailed computational of model of reasoning about climate change, see Paul Thagard and Scott D. Findlay, "Changing Minds About Climate Change: Belief Revision, Coherence, and Emotion," in *Belief Revision Meets Philosophy of Science*, ed. E. J. Olsson and S. Enqvist (Berlin: Springer, 2011), 329–45.
11. Modeling prediction: https://report.ipcc.ch/ar6/wg1/IPCC_AR6_WGI_FullReport.pdf, chap. 4.
12. Tipping elements: David I. Armstrong McKay, Arie Staal, Jesse F. Abrams, Ricarda Winkelmann, Boris Sakschewski, Sina Loriani, Ingo Fetzer, et al., "Exceeding 1.5°C Global Warming Could Trigger Multiple Climate Tipping Points," *Science* 377, no. 6611 (2022): eabn7950.
13. Values: https://report.ipcc.ch/ar6/wg1/IPCC_AR6_WGI_FullReport.pdf, p. 171.
14. Emotion: Paul Thagard, Laurette Larocque, and Ivana Kajić, "Emotional Change: Neural Mechanisms Based on Semantic Pointers," *Emotion* 23 (2023): 182-93.
15. Heartland Institute: https://heartland.org/topics/environment-energy/.

16. Republicans: https://www.businessinsider.com/climate-change-and-republicans-congress-global-warming-2019-2. 2021: https://www.americanprogress.org/article/climate-deniers-117th-congress/.
17. Trump on climate change: Jeremy Schulman, "Every Insane Thing Donald Trump Has Said About Global Warming," *Mother Jones*, December 12, 2018, https://www.motherjones.com/environment/2016/12/trump-climate-timeline/.
18. Dismissing the IPCC: https://heartland.org/opinion/ipcc-report-shows-desperation-not-climate-catastrophe/.
19. *Merchants of Doubt*: Naomi Oreskes and Erik M. Conway, *Merchants of Doubt: How a Handful of Scientists Obscured the Truth on Issues from Tobacco Smoke to Global Warming* (New York: Bloomsbury, 2011).
20. Singer: Rachel White Scheuering, *Shapers of the Great Debate on Conservation: A Biographical Dictionary* (Westport, CT: Greenwood Publishing, 2004), 125.
21. Willie Soon: Justin Gillis and John Schwartz, "Deeper Ties to Corporate Cash for Climate Change Researcher," *New York Times*, February 22, 2015, https://www.nytimes.com/2015/02/22/us/ties-to-corporate-cash-for-climate-change-researcher-Wei-Hock-Soon.html.
22. Oil companies fund Republicans: Nicholas Kusnetz, "The Oil Market May Have Tanked, but Companies Are Still Giving Plenty to Keep Republicans in Office," Inside Climate News, October 28, 2020, https://insideclimatenews.org/news/28102020/oil-campaign-contributions-republicans-trump/.
23. Conflicts of interest: Paul Thagard, "The Moral Psychology of Conflicts of Interest: Insights from Affective Neuroscience," *Journal of Applied Philosophy* 24 (2007): 367–80.
24. Map of values opposed to climate change: Thomas Homer-Dixon, Manjana Milkoreit, Stehemn J Mock, Tobias Schröder, and Paul Thagard, "The Conceptual Structure of Social Disputes: Cognitive-Affective Maps as a Tool for Conflict Analysis and Resolution," *SAGE Open* 4 (2014): 1–20. See also Manjana Milkoreit, *Mindmade Politics: The Cognitive Roots of International Climate Governance* (Cambridge, MA: MIT Press), 2017; and Julius Fenn, Jessica F. Helm, Philipp Höfele, Lars Kulbe, Andreas Ernst, and Andrea Kiesel, "Identifying Key-Psychological Factors Influencing the Acceptance of yet Emerging Technologies–a Multi-Method-Approach to Inform Climate Policy," *PLOS Climate* 2, no. 6 (2023).
25. Motivated ignorance: Brad Plumer, "Florida Isn't the Only State Trying to Shut Down Discussion of Climate Change," Vox, March 10, 2015, https://www.vox.com/2015/3/10/8182513/florida-ban-climate-change.
26. Fear and climate change: Jillian Ambrose, " 'Hijacked by Anxiety': How Climate Dread Is Hindering Climate Action," *Guardian*, October 8, 2020, https://www.theguardian.com/environment/2020/oct/08/anxiety-climate-crisis-trauma-paralysing-effect-psychologists?CMP=Share_iOSApp_Other.
27. Hope: Jacob B. Rode, Amy L. Dent, Caitlin N. Benedict, Daniel B. Brosnahan, Ramona L. Martinez, and Peter H. Ditto, "Influencing Climate Change Attitudes in the United States: A Systematic Review and Meta-Analysis," *Journal of Environmental Psychology* 76 (2021): 101623.
28. Climate change and pandemics: Damian Carrington, "Protecting Nature Is Vital to Escape 'Era of Pandemics': Report," *Guardian*, October 29, 2020, https://www.theguardian

.com/environment/2020/oct/29/protecting-nature-vital-pandemics-report-outbreaks-wild?CMP=Share_iOSApp_Other.

29. Books denying climate change: Riley E. Dunlap and Peter J. Jacques, "Climate Change Denial Books and Conservative Think Tanks: Exploring the Connection," *American Behavioral Scientist* 57, no. 6 (June 2013): 699–731.
30. Videos denying climate change: Joachim Allgaier, "Science and Environmental Communication on YouTube: Strategically Distorted Communications in Online Vvdeos on Climate Hange and Clmate Engineering," *Frontiers in Communication* 4 (2019).
31. Internet spread: Kathie M. d'I, Treen, Hywel T. P. Williams, and Saffron J. O'Neill, "Online Misinformation About Climate Change," *WIREs Climate Change* 11, no. 5 (2020): e665.
32. Social media: Jessica Guynn, "Climate Change Denial on Facebook, YouTube, Twitter and TikTok Is 'As Bad As Ever,' " *USA Today*, January 21, 2022, https://www.usatoday.com/story/tech/2022/01/21/climate-change-misinformation-facebook-youtube-twitter/6594691001/.
33. Twitter: Ramishah Maruf, "Twitter Bans 'Misleading' Climate Change Ads," CNN, April 23, 2022, https://www.cnn.com/2022/04/23/tech/twitter-climate-change-policy/index.html.
34. Facebookposts: Kari Paul, "Climate Misinformation on Facebook 'Increasing Substantially,' Study Says," *Guardian*, November 4, 2021, https://www.theguardian.com/technology/2021/nov/04/climate-misinformation-on-facebook-increasing-substantially-study-says.
35. Spread of climate denial values: Thomas, Marlow, Sean Miller, and J. Timmons Roberts. "Bots and Online Climate Discourses: Twitter Discourse on President Trump's Announcement of US Withdrawal from the Paris Agreement," *Climate Policy* 21, no. 6 (2021): 765–77.
36. Main denier sites: Kari Paul, " 'Super Polluters': The Top 10 Publishers Denying the Climate Crisis on Facebook," *Guardian*, November 2, 2021, https://www.theguardian.com/technology/2021/nov/02/super-polluters-the-top-10-publishers-denying-the-climate-crisis-on-facebook?CMP=Share_iOSApp_Other.
37. Trump's falsehoods: Glenn Kessler, Salvador Rizzo, and Meg Kelly, "Trump's False or Misleading Claims Total 30,573 over 4 Years" *Washington Post*, January 24, 2021, https://www.washingtonpost.com/politics/2021/01/24/trumps-false-or-misleading-claims-total-30573-over-four-years/.
38. Computational methods: Travis G. Coan, Constantine Boussalis, John Cook, and Mirjam O Nanko, "Computer-Assisted Classification of Contrarian Claims About Climate Change," *Scientific Reports* 11, no. 1 (November 16, 2021): 22320.
39. Default assumption: Paul Thagard, "Testimony, Credibility, and Explanatory Coherence," *Erkenntnis* 63 (2005): 295–316; Daniel T. Gilbert, "How Mental Systems Believe," *American Psychologist* 46, no. 2 (1991): 107–19. See also chap. 8 in this book on avoiding gullibility.
40. Exxon: Neela Banerjee, Lisa Song, and David Hasemyer, "Exxon: The Road Not Taken," Inside Climate News, January 12, 2023, https://insideclimatenews.org/project/exxon-the-road-not-taken/; Hiroko Tabuchi, "Exxon Scientists Predicted Global Warming, Even as Company Cast Doubts, Study Finds," *New York Times*, January 12, 2023, https://www.nytimes.com/2023/01/12/climate/exxon-mobil-global-warming-climate-change.html?smid=nytcore-ios-share&referringSource=articleShare.

41. Berkeley researchers: Michael Andrew Ranney and Leela Velautham, "Climate Change Cognition and Education: Given No Silver Bullet for Denial, Diverse Information-Hunks Increase Global Warming Acceptance," *Current Opinion in Behavioral Sciences* 42 (2021): 139–46. See also Junho Lee, Emily F. Wong, and Patricia W. Cheng, "Promoting Climate Actions: A Cognitive-Constraints Approach," *Cognitive Psychology* 143 (Jun 2023): 101565.
42. Cook: John Cook, "Understanding and Countering Misinformation About Climate Change," in *Handbook of Research on Deception, Fake News, and Misinformation Online*, ed. I. Chiluwa and S. Samoilenko (Hershey, PA: IGI-Global, 2019), 281–306; John Cook, Peter Ellerton, and David Kinkead, "Deconstructing Climate Misinformation to Identify Reasoning Errors," *Environmental Research Letters* 13, no. 2 (2018).
43. Climate conspiracy: Stephan Lewandowsky and John Cook, "The Conspiracy Theory Handbook," 2020, http://sks.to/conspiracy.
44. Chemtrails: Allgaier, "Science and Environmental Communication on YouTube."
45. Inference to the best plan: Paul Thagard and Elijah Millgram, "Inference to the Best Plan: A Coherence Theory of Decision., in *Goal-Driven Learning*, ed. A. Ram and D. B. Leake (Cambridge, MA: MIT Press, 1995), 439–54; Paul Thagard, *Coherence in Thought and Action* (Cambridge, MA: MIT Press, 2000).
46. IPCC probabilities: https://report.ipcc.ch/ar6/wg1/IPCC_AR6_WGI_FullReport.pdf, p. 4.
47. Limits of probability: Paul Thagard, *Natural Philosophy: From Social Brains to Knowledge, Reality, Morality, and Beauty* (New York: Oxford University Press, 2019), chap. 3.
48. Inoculation: John Cook, Stephan Lewandowsky, and Ullrich K. H. Ecker, "Neutralizing Misinformation Through Inoculation: Exposing Misleading Argumentation Techniques Reduces Their Influence," *PLoS One* 12, no. 5 (2017): e0175799; Sander van der Linden, *Foolproof: Why Misinformation Infects Our Minds and How to Build Immunity* (New York: Norton, 2023).
49. Priming critical thinking: Lauren Lutzke, Caitlin Drummond, Paul Slovic, and Joseph Árvai, "Priming Critical Thinking: Simple Interventions Limit the Influence of Fake News About Climate Change on Facebook," *Global Environmental Change* 58 (2019): 101964.
50. Motivational interviewing: Emily Senay, Mona Sarfaty, and Mary B. Rice, "Strategies for Clinical Discussions About Climate Change," *Annals of Internal Medicine* 174, no. 3 (March 2021): 417–18.
51. School strike: "School Strike for Climate," Wikipedia, April 27, 2023, https://en.wikipedia.org/wiki/School_Strike_for_Climate.
52. Twitter coronavirus: Taylor Lorenz, "Twitter Ends Its Ban on COVID Misinformation," *Washington Post*, November 29, 2022, https://www.washingtonpost.com/technology/2022/11/29/twitter-covid-misinformation-policy/.
53. Ethical principles: Tom L. Beauchamp and James F. Childress, *Principles of Biomedical Ethics*, 7th ed. (New York: Oxford University Press, 2013).
54. Demarcation doubts: Michael D. Gordin, *On the Fringe: Where Science Meets Pseudoscience* (Oxford: Oxford University Press, 2021). Paul Thagard, *Computational Philosophy of Science* (Cambridge, MA: MIT Press, 1988), shows that verifiability and falsifiability do not work as demarcation criteria.
55. Science versus pseudoscience: Paul Thagard, *Balance: How It Works and What It Means* (New York: Columbia University Press, 2022); Paul Thagard, "Evolution, Creation, and

the Philosophy of Science," in *Evolution, Epistemology, and Science Education*, ed. R. Taylor and M. Ferrari (Milton Park: Routledge, 2010), 20–37.

56. Creationism: Paul Thagard and Scott Findlay, "Getting to Darwin: Obstacles to Accepting Evolution by Natural Selection," *Science & Education* 19 (2010): 625–36; Michael Ruse, "Creationism," *Stanford Encyclopedia of Philosophy* (Stanford, CA: Stanford University, 2018), https://plato.stanford.edu/entries/creationism/.
57. Astrology as pseudoscience: Paul Thagard, "Why Astrology Is a Pseudoscience," in *PSA 1978*, ed. P. Aquith and I. Hacking (East Lansing MI: Philosophy of Science Association, 1978), 223–34.
58. Psychics: Paul Thagard, "Should You Believe in Psychics?," *Psychology Today*, April 24, 2019, https://www.psychologytoday.com/ca/blog/hot-thought/201904/should-you-believe-in-psychics.
59. Flat earth: Lee McIntyre, *How to Talk to a Science Denier: Conversations with Flat Earthers, Climate Deniers, and Others Who Defy, Reason* (Cambridge, MA: MIT Press, 2021); Steve Mirsky, "Flat Earthers: What They Believe and Why," Scientific American, March 27, 2020, https://www.scientificamerican.com/podcast/episode/flat-earthers-what-they-believe-and-why/. See also the documentary film *Behind the Curve*, directed by Daniel J. Clark (2018; London: Delta-v Productions), documentary.
60. Mistakes of science deniers: McIntyre, *How to Talk to a Science Denier*; Matthew J. Hornsey, "Why Facts Are Not Enough: Understanding and Managing the Motivated Rejection of Science," *Current Directions in Psychological Science* 29, no. 6 (2020): 583–91, has a similar list of the roots of science denial: conspiracist worldviews, vested interests, ideologies, anxieties, personal identities, and social identities.
61. Agricultural industry: Ignacio Calderon, "Once Climate Change Deniers, the Agriculture Industry Positions Itself As Part of the Solution," Investigate Midwest, April 1, 2021, https://investigatemidwest.org/2021/04/01/once-climate-change-deniers-the-agriculture-industry-positions-itself-as-part-of-the-solution/.
62. Oil companies: Chris McGreal, "How a Powerful US Lobby Group Helps Big Oil to Block Climate Action," *Guardian*, July 19, 2021, https://www.theguardian.com/environment/2021/jul/19/big-oil-climate-crisis-lobby-group-api?CMP=Share_iOSApp_Other.

6. PLOTS: CONSPIRACY THEORIES AND POLITICAL MISINFORMATION

1. Conspiracy theories: More than 100 real and fake conspiracies are listed by Richard M. Bennett, *Conspiracy: Plots, Lies, and Cover-Ups* (London: Virgin Books, 2003). Wikipedia lists more than fifty conspiracy theories at https://en.wikipedia.org/wiki/List_of_conspiracy_theories. A comprehensive classification is in Michael Shermer, *Conspiracy: Why the Rational Believe the Irrational* (Baltimore, MD: Johns Hopkins University Press, 2022). For more examples, see "The Conspiracy Chart" at https://conspiracychart.com.
2. Social mechanisms: Paul Thagard, *Mind-Society: From Brains to Social Sciences and Professions* (New York: Oxford University Press, 2019), calls this approach a social cognitive-emotional workup.

3. Caesar: Barry Strauss, *The Death of Caesar: The Story of History's Most Famous Assassination* (New York: Simon & Schuster, 2015).
4. Motives of assassins: Strauss, *The Death of Caesar*, 135.
5. Plead guilty: U.S. Department of Justice, Office of Public Affairs, "Leader of North Caroline Chapter of Oath Keepers Pleads Guilty to Seditious Conspiracy and Obstruction of Congress for Efforts to Stop Transfer of Power Following 2020 Presidential Election," Press Release, May 4, 2022, https://www.justice.gov/opa/pr/leader-north-carolina-chapter-oath-keepers-pleads-guilty-seditious-conspiracy-and-obstruction.
6. Convicted: Hannah Rabinowitz, Holmes Lybrand, and Sonnet Swire, "Oath Keepers Leader and Associates Convicted of Multiple Charges in Seditious Conspiracy Case," CNN, November 30, 2022, https://www.cnn.com/2022/11/29/politics/oath-keepers-convicted-verdict-charges-january-6-seditious-conspiracy/index.html.
7. More convicted: Manisha Ganguly, "'Aims': The Software for Hire That Can Control 30,000 Fake Online Profiles," *Guardian*, February 14, 2023, https://www.theguardian.com/world/2023/feb/15/aims-software-avatars-team-jorge-disinformation-fake-profiles.
8. Oath Keepers: Sam Jackson, *Oath Keepers: Patriotism and the Edge of Violence in a Right-Wing Antigovernment Group* (New York: Columbia University Press, 2020).
9. Indictment of Oath Keepers: U.S. Department of Justice, Office of Public Affairs, "Leader of Oath Keepers and 10 Other Individuals Indicted in Federal Court for Seditious Conspiracy and Other Offenses Related to U.S. Capitol Breach," Press Release, January 13, 2022, https://www.justice.gov/opa/pr/leader-oath-keepers-and-10-other-individuals-indicted-federal-court-seditious-conspiracy-and.
10. Proud Boys: Hannah Rabinowitz and Katelyn Polantz, "Proud Boys Leader and Top Members Charged with Seditious Conspiracy over January 5," CNN, June 7, 2022, https://www.cnn.com/2022/06/06/politics/tarrio-proud-boys-charged-seditious-conspiracy-january-6/index.html.
11. Proud Boys: U.S. Department of Justice, Office of Public Affairs, "Former Leader of Proud Boys Pleads Guilty to Seditious Conspiracy for Efforts to Stop Transfer of Power Following 2020 Presidential Election," Press Release, October 6, 2022, https://www.justice.gov/opa/pr/former-leader-proud-boys-pleads-guilty-seditious-conspiracy-efforts-stop-transfer-power.
12. January 6 committee: https://www.govinfo.gov/committee/house-january6th?path=/browsecommittee/chamber/house/committee/january6th/collection/CRPT.
13. Special counsel: Josh Gerstein and Kyle Cheney, "New Trump Special Counsel Launches Investigation in Mueller's Shadow," Politico, November 18, 2022, https://www.politico.com/news/2022/11/18/trump-special-counsel-investigation-mueller-00069578; Trump's indictment for conspiracy: Allen Feuer and Maggie Haberman, "Trump Indictment," *New York Times*, August 1, 2023, https://www.nytimes.com/live/2023/08/01/us/trump-indictment-jan-6.
14. Chile: Bennett, *Conspiracy*.
15. QAnon: Paul Bleakley, "Panic, Pizza and Mainstreaming the Alt-Right: A Social Media Analysis of Pizzagate and the Rise of the QAnon Conspiracy," *Current Sociology* (2021): 1–17; Amanda Garry, Samantha Walther, Rukaya Rukaya, and Ayan Mohammed, "QAnon Conspiracy Theory: Examining Its Evolution and Mechanisms of Radicalization,"

Journal for Deradicalization 26, no. 3 (2021): 152–216; Charles Kaiser, "*The Storm Is Upon Us* Review: Indispensable QAnon History, Updated," *Guardian*, August 21, 2022, https://www.theguardian.com/books/2022/aug/21/the-storm-is-upon-us-review-qanon-history-updated-trump-january-6?CMP=Share_iOSApp_Other; Edward Tian, "The QAnon Timeline: Four Years, 5,000 Drops and Countless Failed Prophecies," Bellingcat, January 29, 2022, https://www.bellingcat.com/news/americas/2021/01/29/the-qanon-timeline/.

16. QAnon Facebook followers: Ari Sen and Brandy Zadrozny, "QAnon Groups Have Millions of Members on Facebook, Documents Show," NBCNews, August 10, 2020, https://www.nbcnews.com/tech/tech-news/qanon-groups-have-millions-members-facebook-documents-show-n1236317.
17. QAnon Republicans: Daniel A. Cox, "After the Ballots Are Counted: Conspiracies, Political Violence, and American Exceptionalism," Survey Center on American Life, February 11, 2021, https://www.americansurveycenter.org/research/after-the-ballots-are-counted-conspiracies-political-violence-and-american-exceptionalism/.
18. QAnon merges with COVID: Marianna Spring and Mike Wendling, "How COVID-19 Myths Are Merging with the QAnon Conspiracy Theory," BBC, September 3, 2020, https://www.bbc.com/news/blogs-trending-53997203.
19. Residential schools: Tim Naumetz, "One in Five Students Suffered Sexual Abuse at Residential Schools, Figures Indicate," *Globe and Mail*, January 17, 2009, https://www.theglobeandmail.com/news/national/one-in-five-students-suffered-sexual-abuse-at-residential-schools-figures-indicate/article20440061/.
20. Pedophilia: Michael C. Seto, "Pedophilia," *Annual Review of Clinical Psychology* 5 (2009): 391–407.
21. Psychological explanations: Rob Brotherton, *Suspicious Minds: Why We Believe Conspiracy Theorie* (London: Bloomsbury Sigma, 2015); Quassim Cassam, *Conspiracy Theories* (Cambridge: Polity, 2019); Karen M. Douglas, Joseph E. Uscinski, Robbie M. Sutton, Aleksandra Cichocka, Turkay Nefes, Chee Siang Ang, and Farzin Deravi, "Understanding Conspiracy Theorie." *Political Psychology* 40, no. S1 (2019): 3–35; Jan-Willem van Prooijen, *The Psychology of Conspiracy Theories* (London: Routledge, 2018); Jan-Willem van Prooijen, "Injustice Without Evidence: The Unique Role of Conspiracy Theories in Social Justice Research," *Social Justice Research* (September 28, 2021): 1–19; Joseph E. Uscinski, ed., *Conspiracy Theories and the People Who Believe Them* (New York: Oxford University Press, 2018).
22. Motivated inference in politics: Joanne M. Miller, Kyle L. Saunders, and Christina E. Farhart, "Conspiracy Endorsement as Motivated Reasoning: The Moderating Roles of Political Knowledge and Trust," *American Journal of Political Science* 60, no. 4 (2016): 824–844; Charles S. Taber and Milton Lodge, "Motivated Skepticism in the Evaluation of Political Beliefs," *American Journal of Political Science* 50, no. 3 (2006): 755–769.
23. Motivated collective cognition: Péter Krekó, "Conspiracy Theory as Collective Motivated Cognition," in *The Psychology of Conspiracy*, ed. Michal Bilewicz, Aleksandra Cichocka and Wiktor Soral (London: Routledge, 2015), 62–77.
24. Disgust: Julia Elad-Strenger, Jutta Proch, and Thomas Kessler, "Is Disgust a 'Conservative' Emotion?," *Personality and Social Psychology Bulletin* 46, no. 6 (2020): 896–912.

25. Religious: Kaleigh Rogers, "Why QAnon Has Attracted So Many White Evangelicals," FiveThirtyEight, March 4, 2021, https://fivethirtyeight.com/features/why-qanon-has-attracted-so-many-white-evangelicals/.
26. Faith: Paul Thagard, *The Brain and the Meaning of Life* (Princeton, NJ: Princeton University Press, 2010).
27. InfoWars wellness: Farhad Manjoo, Alex Jones and the Wellness-Conspiracy Industrial Complex, *New York Times*, August 11, 2022, https://www.nytimes.com/2022/08/11/opinion/alex-jones-wellness-conspiracy.html?referringSource.
28. Entrepreneurs: Cass Robert Sunstein and Cornelius Adrian Vermeule, "Conspiracy Theories: Causes and Cures," *Journal of Political Philosophy* 17 (2009): 202–227.
29. Trump on repetition: Chris Cilizza, "Donald Trump Just Accidentally Told the Truth About His Disinformation Strategy," CNN, July 5, 2021, https://www.cnn.com/2021/07/05/politics/trump-disinformation-strategy/index.html.
30. Bannon: Brian Stelter, "This Infamous Steve Bannon Quote Is Key to Understanding America's Crazy Politics," CNN, November 16, 2021, https://www.cnn.com/2021/11/16/media/steve-bannon-reliable-sources/index.html.
31. Great replacement: Renaud Camus, *Le Grand Remplacement: Introduction au Remplacisme Global* (Paris: La Nouvelle Librairie, 2021).
32. Carlson: John Haltiwanger, "Tucker Carlson Peddled a White Supremacist Conspiracy Theory While Attacking Biden over the Haitian Migrant Crisis," Insider, September 23, 2021, https://www.businessinsider.com/tucker-carlson-again-pushes-white-supremacist-conspiracy-theory-2021-9.
33. Canada immigration: "Distribution of New Immigrants in Canada in 2021, by Country of Origin," Statista, https://www.statista.com/statistics/1171597/new-immigrants-canada-country/.
34. Canada survey: David Coletto, "Millions Believe in Conspiracy Theories in Canada, Abacus Data, June 12, 2022, https://abacusdata.ca/conspiracy-theories-canada/.
35. Republicans on replacement: Philip Bump, "Nearly Half of Republicans Agree with 'Great Replacement Theory,'" *Washington Post*, May 9, 2022, https://www.washingtonpost.com/politics/2022/05/09/nearly-half-republicans-agree-with-great-replacement-theory/.
36. Existence: Paul Thagard, *Natural Philosophy: From Social Brains to Knowledge, Reality, Morality, and Beauty* (New York: Oxford University Press, 2019), 115.
37. Handbook: Stephan Lewandowsky and John Cook, "The Conspiracy Theory Handbook," https://sks.to/conspiracy.
38. New conspiracism: Russell Muirhead and Nancy L Rosenblum, *A Lot of People Are Saying: The New Conspiracism and the Assault on Democracy* (Princeton, NJ: Princeton University Press, 2020).
39. Error tendencies: Paul Thagard, "Critical Thinking and Informal Logic: Neuropsychological Perspectives," *Informal Logic* 31 (2011): 152–70.
40. Empathy: Lewandowsky and Cook, "The Conspiracy Theory Handbook."
41. Powerlessness: Lewandowsky and Cook, "The Conspiracy Theory Handbook."
42. Multilingual: Tiffany Hsu, "Misinformation Swirls in Non-English Languages Ahead of Midterms," *New York Times*, October 12, 2022, https://www.nytimes.com/2022/10/12/business/media/midterms-foreign-language-misinformation.html.

43. Alex Jones lawsuits: Associated Press, "Nova Scotia Offers Prescription for Calgary 'Ailment,'" CBC News, March 26, 2007, https://www.cbc.ca/news/world/alex-jones-verdict-damages-hoax-1.6647707.
44. Digital safety: Cristriano Lima, "A New Road Map for Reining in Social Media Companies Is Gaining Steam," *Washington Post*, February 23, 2022, https://www.washingtonpost.com/politics/2022/02/23/new-roadmap-reining-social-media-companies-is-gaining-steam/.
45. Child pornography: Roger Fingas, "Telegram Was Pulled Because of Child Pornography, Says Apple's Phil Schiller," AppleInsider, February 5, 2018, https://appleinsider.com/articles/18/02/05/telegram-was-pulled-because-of-child-pornography-says-apples-phil-schiller.
46. New platforms: Steven Lee Myers and Sheera Frenkel, "How Disinformation Splintered and Became More Intractable," *New York Times*, October 20, 2022, https://www.nytimes.com/2022/10/20/technology/disinformation-spread.html.
47. QAnon on Truth Social: Tiffany Hsu, "QAnon Accounts Found a Home, and Trump's Support, on Truth Social," *New York Times*, August 29, 2022, https://www.nytimes.com/2022/08/29/technology/qanon-truth-social-trump.html.
48. Cognitive infiltration: Sunstein and Vermeule, "Conspiracy Theories;" critique by David Coady, "Cass Sunstein and Adrian Vermeule on Conspiracy Theories," *Argumenta* 3, no. 2 (2018): 291–302.
49. Lies: Bill Fawcett, ed., *You Said What? Lies and Propaganda Throughout History* (New York: Harper, 2007).
50. Military deception: Paul Thagard, "Adversarial Problem Solving: Modelling an Opponent Using Explanatory Coherence," *Cognitive Science* 16 (1992): 123–49; Jon Latimer, *Deception in War* (New York: Overlook Press, 2001).
51. Disinformation campaigns: Thomas Rid, *Active Measures: The Secret History of Disinformation and Political Warfare* (New York: Farrar, Straus and Giroux, 2020).
52. Trump's deceptions: Sophie Van Der Zee, Ronald Poppe, Alice Havrileck, and Aurelien Baillon, "A Personal Model of Trumpery: Linguistic Deception Detection in a Real-World High-Stakes Setting," *Psychological Science* 33, no. 1 (January 2022): 3–17.
53. Propaganda: Oliver Thomson, *Easily Led: A History of Propaganda* (Phoenix Mill, UK: Sutton, 1999); Jason Stanley, *How Propaganda Works* (Princeton, NJ: Princeton University Press, 2015).
54. Russian trolls on QAnon: Ben Collins and Joe Murphy, "Russian Troll Accounts Purged by Twitter Pushed QAnon and Other Conspiracy Theories," NBCNews, February 2, 2019, https://www.nbcnews.com/tech/social-media/russian-troll-accounts-purged-twitter-pushed-qanon-other-conspiracy-theories-n966091.
55. Media against Russian influence: Shannon Bond, "Tech's Crackdown on Russian Propaganda Is a Geopolitical High-Wire Act," NPR, March 2, 2022, https://www.npr.org/2022/03/01/1083824030/techs-crackdown-on-russian-propaganda-is-a-geopolitical-high-wire-act.
56. AIMS: My AIMS theory is the opposite of the "Aims" disinformation software that controls 30,000 fake social media profiles; see Ganguly, "Aims," https://www.theguardian.com/world/2023/feb/15/aims-software-avatars-team-jorge-disinformation-fake-profiles.

7. EVILS: INEQUALITY AND SOCIAL MISINFORMATION

1. Equality history: George L. Abernethy, ed., *The Idea of Equality: An Anthology* (Richmond, VA: John Knox, 1959).
2. Analysis of equality and inequality: Paul Thagard, "Social Equality: Cognitive Modeling Based on Emotional Coherence Explains Attitude Change," *Policy Insights from Behavioral and Brain Sciences* 5, no. 2 (2018): 247–256.
3. Equal needs: Michael Walzer, *Spheres of Justice: A Defense of Pluralism and Equality* (New York: Basic Books, 1983).
4. Needs: Richard M. Ryan and Edward L. Deci, *Self-Determination Theory: Basic Psychological Needs in Motivation, Development, and Wellness* (New York: Guilford, 2017); Paul Thagard, *The Brain and the Meaning of Life* (Princeton, NJ: Princeton University Press, 2010); Paul Thagard, *Natural Philosophy: From Social Brains to Knowledge, Reality, Morality, and Beauty* (New York: Oxford University Press, 2019).
5. Arguments against equality: William Letwin, ed. *Against Equality: Readings on Economic and Social Policy* (London: Macmillan, 1983).
6. Inequality harms: Nabil Ahmed, Anna Marriott, Nafkote Dabi, Megan Lowthers, Max Lawson, and Leah Mugehera, "Inequality Kills: The Unparalleled Action Needed to Combat Unprecedented Inequality in the Wake of Covid-19," Oxfam, 2022, https://oxfamilibrary.openrepository.com/handle/10546/621341; Anthony B. Atkinson, *Inequality: What Can Be Done?* (Cambridge, MA: Harvard University Press, 2015); Branko Milanovic, *The Haves and the Have-Nots: A Brief and Idiosyncratic History of Global Inequality* (New York: Basic Books, 2010): Kate E. Pickett and Richard G. Wilkinson, "Income Inequality and Health: A Causal Review," *Social Science & Medicine* 128 (2015): 316–326; Joseph E. Stiglitz, *The Price of Inequality: How Today's Divided Society Endangers Our Future* (New York: Norton, 2013); Richard G. Wilkinson and Kate Pickett, *The Spirit Level: Why Greater Equality Makes Societies Stronger* (New York: Penguin, 2010).
7. Stress: Richard G. Wilkinson and Kate Picket, *The Inner Level: How More Equal Societies Reduce Stress, Restore Sanity and Improve Everyone's Well-Being* (New York: Penguin, 2020); Richard Sennett, *Respect in a World of Inequality* (New York: Norton, 2003).
8. Health: Pickett and Wilkinson, "Income Inequality and Health."
9. Wealth tax: Thomas Piketty, *Capital in the Twenty-First Century* (Cambridge, MA: Harvard University Press, 2014).
10. Basic income: Phillippe Van Parijs and Yannick Vanderborght, *Basic Income: A Radical Proposal for a Free Society and a Sane Economy* (Cambridge, MA: Harvard University Press, 2017); Hilary W. Hoynes and Jesse Rothstein, "Universal Basic Income in the US and Advanced Countries," *Annual Review of Economics* 11 (2019): 929–958.
11. Basic income experiments: Evelyn L. Forget, "The Town with No Poverty: The Health Effects of a Canadian Guaranteed Annual Income Field Experiment," *Canadian Public Policy* 37, no. 3 (2011): 283–305; Roberto Merrill, Catarina Neves, and Bru Laín, *Basic Income Experiments: A Critical Examination of Their Goals, Contexts, and Methods* (London: Palgrave Macmillan, 2021).
12. Minimum wage: David Card and Alan B. Krueger, *Myth and Measurement: The New Economics of the Minimum Wage* (Princeton, NJ: Princeton University Press, 1997).

13. Canadian Income Survey: "Canadian Income Survey: 2021 (CIS)," Statistics Canada, May 2, 2023, https://www23.statcan.gc.ca/imdb/p2SV.pl?Function=getSurvey&SDDS=5200.
14. Gini limitations: Lars Osberg, *The Age of Increasing Inequality: The Astonishing Rise of Canada's 1 percent* (Toronto: James Lorimer, 2018).
15. Consumer surveys in the United States: U.S. Bureau of Labor Statistics, "Consumer Expenditure Surveys," U.S. Department of Labor, https://www.bls.gov/cex/.
16. Larry Elliott, "World's Eight Richest People Have Same Wealth as Poorest 50%," Guardian, January 15, 2017, https://www.theguardian.com/global-development/2017/jan/16/worlds-eight-richest-people-have-same-wealth-as-poorest-50.
17. World inequality: Lucas Chancel, Thomas Piketty, Emmanuel Saez, and Gabriel Zucman, "World Inequality Report 2022," World Inequlity Lab, 2022, https://wir2022.wid.world.
18. World Inequality Database: World Inequality Database, https://wid.world/.
19. Argument against equality: compare Letwin, *Against Equality*.
20. Genesis 3:16, English Standard Version.
21. Ephesians 5:22–24, English Standard Version.
22. Myth of the metals: Plato, *The Collected Dialogues*, ed. E. Hamilton and E. Cairns (Princeton, NJ: Princeton University Press, 1961); Plato, *The Republic*, Book III, 414e.
23. Slavery: Aristotle, *The Complete Works of Aristotle.*, ed. J. Barnes, 2 vols. (Princeton, NJ: Princeton University Press, 1984); *Politics*, 1254b.
24. Toxic analogies: Paul Thagard, *Balance: How It Works and What It Means* (New York: Columbia University Press, 2022).
25. Divine right: Glenn Burgess, "The Divine Right of Kings Reconsidered," *The English Historical Review* 107, no. 425 (1992): 837–861.
26. Monarchy: John Locke, *John Locke: Two Treatises of Government* (Cambridge: Cambridge University Press, 1967).
27. Existence of God: Thagard, *The Brain and the Meaning of Life*; Thagard, *Natural Philosophy*.
28. Religion as motivated reasoning: Paul Thagard, "The Emotional Coherence of Religion," *Journal of Cognition and Culture* 5 (2005): 58–74.
29. Apostles: Acts of the Apostles, 4:32–35.
30. Leviathan: Thomas Hobbes, *Leviathan* (Oxford: Basil Blackwell, 1947).
31. Philosophical thought experiments: Thagard, The *Brain and the Meaning of Life*; Paul Thagard, "Thought Experiments Considered Harmful," *Perspectives on Science* 22 (2014): 288–305; Thagard, *Natural Philosophy*.
32. Born free: Jean-Jacques Rousseau, *Discourse on the Origin of Inequality* (Oxford: Oxford University Press, 1999), 165.
33. Rawls: John Rawls, *A Theory of Justice* (Cambridge, MA: Harvard University Press, 1971).
34. Nozick: Robert Nozick, *Anarchy, State, and Utopia* (New York: Basic Books, 1974).
35. Sex: Carole Pateman, *The Sexual Contract* (Cambridge: Polity Press, 1988). Race: Charles W. Mills, *The Racial Contract* (Ithaca, NY: Cornell University Press, 2014).
36. Creation of inequality: Kent Flannery and Joyce Marcus, *The Creation of Inequality: How Our Prehistoric Ancestors Set the Stage for Monarchy, Slavery, and Empire* (Cambridge, MA: Harvard University Press, 2012); Kim Sterelny, *The Pleistocene Social Contract: Culture and Cooperation in Human Evolution* (Oxford: Oxford University Press, 2021).
37. Origin of inequality: David Graeber and David Wengrow, *The Dawn of Everything: A New History of Humanity* (London: Allen Lane, 2021), 11.

38. Challenge: Kwame Anthony Appiah, "Digging for Utiopia," *New York Review of Books*, December 16, 2021.
39. Fascism: Roger Eatwell, *Fascism: A History* (London: Pimlico, 2003); Madeline Albright, *Fascism: A Warning* (New York: HarperCollins, 2018); Jason Stanley, *How Fascism Works: The Politics of Us and Them* (New York: Random House, 2020).
40. Virility: Ruth Ben-Ghiat, *Strongmen: Mussolini to the Present* (New York: Norton, 2020).
41. Social Darwinism: Jeffrey O'Connell and Michael Ruse, *Social Darwinism* (Cambridge: Cambridge University Press, 2021).
42. Peterson: Jordan B. Peterson, *12 Rules for Life: An Antidote to Chaos* (Toronto: Random House Canada, 2018). For critique, see Paul Thagard, "Jordan Peterson's Flimsy Philosophy of Life," *Psychology Today*, February 14, 2018, https://www.psychologytoday.com/ca/blog/hot-thought/201802/jordan-peterson-s-flimsy-philosophy-life; Paul Thagard, "Jordan Peterson's Murky Maps of Meaning," *Psychology Today*, March 12, 2018; https://www.psychologytoday.com/ca/blog/hot-thought/201803/jordan-petersons-murky-maps-meaning.
43. Rohingya: "Rohingya Sue Facebook for $150 Billion over Myanmar Hate Speech," BBC, December 7, 2021, https://www.bbc.com/news/world-asia-59558090.
44. Facebook white supremacy: Naomi Nix, "Facebook Bans Hate Speech but Still Makes Money from White Supremacists," *Washington Post*, August 10, 2022, https://www.washingtonpost.com/technology/2022/08/10/facebook-white-supremacy-ads/.
45. Motivational interviewing and culture: Kamilla L. Venner and Steven P Verney, "Motivational Interviewing: Reduce Student Reluctance and Increase Engagement in Learning Multicultural Concepts," *Professional Psychology: Research and Practice* 46, no. 2 (2015): 116.
46. Economic myths: Stiglitz, *The Price of Inequality*.
47. Zombie ideas: Paul Krugman, *Arguing with Zombies: Economics, Politics, and the Fight for a Better Future* (New York: Norton, 2020), 3.
48. Recession: Paul Thagard, *Mind-Society: From Brains to Social Sciences and Professions* (New York: Oxford University Press, 2019), chap. 7.
49. Cato Institute and right-wing myths: Michael D. Tanner, "Five Myths About Economic Inequality in America: Myth 4. More Inequality Means More Poverty," Cato Institute, September 7, 2016, https://www.cato.org/policy-analysis/five-myths-about-economic-inequality-america#myth-4-more-inequality-means-more-poverty-nbsp.
50. Psychological support of inequality: Hans Jurgen Eysenck, *The Inequality of Man* (London: Temple Smith, 1973); Richard J. Herrnstein and Charles Murray. *The Bell Curve: Intelligence and Class Structure in American Life* (New York: Free Press, 1994); J. Philippe Rushton and Arthur R. Jensen, "Thirty Years of Research on Race Differences in Cognitive Ability," *Psychology, Public Policy, and Law* 11, no. 2 (2005): 235–294.
51. Nisbett: Richard E. Nisbett, *Intelligence and How We Get It* (New York: Norton, 2009); Richard E. Nisbett, Joshua Aronson, Clancy Blair, William Dickens, James Flynn, Diane F. Halpern, and Eric Turkheimer, "Intelligence: New Findings and Theoretical Developments," *American Psychologist* 67, no. 2 (February–March 2012): 130–159.
52. Poverty affects cognitive function: Sendhil Mullainathan and Eldar Shafir, *Scarcity: Why Having Too Little Means So Much* (New York: Macmillan, 2013).
53. Pioneer fund: William H. Tucker, *The Funding of Scientific Racism: Wickliffe Draper and the Pioneer Fund* (Champaign: University of Illinois Press, 2002).

54. Gender and the brain: Gina Rippon, *Gender and Our Brains: How New Neuroscience Explodes the Myths of the Male and Female Mind* (New York: Vintage, 2020); Agustin Fuentes, "Busting Myths About Sex and Gender," *Sapiens*, May 11, 2022, https://www.sapiens.org/biology/busting-myths-about-sex-and-gender/.
55. Second sex: Simone de Beauvoir, *The Second Sex*, trans. C. Borde and S. Malovany-Chevallier (New York: Alfred A. Knopf, 2010).
56. Raw talent: Melis Muradoglu, Zachary Horne, Matthew D. Hammond, Sarah-Jane Leslie, and Andrei Cimpian, "Women—Particularly Underrepresented Minority Women—and Early-Career Academics Feel Like Impostors in Fields That Value Brilliance," *Journal of Educational Psychology* 114 (2021): 1086–1100.
57. COVID-19 and women: Maria Giovanna Sessa, "Misogyny and Misinformation: An Analysis of Gendered Disinformation Tactics During the COVID-19 Pandemic," EU DisinfoLab, December 4, 2020, https://www.disinfo.eu/publications/misogyny-and-misinformation:-an-analysis-of-gendered-disinformation-tactics-during-the-covid-19-pandemic/.
58. Lies about women: Nina Jankowicz, Jillian Hunchak, Alexandra Pavliuc, et al., "Malign Creativity: How Gender, Sex, and Lies are Weaponized Against Women Online," accessed July 19, 2023, https://www.wilsoncenter.org/publication/malign-creativity-how-gender-sex-and-lies-are-weaponized-against-women-online; EU DisinfoLab, October 20, 2021, "Gender-Based Disinformation: Advancing Our Understanding and Response," https://www.disinfo.eu/publications/gender-based-disinformation-advancing-our-understanding-and-response/.
59. False consciousness: Denise Meyerson, *False Consciousness* (Oxford: Clarendon Press, 1991).
60. System justification: John T. Jost, *A Theory of System Justification* (Cambridge, MA: Harvard University Press, 2020).
61. Self-deception: Sahdra Baljinder and Paul Thagard, "Self-Deception and Emotional Coherence," *Minds and Machines* 15 (2003): 213–231.
62. Systemic racism: Paul Thagard, "What Systemic Racism Is and How to Overcome It," *Psychology Today*, January 9, 2023, https://www.psychologytoday.com/ca/blog/hot-thought/202301/what-systemic-racism-is-and-how-to-overcome-it. See also the online supplemental material.
63. Misperception of inequality: Vladimir Gimpelson and Daniel Treisman, "Misperceiving Inequality," *Economics & Politics* 30, no. 1 (2018): 27–54; Oliver P. Hauser and Michael I. Norton, "(Mis)Perceptions of Inequality," *Current Opinion in Psychology* 18 (2017): 21–25.
64. Equality harms: N. Derek Brown, Drew S. Jacoby-Senghor, and Isaac Raymundo, "If You Rise, I Fall: Equality Is Prevented by the Misperception That It Harms Advantaged Groups," *Science Advances* 8, no. 18 (2022): eabm2385.
65. Misperceptions from motivated reasoning: Michael W. Kraus, Sa-kiera T. J. Hudson, and Jennifer A. Richeson, "Framing, Context, and the Misperception of Black–White Wealth Inequality," *Social Psychological and Personality Science* 13, no. 1 (2022): 4–13.
66. Race: Rashawn Ray and Alexandra Gibbons, "Why Are States Banning Critical Race Theory?," Brookings Institution, November 20221, https://www.brookings.edu/blog/fixgov/2021/07/02/why-are-states-banning-critical-race-theory/.

67. Relationship myths: John Gottman, "Debunking 12 Myths About Relationships," Gottman Institute, https://www.gottman.com/blog/debunking-12-myths-about-relationships/; John M. Gottman, *Principia Amoris: The New Science of Love* (New York: Routledge, 2015).
68. Relationship metaphors (longer list): Paul Thagard, "Relationship Metaphors: Helpful or Toxic?," *Psychology Today*, August 19, 2021, https://www.psychologytoday.com/ca/blog/hot-thought/202108/relationship-metaphors-helpful-or-toxic.
69. Metaphor therapy: Donald Meichenbaum, *A Clinical Handbook/Practical Therapist Manual for Assessing and Treating Adults with Post-Traumatic Stress Disorder (PTSD)* (Waterloo, Ontario: Institute Press, 1994).
70. Learning styles: Harold Pashler, Mark McDaniel, Doug Rohrer, and Robert Bjork, "Learning Styles: Concepts and Evidence," *Psychological Science in the Public Interest* 9, no. 3 (2008): 105–119.
71. Reading: James S. Kim, Mary A. Burkhauser, Laura M. Mesite, Catherine A. Asher, Jackie Eunjung Relyea, Jill Fitzgerald, and Jeff Elmore, "Improving Reading Comprehension, Science Domain Knowledge, and Reading Engagement Through a First-Grade Content Literacy Intervention," *Journal of Educational Psychology* 113, no. 1 (2021): 3–26. Mathematics: Bethany Rittle-Johnson, Jon R. Star, and Kelley Durkin, "How Can Cognitive-Science Research Help Improve Education? The Case of Comparing Multiple Strategies to Improve Mathematics Learning and Teaching," *Current Directions in Psychological Science* 29, no. 6 (2020): 599–609.
72. Zombie ideas in education: Panayiota Kendeou, Daniel H. Robinson, and Matthew T. McCrudden, *Misinformation and Fake News in Education* (Charlotte, NC: Information Age Publishing, 2019).
73. Residential schools: J. R. Miller, "Residential Schools in Canada," The Canadian Encyclopedia, Historica Canada, article published October 10, 2012, last edited January 6, 2023; https://www.thecanadianencyclopedia.ca/en/article/residential-schools.
74. Davin: E. McColl, "Report on Industrial Schools for Indians and Half Breeds," HathiTrust Digital Library, March 14, 1879, digitized September 16, 2014, https://babel.hathitrust.org/cgi/pt?id=aeu.ark:/13960/t7np2g21f&view=1up&seq=6.

8. MISINFORMATION SELF-DEFENSE: A MANUAL ILLUSTRATED BY THE RUSSIA-UKRAINE WAR

1. Theory: Kurt Lewin, *Field Theory in Social Science* (New York: Harper & Row, 1951), 169.
2. Russian misinformation: Dan Milmo, "Analysts Identify Top 10 'War Myths' of Russia-Ukraine Conflict," *Guardian*, March 3, 2022, https://www.theguardian.com/world/2022/mar/03/russia-ukraine-conflict-top-10-war-myths-newsguard.
3. Neo-nazis: "Ukraine Compares its Struggles to Nazi Germany," *RT*, March 23, 2022, https://www.rt.com/russia/552585-ukraine-nazi-germany-post/; Anton Troianovski, "Why Vladimir Putin Invokes Nazis to Justify His Invasion of Ukraine," *New York Times*, March 3, 2022, https://www.nytimes.com/2022/03/17/world/europe/ukraine-putin-nazis.html.
4. Ukrainian refuges: Loveday Morris and Will Oremus, "Russian Disinformation Is Demonizing Ukrainian Refugees," *Washington Post*, December 8, 2022, https://www.washingtonpost.com/technology/2022/12/08/russian-disinfo-ukrainian-refugees-germany/.

5. Zelensky quote: Glenn Kessler, "Zelensky's Famous Quote 'Need Ammo, Not a Ride' Not Easily Confirmed," *Washington Post*, March 6, 2022, https://www.washingtonpost.com/politics/2022/03/06/zelenskys-famous-quote-need-ammo-not-ride-not-easily-confirmed/.
6. Ghost of Kyiv: Ines Eisele, "Fact Check: The 'Ghost of Kyiv' Fighter Pilot," DW, May 4, 2022, https://www.dw.com/en/fact-check-ukraines-ghost-of-kyiv-fighter-pilot/a-60951825; Stuart A. Thompson and Davey Alba, "Fact and Mythmaking Blend in Ukraine's Information War," *New York Times*, March 3, 2022, https://www.nytimes.com/2022/03/03/technology/ukraine-war-misinfo.html; Lateshia Beachum, "The Ghose of Kyiv Was Never Alive, Ukrainian Air Force Says" *Washington Post*, May 1, 2022, https://www.washingtonpost.com/world/2022/05/01/ghost-of-kyiv-propaganda/.
7. Fake pictures: Kate O'Flaherty, "Russia-Ukraine Conflict—How to Tell if Pictures and Videos Are Fake," Forbes, February 25, 2022, https://www.forbes.com/sites/kateoflahertyuk/2022/02/25/russia-ukraine-crisis-how-to-tell-if-pictures-and-videos-are-fake/?sh=2e657f9734ad.
8. TikTok: Kari Paul, "TikTok Was 'Just a Dancing App,' Then the Ukraine War Started," *Guardian*, March 20, 2022, https://www.theguardian.com/technology/2022/mar/19/tiktok-ukraine-russia-war-disinformation; Dan Milmo and Hibaq Farah Sat, "Wat as Seen on TikTok: Ukraine Clips Get Views Whether True of Not," *Guardian*, March 5, 2022, https://www.theguardian.com/technology/2022/mar/05/tiktok-ukraine-russia-invasion-clips-get-views-whether-true-or-not?CMP=Share_iOSApp_Other; Robert Booth, "Russia's Trolling on Ukraine Gets 'Incredible Traction' on TikTok," *Guardian*, May 1, 2022, https://www.theguardian.com/world/2022/may/01/russia-trolling-ukraine-traction-tiktok.
9. Deepfake: Rachel Metz, "Deepfakes Are Now Trying to Change the Course of War," CNN, March 25, 2022, https://www.cnn.com/2022/03/25/tech/deepfakes-disinformation-war/index.html.
10. Identify fakes: Elyse Samuels, Sarah Cahlan, Emily Sabens, "Fact Checker: How to Spot a Fake Video," *Washington Post*, March 19, 2021, https://www.washingtonpost.com/politics/2021/03/19/how-spot-fake-video/?itid=lk_inline_manual_26; Katie Nicholson, "There's a Flood of Disinformation About Russia's Invasion of Ukraine: Here's Who's Sorting It Out," CBC, February 27, 2022, https://www.cbc.ca/news/world/fact-checkers-ukraine-1.6365682; Bellingcat Investigation Team, "Documenting and Debunking Dubious Footage from Ukraine's Frontlines," Bellingcat, February 23, 2022, https://www.bellingcat.com/news/2022/02/23/documenting-and-debunking-dubious-footage-from-ukraines-frontlines/.
11. Russian censorship: Anton Troianovski and Valeriya Safronova, "Russia Takes Censorship to Extremes, Stifling War Coverage," *New York Times*, March 4, 2022, https://www.nytimes.com/2022/03/04/world/europe/russia-censorship-media-crackdown.html.
12. Conspiracies: Ilya Yablokov, "The Five Conspiracy Theories That Putin Has Weaponized," *New York Times*, April 25, 2022, https://www.nytimes.com/2022/04/25/opinion/putin-russia-conspiracy-theories.html.
13. Troll farm: Staff and Agencies, " 'Troll Factory' Spreading Russian Pro-War Lies Online, Says UK," *Guardian*, April 30, 2022, https://www.theguardian.com/world/2022/may/01/troll-factory-spreading-russian-pro-war-lies-online-says-uk.

14. Russian TV: Brian Stelter, "Why Russian TV Propaganda Is Crucial to Understanding the War in Ukraine," CNN, April 12, 2022, https://www.cnn.com/2022/04/12/media/russian-tv-propaganda-reliable-sources/index.html.
15. Firehose of falsehood: Christopher Paul and Miriam Matthews, "The Russian 'Firehose of Falsehoods' Propaganda Model: Why It Might Work and Options to Counter It," Rand, https://www.rand.org/pubs/perspectives/PE198.html.
16. Putin's misinformation: Julian E. Barnes, Lara Jakes, and John Ismay, "U.S. Intelligence Suggests That Putin's Advisers Misinformed Him on Ukraine," *New York Times*, March 30, 2022, https://www.nytimes.com/2022/03/30/world/europe/putin-advisers-ukraine.html.
17. Ukraine disinformation: Melissa de Witte, "Seven Tips for Spotting Disinformation Related to the Russia-Ukraine Conflict," *Stanford News*, March 3, 2022, https://news.stanford.edu/2022/03/03/seven-tips-spotting-disinformation-russia-ukraine-war/,
18. U.S. right: Margaret Sullivan, "Putin's Full Scale Information War Got a Key Assist from Donald Trump and Right-Wing Media," *Washington Post*, March 6, 2022, https://www.washingtonpost.com/media/2022/03/06/putin-information-war-trump/; Sheera Frenkel and Stuart A. Thompson, "How Russia and Right-Wing Americans Converged on War in Ukraine," *New York Times*, March 23, 2022, https://www.nytimes.com/2022/03/23/technology/russia-american-far-right-ukraine.html; Donie O'Sullivan, "Analysis: Russia and QAnon Have the Same False Conspiracy Theory About Ukraine," CNN, March 10, 2022, https://www.cnn.com/2022/03/09/media/biolab-ukraine-russia-qanon-false-conspiracy-theory/index.html.
19. Russian censorship: Frank Bruni, "Opinion: Russia, Where All the News Is Fake," *New York Times*, March 10, 2022, https://www.nytimes.com/2022/03/10/opinion/russia-ukraine-fake-news.html.
20. The Grayzone: Alexander Rubinstein and Max Blumenthal, "How Ukraine's Jewish President Zelensky Made Peace with Neo-Nazi Paramilitaries on Front Lines of War with Russia," Grayzone, March 4, 2022, https://thegrayzone.com/2022/03/04/nazis-ukrainian-war-russia/.
21. Putin's motives: Bret Stephens, "Opinion: What if Putin Didn't Miscalculate?," *New York Times*, March 29, 2022, https://www.nytimes.com/2022/03/29/opinion/ukraine-war-putin.html.
22. Snyder on Putin: Timothy Snyder, "Ivan Ilyin, Putin's Philospher of Russian Fascism," *New York Review of Books*, March 16, 2018, https://www.nybooks.com/daily/2018/03/16/ivan-ilyin-putins-philosopher-of-russian-fascism/; Roger Cohen, "The Making of Vladimir Putin: Tracing Putin's 22-Year Slide from Statesman to Tyrant," *New York Times*, March 26, 2022, https://www.nytimes.com/2022/03/26/world/europe/vladimir-putin-russia.html.
23. Putin on Ukraine: Vladimir Putin, "On the Historical Unity of Russians and Ukrainians," Wikisource,2021,https://en.wikisource.org/wiki/On_the_Historical_Unity_of_Russians_and_Ukrainians.
24. Putin as fascist: Jason Stanley, "The Antisemitism Animating Putin's Claim to 'Denazify' Ukraine," *Guardian*, February 25, 2022, https://www.theguardian.com/world/2022/feb/25/vladimir-putin-ukraine-attack-antisemitism-denazify.
25. American spreaders of Russian propaganda: Philip Bump, "A Layer Cake of Misinformation, Dishonesty, and Pro-Russia Apologism" *Washington Post*, March 17, 2022,

https://www.washingtonpost.com/politics/2022/03/17/layer-cake-misinformation-dishonesty-pro-russia-apologism/.

26. Instruments: Nic Robertson, "Drones, Phones and Satellite Technology Are Exposing the Truth About Russia's War in Ukraine in Near Real-Time," CNN, April 7, 2022, https://www.cnn.com/2022/04/06/europe/ukraine-russia-war-technology-intl-cmd/index.html.
27. Azov regiment: Tara John and Tim Lister, "A Far-Right Battalion Has a Key Role in Ukraine's Resistance: It's Neo-Nazi History Has Been Exploited by Putin," CNN, March 30, 2022, https://www.cnn.com/2022/03/29/europe/ukraine-azov-movement-far-right-intl-cmd/index.html; Vice, https://www.vice.com/en/article/3ab7dw/azov-battalion-ukraine-far-right.
28. Bad analogies: Paul Thagard, *Balance: How It Works and What It Means* (New York: Columbia University Press, 2022).
29. Right wing versus Ukraine: Peter Stone, "Top US Conservatives Pushing Russia's Spin on Ukraine War, Experts Say" *Guardian*, December 6, 2022, https://www.theguardian.com/world/2022/dec/06/us-conservatives-pushing-russian-spin-ukraine-war.
30. CBC: Brodie Fenlon, "As Russia's War with Ukraine Continues, Here Are Some of the Decisions We've Made at CBC News," CBC, March 13, 2022, https://www.cbc.ca/news/editorsblog/editor-s-blog-russia-ukraine-journalism-1.6382891.
31. NATO response: Raisa Patel, *Toronto Star*, "Justin Trudeau Says He Has a Plan to Fight Russian Disinformation," March 13, 2022, https://www.thestar.com/politics/federal/2022/03/13/justin-trudeau-says-he-has-a-plan-to-fight-russian-disinformation.html.
32. Powerlessness: Jan-Willem van Prooijen, *The Psychology of Conspiracy Theories* (London: Routledge, 2018).
33. Social media: Kari Paul, " 'Game of Whac-a-Mole': Why Russian Disinformation Is Still Running Amok on Social Media," *Guardian*, March 16, 2022, https://www.theguardian.com/media/2022/mar/15/russia-disinformation-social-media-ukraine; Kari Paul, "Flood of Russian Misinformation Puts Tech Companies in the Hot Seat," *Guardian*, March 1, 2022, https://www.theguardian.com/media/2022/feb/28/facebook-twitter-ukraine-russia-misinformation?CMP=Share_iOSApp_Other.
34. Facebook outside the United States: Dan Milmo, "Frances Haugen: 'I Never Wanted to Be a Whistleblower. But Lives Were in Danger,' " *Guardian*, October 24, 2021, https://www.theguardian.com/technology/2021/oct/24/frances-haugen-i-never-wanted-to-be-a-whistleblower-but-lives-were-in-danger; Cat Zakrzewski, Gerrit De Vynck, Niha Misah, and Shibani Mitanhi, "How Facebook Neglected the Rest of the World, Fueling Hate Speech and Violence in India," *Washington Post*, October 24, 2021, https://www.washingtonpost.com/technology/2021/10/24/india-facebook-misinformation-hate-speech/; Stephanie Valencia, "Misinformation Online Is Bad In English. But It's Far Worse in Spanish," *Washington Post*, October 29, 2021, https://www.washingtonpost.com/outlook/2021/10/28/misinformation-spanish-facebook-social-media/.
35. Twitter: Will Oremus, "To Fight Misinformation, Twitter Expands Project to Let Users Fact-Check Each Other's Tweets," *Washington Post*, March 3, 2022, https://www.washingtonpost.com/technology/2022/03/03/twitter-birdwatch-fact-check-misinfo-test/.
36. Gullibility: Joseph P. Forgas and Roy Baumeister, *The Social Psychology of Gullibility: Conspiracy Theories, Fake News and Irrational Beliefs* (London: Routledge, 2019).

37. Automatic belief: Daniel T. Gilbert, "How Mental Systems Believe," *American Psychologist* 46, no. 2 (1991): 107–119.
38. Testimony: Paul Thagard, "Testimony, Credibility, and Explanatory Coherence," *Erkenntnis* 63 (2005): 295–316.
39. Critical thinking: Paul Thagard, "Critical Thinking," Winter 2003 course at University of Waterloo, Canada, http://cogsci.uwaterloo.ca/courses/phil145.html; Paul Thagard, "Critical Thinking and Informal Logic: Neuropsychological Perspectives," *Informal Logic* 31 (2011): 152–70.
40. Schools: Jay Caspian Kang, "Fighting Disinformation Can Feel Like a Lost Cause. It Isn't," *New York Times*, March 7, 2022, https://www.nytimes.com/2022/03/07/opinion/fighting-disinformation-education.html.
41. Join in: H. Collen Sinclair, "7 Ways to Avoid Becoming a Misinformation Superspreader," The Conversation, March 18, 2021, https://theconversation.com/7-ways-to-avoid-becoming-a-misinformation-superspreader-157099.
42. Empathy and psychotherapy: Leslie S. Greenberg, *Emotion-Focused Therapy: Coaching Clients to Work Through Their Feelings* (Washington, DC: American Psychological Association, 2015).
43. Roots of empathy: https://rootsofempathy.org.
44. Prebunking: Li Qian Tay, Mark J. Hurlstone, Tim Kurz, and Ullrich K. H. Ecker, "A Comparison of Prebunking and Debunking Interventions for Implied Versus Explicit Misinformation," *British Journal of Psychology* 113 (2021): 591–607; Stephan Lewandowsky and Sander Van Der Linden, "Countering Misinformation and Fake News Through Inoculation and Prebunking," *European Review of Social Psychology* 32, no. 2 (2021): 348–84.
45. Ukraine prebunking: Farhad Manjoo, "Opinion: Putin No Longer Seems Like a Master of Disinformation," *New York Times*, March 2, 2022, https://www.nytimes.com/2022/03/02/opinion/putin-disinformation-social-media.html; Dan Milmo and Pjotr Sauer Sat, "Deepfakes v. Pre-Bunking: Is Russia Losing the Infowar?," *Guardian*, March 19, 2022, https://www.theguardian.com/world/2022/mar/19/russia-ukraine-infowar-deepfakes.
46. Inoculation: Jon Roozenbeek, Sander Van Der Linden, Beth Goldberg, Steve Rathje, and Stephan Lewandowsky, "Psychological Inoculation Improves Resilience Against Misinformation on Social Media," *Science Advances* 8, no. 34 (2022): eabo6254; Sander van der Linden, *Foolproof: Why Misinformation Infects Our Minds and How to Build Immunity* (New York: Norton, 2023).
47. Critical ignoring: Anastasia Kozyreva, Sam Wineburg, Stephan Lewandowsky, and Ralph Hertwig, "Critical Ignoring as a Core Competence for Digital Citizens," *Current Directions in Psychological Science* (2022): 09637214221121570.
48. Gatekeeping: Pamela J. Shoemaker and Timothy Vos, *Gatekeeping Theory* (London: Routledge, 2009), 1.
49. Journalistic ethics: Stephen J. A. Ward, *Disrupting Journalism Ethics: Radical Change on the Frontier of Digital Media* (London: Routledge, 2019).
50. Peer review: Tom, Jefferson, Melanie Rudin, Suzanne Brodney Folse, and Frank Davidoff, "Editorial Peer Review for Improving the Quality of Reports of Biomedical Studies," *Cochrane Database of Systematic Reviews*, no. 2 (2007): 1–35.
51. Evidence-based medicine: Jeremy H. Howick, *The Philosophy of Evidence-Based Medicine* (Oxford: Wiley-Blackwell, 2011).

52. Legal evidence: Michael H. Graham, *Federal Rules of Evidence in a Nutshell* (St. Paul, MN: West, 1987).
53. Pause and care before you share: "Pause before sharing, to help stop viral spread of COVID-19 information," UN News, June 30, 2020, https://news.un.org/en/story/2020/06/1067422.
54. Gatekeeping on Russia: Ryan Mac, Mike Isaac, and Sheera Frenkel, "How War in Ukraine Roiled Facebook and Instagram," *New York Times*, March 30, 2022, https://www.nytimes.com/2022/03/30/technology/ukraine-russia-facebook-instagram.html.
55. Deception: Thomas Rid, *Active Measures: The Secret History of Disinformation and Political Warfare* (New York: Farrar, Straus and Giroux, 2020).
56. Lie machines: Philip N. Howard, *Lie Machines: How to Save Democracy from Troll Armies, Deceitful Robots, Junk News Operations, and Political Operatives* (New Haven, CT: Yale University Press, 2020), 4.
57. Cheap speech: Richard L. Hasen, *Cheap Speech: How Disinformation Poisons Our Politics—and How to Cure It* (New Haven, CT: Yale University Press, 2022).
58. Free speech: Peter Warren Singer and Emerson T. Brooking, *LikeWar: The Weaponization of Social Media* (Boston: Houghton Mifflin Harcourt, 2018).
59. Misinformation tolerance: Kaveh Waddell, "On Social Media, Only Some Lies Are Against the Rules," *Consumer Reports*, August 13, 2020, https://www.consumerreports.org/social-media/social-media-misinformation-policies.
60. Wikipedia: Heather Kelly, "On Its 20th Birthday, Wikipedia May Be the Safest Place Online," *Washington Post*, January 15, 2021, https://www.washingtonpost.com/technology/2021/01/15/wikipedia-20-year-anniversary/.
61. European Union: "The 2022 Code of Practice on Disinformation," European Commission, June 16, 2022, https://digital-strategy.ec.europa.eu/en/policies/code-practice-disinformation; "The Digital Services Act Package," European Commission, October 27, 2022, https://digital-strategy.ec.europa.eu/en/policies/digital-services-act-package.
62. VR chat: "Facebook's Metaverse: One Incident of Abuse and Harassment Every 7 Minutes," Center for Countering Digital Hate, https://counterhate.com/research/facebooks-metaverse/.
63. Alignment problem: Paul Thagard, *Bots and Beasts: What Makes Machines, Animals, and People Smart?* (Cambridge, MA: MIT Press, 2021).

9. REALITY RESCUED: BEYOND POST-TRUTH

1. Giuliani: Rebecca Morin and David Cohen, "Giuliani: 'Truth Isn't Truth,'" Politico, August 19, 2018, https://www.politico.com/story/2018/08/19/giuliani-truth-todd-trump-788161. Post-truth: Lee McIntyre, *Post-Truth* (Cambridge, MA: MIT Press, 2018).
2. Truth as correspondence: David, Marian, "The Correspondence Theory of Truth," in The Stanford Encyclopedia of Philosophy (Summer 2022 ed.), ed. Edward N. Zalta, https://plato.stanford.edu/archives/sum2022/entries/truth-correspondence/; Paul Thagard, *Natural Philosophy: From Social Brains to Knowledge, Reality, Morality, and Beauty* (New York: Oxford University Press, 2019), chap. 4.
3. Idealism: Paul Guyer and Rolf-Peter Horstmann, "Idealism," in The Stanford Encyclopedia of Philosophy (Spring 2023 ed.), ed. Edward N. Zalta and Uri Nodelman, https://plato.stanford.edu/archives/spr2023/entries/idealism.

4. Quantum idealism: Deepak Chopra, *Metahuman: Unleashing Your Infinite Potential* (New York: Harmony, 2019).
5. Empiricism: Peter Markie and M. Folescu, "Rationalism vs. Empiricism," in The Stanford Encyclopedia of Philosophy (Spring 2023 ed. Edward N. Zalta and Uri Nodelman https://plato.stanford.edu/archives/spr2023/entries/rationalism-empiricism/.
6. Reality is mathematical: Carl Huffman, "Pythagoreanism," in The Stanford Encyclopedia of Philosophy (Fall 2019 ed.), ed. Edward N. Zalta, https://plato.stanford.edu/archives/fall2019/entries/pythagoreanism/; Max Tegmark, *Our Mathematical Universe: My Quest for the Ultimate Nature of Reality* (New York: Knopf, 2014).
7. It from bit: David J. Chalmers, *Reality+: Virtual Worlds and the Problems of Philosophy* (New York: Norton, 2022).
8. Social constructionism: Bruno Latour and Steve Woolgar, *Laboratory Life: The Construction of Scientific Facts* (Princeton, NJ: Princeton University Press, 1986).
9. Ideology: Loren R. Graham, *Science in Russia and the Soviet Union: A Short History* (Cambridge: Cambridge University Press, 1993).
10. Opinions/facts: Bernard Baruch, Daniel Patrick Moynihan, Rayburn H. Carrell, James R. Schlesinger, and Alan Greenspan, "People Are Entitle to Their Own Opinions but Not to Their Own Facts," Quote Investigator®, March 17, 2020, https://quoteinvestigator.com/2020/03/17/own-facts/.
11. Controlled hallucination: Anil Seth, *Being You: A New Science of Consciousness* (New York: Dutton 2021), 87.
12. Conscious experiences explained: Paul Thagard, *Brain-Mind: From Neurons to Consciousness and Creativity* (New York: Oxford University Press, 2019); Thagard, *Natural Philosophy*; Paul Thagard, *Balance: How It Works and What It Means* (New York: Columbia University Press, 2022).
13. Critique of Bayesian cognition: Max Jones and Bradley C. Love, "Bayesian Fundamentalism or Enlightenment: On the Explanatory Status and Theoretical Contributions of Bayesian Models of Cognition," *Behavioral and Brain Sciences* 34 (2011): 169–231; Thagard, *Natural Philosophy*.
14. Constraint satisfaction: James L. McClelland, Daniel Mirman, Donald J. Bolger, and Pranav Khaitan, "Interactive Activation and Mutual Constraint Satisfaction in Perception and Cognition," *Cognitive Science* 38, no. 6 (2014): 1139–1189; Thagard, *Brain-Mind*; Thagard, *Natural Philosophy*.
15. Recurrent: Mika Koivisto, Henry Railo, Antti Revonsuo, Simo Vanni, and Niina Salminen-Vaparanta, "Recurrent Processing in V1/V2 Contributes to Categorization of Natural Scenes," *Journal of Neuroscience* 31, no. 7 (2011): 2488–2492.
16. Always hallucinating: Seth, *Being You*, 92.
17. Measuring instruments: "Measuring Instrument: Alphabetical Listing," Wikipedia, May 11, 2023, https://en.wikipedia.org/wiki/Measuring_instrument#Alphabetical_listing.
18. Against reality: Donald Hoffman, *The Case Against Reality: Why Evolution Hid the Truth from Our Eyes* (New York: Norton, 2019).
19. Fitness-beats-truth theorem: Chetan Prakash, Kyle D. Stephens, Donald D. Hoffman, Manish Singh, and Chris Fields, "Fitness Beats Truth in the Evolution of Perception," *Acta Biotheoretica* 69, no. 3 (2021): 319–341.

20. Temperature: Hasok Chang, *Inventing Temperature: Measurement and Scientific Progress* (Oxford: Oxford University Press, 2004).
21. Correct sensation: Chang, *Inventing Temperature*, 43.
22. Coherence as constraint satisfaction: Paul Thagard and Karstn Verbeurgt, "Coherence as Constraint Satisfaction," *Cognitive Science* 22 (1998): 1–24; Paul Thagard, *Coherence in Thought and Action* (Cambridge, MA: MIT Press, 2000).
23. Wastewater: "National Wastewater Surveillance System (NWSS) Centers for Disease Control and Prevention, March 14, 2023, https://www.cdc.gov/healthywater/surveillance/wastewater-surveillance/wastewater-surveillance.html.
24. Replication crisis: Fiona Fidler and John Wilcox, "Reproducibility of Scientific Results," in The Stanford Encyclopedia of Philosophy (Summer 2021 ed.), ed. Edward N. Zalta, https://plato.stanford.edu/archives/sum2021/entries/scientific-reproducibility/.
25. Empiricism: Bas van Fraassen, *The Scientific Image* (Oxford: Clarendon Press, 1980).
26. Pessimistic induction: Larry Laudan, "A Confutation of Convergent Realism," *Philosophy of Science* 48 (1981): 19–49.
27. Underlying mechanisms: Paul Thagard, *The Cognitive Science of Science: Explanation, Discovery, and Conceptual Change* (Cambridge, MA: MIT Press, 2012).
28. Balance: Thagard, *Balance*.
29. Moral realism: Geoff Sayre-McCord, "Moral Realism," in The Stanford Encyclopedia of Philosophy (Summer 2021 ed.), ed. Edward N. Zalta, https://plato.stanford.edu/archives/sum2021/entries/moral-realism/.
30. Railton: Peter Railton, *Facts, Values, and Norms: Essays Toward a Morality of Consequence* (Cambridge: Cambridge University Press, 2003).
31. Psychological needs: Richard M. Ryan and Edward L. Deci, *Self-Determination Theory: Basic Psychological Needs in Motivation, Development, and Wellness* (New York: Guilford, 2017).
32. Needs-based ethics: Paul Thagard, *The Brain and the Meaning of Life* (Princeton, NJ: Princeton University Press, 2010); Thagard, *Natural Philosophy*. These books explain why needs are a better approach to morality than the capabilities advocated by Amartya Sen and Martha Nussbaum.
33. Universal needs: A. Timothy Church, Marcia S. Katigbak, Kenneth D. Locke, Hengsheng Zhang, Jiliang Shen, José de Jesús Vargas-Flores, Joselina Ibáñez-Reyes, et al., "Need Satisfaction and Well-Being: Testing Self-Determination Theory in Eight Cultures," *Journal of Cross-Cultural Psychology* 44, no. 4 (2013): 507–534.
34. Moral coherence: Thagard, *Balance*.
35. Simulation hypothesis: Mike Wall, "We're Probably Living in a Simulation, Elon Musk Says," Space.com, September 7, 2018, https://www.space.com/41749-elon-musk-living-in-simulation-rogan-podcast.html; Nick Bostrom, "Are We Living in a Computer Simulation?," *The Philosophical Quarterly* 53 (2003): 243–255; Preston Greene, "The Termination Risks of Simulation Science," *Erkenntnis* 85 (2018): 485–509; Chalmers, *Reality+*.
36. Conclusions: Chalmers, *Reality+*, xvii, 27.
37. Argument: Chalmers, *Reality+*, 101.
38. *Consciousness:* Paul Thagard, *Bots and Beasts: What Makes Machines, Animals, and People Smart?* (Cambridge, MA: MIT Press, 2021).

39. Energy and substrate independence: Paul Thagard, "Energy Requirements Undermine Substrate Independence and Mind-Body Functionalism," *Philosophy of Science* 89 (2022): 70–88.
40. Simulate universe: Dom Galeon, "Sorry, Elon. Physicists Say We Definitely Aren't Living in a Computer Simulation," Futurism, October 3, 2017, https://futurism.com/sorry-elon-physicists-say-we-definitely-arent-living-in-a-computer-simulation.
41. Reality as bits: Chalmers, *Reality+*, 105. Virtual objects are real: Chalmers, *Reality+*, 116.
42. Objectivism: George Lakoff and Mark Johnson, *Philosophy in the Flesh: The Embodied Mind and Its Challenge to Western Thought* (New York: Basic Books, 1999). Response: Thagard, *Balance*, chap. 5.

BIBLIOGRAPHY

Abaluck, Jason, Laura H. Kwong, Ashley Styczynski, Ashraful Haque, Md Alagir Kabir, Ellen Bates-Jefferys, Emily Crawford, et al. "Impact of Community Masking on Covid-19: A Cluster-Randomized Trial in Bangladesh." *Science* 375, no. 6577 (January 14, 2022): eabi9069.

Abernethy, George L., ed. *The Idea of Equality: An Anthology*. Richmond, VA: John Knox, 1959.

Adam, David. "Special Report: The Simulations Driving the World's Response to Covid-19." *Nature* 580, no. 7802 (2020): 316–19.

Agley, Jon, and Yunyu Xiao. "Misinformation About Covid-19: Evidence for Differential Latent Profiles and a Strong Association with Trust in Science." *BMC Public Health* 21, no. 1 (January 7, 2021): 89.

Ahmed, Nabil, Anna Marriott, Nafkote Dabi, Megan Lowthers, Max Lawson, and Leah Mugehera. "Inequality Kills: The Unparalleled Action Needed to Combat Unprecedented Inequality in the Wake of Covid-19." (2022). https://oxfamilibrary.openrepository.com/handle/10546/621341.

Albright, Madeline. *Fascism: A Warning*. New York: HarperCollins, 2018.

Albright, Thomas D. "Why Eyewitnesses Fail." *Proceedings of the National Academy of Sciences* 114, no. 30 (July 25, 2017): 7758–64.

Allgaier, Joachim. "Science and Environmental Communication on YouTube: Strategically Distorted Communications in Online Vvdeos on Climate Hange and Clmate Engineering." *Frontiers in Communication* 4 (2019).

Anderson, Michael C., and Simon Hanslmayr. "Neural Mechanisms of Motivated Forgetting." *Trends in Cognitive Sciences* 18, no. 6 (2014): 279–92.

Andreu-Sánchez, Celia, and Miguel-Ángel Martín-Pascual. "Fake Images of the Sars-Cov-2 Coronavirus in the Communication of Information at the Beginning of the First Covid-19 Pandemic." *El Profesional de la Información* 29, no. 3 (2020): e290309.

Appiah, Kwame Anthony. "Digging for Utopia." *New York Review of Books*, December 16, 2021, 80.

Aristotle. *The Complete Works of Aristotle*. Ed. J. Barnes. 2 vols. Princeton, NJ: Princeton University Press, 1984.

——. *Ethics*. Trans. J. A. K. Thomson. Harmondsworth, Middlesex: Penguin, 1955.

Armstrong McKay, David I, Arie Staal, Jesse F Abrams, Ricarda Winkelmann, Boris Sakschewski, Sina Loriani, Ingo Fetzer, et al. "Exceeding 1.5 C Global Warming Could Trigger Multiple Climate Tipping Points." *Science* 377, no. 6611 (2022): eabn7950.

Atkinson, Anthony B. *Inequality: What Can Be Done?* Cambridge, MA: Harvard University Press, 2015.

Bacon, Francis. *The New Organon and Related Writings*. Ed. F. Anderson. Indianapolis, IN: Bobbs-Merrill, 1960.

Baird, Davis. *Thing Knowledge: A Philosophy of Scientific Instruments*. Berkeley: University of California Press, 2004.

Banich, Marie T., and Rebecca J. Compton. *Cognitive Neuroscience*. 4th ed. Cambridge: Cambridge University Press, 2018.

Basol, Melisa, Jon Roozenbeek, Manon Berriche, Fatih Uenal, William P. McClanahan, and Sander van der Linden. "Towards Psychological Herd Immunity: Cross-Cultural Evidence for Two Prebunking Interventions Against Covid-19 Misinformation." *Big Data & Society* 8, no. 1 (2021).

Bazerman, Max H., and Don A. Moore. *Judgment in Managerial Decision Making*. 8th ed. New York: Wiley, 2012.

Beauchamp, Tom L., and James F. Childress. *Principles of Biomedical Ethics*. 7th ed. New York: Oxford University Press, 2013.

Bechtel, William. *Mental Mechanisms: Philosophical Perspectives on Cognitive Neuroscience*. New York: Routledge, 2008.

Ben-Ghiat, Ruth. *Strongmen: Mussolini to the Present*. New York: Norton, 2020.

Bennett, Richard M. *Conspiracy: Plots, Lies, and Cover-Ups*. London: Virgin, 2003.

Bleakley, Paul. "Panic, Pizza and Mainstreaming the Alt-Right: A Social Media Analysis of Pizzagate and the Rise of the Qanon Conspiracy." *Current Sociology* 71, no. 3 (2021): 1–17.

Boness, Cassandra L, Mackenzie Nelson, and Antoine B Douaihy. "Motivational Interviewing Strategies for Addressing Covid-19 Vaccine Hesitancy." *Journal of the American Board of Family Medicine* 35, no. 2 (March–April 2022): 420–26.

Bostrom, Nick. "Are We Living in a Computer Simulation?" *The Philosophical Quarterly* 53 (2003): 243–55.

Brashier, Nadia M., and Elizabeth J Marsh. "Judging Truth." *Annual Review of Psychology* 71 (January 4, 2020): 499–515.

Britt, Robert Roy. "What the Coronavirus Image You've Seen a Million Times Really Shows." *Elemental*. (2020). https://elemental.medium.com/what-the-coronavirus-image-youve-seen-a-million-times-really-shows-3d8de7e3eb1f.

Brotherton, Rob. *Suspicious Minds: Why We Believe Conspiracy Theories*. London: Bloomsbury Sigma, 2015.

Brown, N. Derek, Drew S. Jacoby-Senghor, and Isaac Raymundo. "If You Rise, I Fall: Equality Is Prevented by the Misperception That It Harms Advantaged Groups." *Science Advances* 8, no. 18 (2022): eabm2385.

Burgess, Glenn. "The Divine Right of Kings Reconsidered." *The English Historical Review* 107, no. 425 (1992): 837–61.

Butler, Heather A., Christopher P. Dwyer, Michael J. Hogan, Amanda Franco, Silvia F. Rivas, Carlos Saiz, and Leandro S. Almeida. "The Halpern Critical Thinking Assessment and Real-World Outcomes: Cross-National Applications." *Thinking Skills and Creativity* 7, no. 2 (2012): 112–21.

Byrne, Rhonda. *The Secret*. New York: Simon and Schuster, 2006.

Camus, Renaud. *Le Grand Remplacement: Introduction Au Remplacisme Global*. Paris: La Nouvelle Librairie, 2021.

Card, David, and Alan B. Krueger. *Myth and Measurement: The New Economics of the Minimum Wage*. Princeton, NJ: Princeton University Press, 1997.

Carlson, Colin J., Gregory F. Albery, Cory Merow, Christopher H. Trisos, Casey M. Zipfel, Evan A. Eskew, Kevin J. Olival, Noam Ross, and Shweta Bansal. "Climate Change Increases Cross-Species Viral Transmission Risk." *Nature* 607 (2022): 555–62.

Cassam, Quassim. *Conspiracy Theories*. Cambridge: Polity, 2019.

Castonguay, Louis G., and Clara E. Hill, eds. *How and Why Are Some Therapists Better Than Others? Understanding Therapist Effects*. Washington, DC: American Psychological Association, 2017.

CDC. "Scientific Brief: Community Use of Cloth Masks to Controll the Spread of Sars-Cov-2." Centers for Disease Control and Prevention. 2021. Accessed March 10, 2023. https://www.cdc.gov/coronavirus/2019-ncov/science/science-briefs/masking-science-sars-cov2.html.

Chalmers, David J. *Reality+: Virtual Worlds and the Problems of Philosophy*. New York: Norton, 2022.

Chancel, Lucas, Thomas Piketty, Emmanuel Saez, and Gabriel Zucman. "World Inequality Report 2022." 2022. https://wir2022.wid.world.

Chang, Hasok. *Inventing Temperature: Measurement and Scientific Progress*. Oxford: Oxford University Press, 2004.

Chopra, Deepak. *Metahuman: Unleashing Your Infinite Potential*. New York: Harmony, 2019.

Church, A. Timothy, Marcia S. Katigbak, Kenneth D. Locke, Hengsheng Zhang, Jiliang Shen, José de Jesús Vargas-Flores, Joselina Ibáñez-Reyes, et al. "Need Satisfaction and Well-Being: Testing Self-Determination Theory in Eight Cultures." *Journal of Cross-Cultural Psychology* 44, no. 4 (2013): 507–34.

Coady, David. "Cass Sunstein and Adrian Vermeule on Conspiracy Theories." *Argumenta* 3, no. 2 (2018): 291–302.

Coan, Travis G., Constantine Boussalis, John Cook, and Mirjam O. Nanko. "Computer-Assisted Classification of Contrarian Claims About Climate Change." *Scientific Reports* 11, no. 1 (November 16, 2021): 22320.

Collins, Randall. *Interaction Ritual Chains*. Princeton, NJ: Princeton University Press, 2004.

Cook, J. "Understanding and Countering Misinformation About Climate Change." In *Handbook of Research on Deception, Fake News, and Misinformation Online*, ed. I. Chiluwa and S. Samoilenko, 281–306. Hershey, PA: IGI-Global, 2019.

Cook, John, Peter Ellerton, and David Kinkead. "Deconstructing Climate Misinformation to Identify Reasoning Errors." *Environmental Research Letters* 13, no. 2 (2018): 024018.

Cook, John, Stephan Lewandowsky, and Ullrich K. H. Ecker. "Neutralizing Misinformation Through Inoculation: Exposing Misleading Argumentation Techniques Reduces Their Influence." *PLoS One* 12, no. 5 (2017): e0175799.

Cook, John, Naomi Oreskes, Peter T. Doran, William R. L. Anderegg, Bart Verheggen, Ed W. Maibach, J. Stuart Carlton, et al. "Consensus on Consensus: A Synthesis of Consensus Estimates on Human-Caused Global Warming." *Environmental Research Letters* 11, no. 4 (2016): 048081.

Craver, Carl F., and Lindley Darden. *In Search of Mechanisms: Discoveries Across the Life Sciences*. Chicago: University of Chicago Press, 2013.

Dammann, Olaf, Ted Poston, and Paul Thagard. "How Do Medical Researchers Make Causal Inferences?" In *What Is Scientific Knowledge? An Introduction to Contemporary Epistemology of Science*, ed. K. McCain and K. Kampourakis, 33–51. New York: Routledge, 2019.

Darden, Lindley, Kunal Kundu, Lipika R. Pal, and John Moult. "Harnessing Formal Concepts of Biological Mechanism to Analyze Human Disease." *PLoS Computational Biology* 14, no. 12 (December 2018): e1006540.

de Beauvoir, Simone. *The Second Sex*. Trans. C. Borde and S. Malovany-Chevallier. New York: Alfred A. Knopf, 2010.

Deci, Edward. L., and Richard. M. Ryan, eds. *Handbook of Self-Determination Research*. Rochester, NY: Univerity of Rochester Press, 2002.

Derreumaux, Yrian, Robin Bergh, and Brent L. Hughes. "Partisan-Motivated Sampling: Re-Examining Politically Motivated Reasoning Across the Information Processing Stream." *Journal of Personality and Social Psychology* 123 (2022): 316–36.

Dobelli, Rolf. *The Art of Thinking Clearly: Better Thinking, Better Decisions*. New York: Harper, 2013.

Douglas, Karen M., Joseph E. Uscinski, Robbie M. Sutton, Aleksandra Cichocka, Turkay Nefes, Chee Siang Ang, and Farzin Deravi. "Understanding Conspiracy Theories." *Political Psychology* 40, no. S1 (2019): 3–35.

Dunlap, Riley E., and Peter J. Jacques. "Climate Change Denial Books and Conservative Think Tanks: Exploring the Connection." *American Behavioral Scientist* 57, no. 6 (June 2013): 699–731.

Dunning, David, and Emily Balcetis. "Wishful Seeing: How Preferences Shape Perception." *Current Directions in Psychological Science* 22 (2013): 33–37.

Dunsmoor, Joseph E., Marijn C. W. Kroes, Vishnu P Murty, Stephen H. Braren, and Elizabeth A. Phelps. "Emotional Enhancement of Memory for Neutral Information: The Complex Interplay Between Arousal, Attention, and Anticipation." *Biological Psychology* 145 (2019): 134–41.

Eatwell, Roger. *Fascism: A History*. London: Pimlico, 2003.

Ecker, Ullrich K. H., Stephan Lewandowsky, John Cook, Philipp Schmid, Lisa K. Fazio, Nadia Brashier, Panayiota Kendeou, Emily K. Vraga, and Michelle A. Amazeen. "The Psychological Drivers of Misinformation Belief and Its Resistance to Correction." *Nature Reviews Psychology* 1, no. 1 (2022): 13–29.

Eitam, Baruch, David B. Miele, and E. Tory Higgins. "Motivated Remembering: Remembering as Accessibility and Accessibility as Motivational Relevance." In *The Oxford Handbook of Social Cognition*, ed. Donal E. Carlston, 463–75. Oxford: Oxford University Press, 2013.

Elad-Strenger, Julia, Jutta Proch, and Thomas Kessler. "Is Disgust a 'Conservative' Emotion?" *Personality and Social Psychology Bulletin* 46, no. 6 (2020): 896–912.

Eliasmith, Chris. *How to Build a Brain: A Neural Architecture for Biological Cognition*. Oxford: Oxford University Press, 2013.

Ellis, Jon. "Motivated Reasoning and the Ethics of Belief." *Philosophy Compass* 17 (2022): e12828.

Elster, Jon. *Explaining Social Behavior*. Cambridge: Cambridge University Press, 2007.

Epley, Nicholas, and Thomas Gilovich. "The Mechanics of Motivated Reasoning." *Journal of Economic Perspectives* 30, no. 3 (2016): 133–40.

Eysenck, Hans Jurgen. *The Inequality of Man*. London: Temple Smith. 1973.

Fawcett, Bill, ed. *You Said What? Lies and Propagand Throughout History*. New York: Harper, 2007.

Fenn, Julius, Jessica F. Helm, Philipp Höfele, Lars Kulbe, Andreas Ernst, and Andrea Kiesel. "Identifying Key-Psychological Factors Influencing the Acceptance of yet Emerging Technologies–a Multi-Method-Approach to Inform Climate Policy." *PLOS Climate* 2, no. 6 (2023).

Flannery, Kent, and Joyce Marcus. *The Creation of Inequality: How Our Prehistoric Ancestors Set the Stage for Monarchy, Slavery, and Empire*. Cambridge, MA: Harvard University Press, 2012.

Floridi, Luciano, and Phyllis Illari. *The Philosophy of Information Quality*. Berlin: Springer, 2014.

Forgas, Joseph P., and Roy Baumeister. *The Social Psychology of Gullibility: Conspiracy Theories, Fake News and Irrational Beliefs*. London: Routledge, 2019.

Forget, Evelyn L. "The Town with No Poverty: The Health Effects of a Canadian Guaranteed Annual Income Field Experiment." *Canadian Public Policy* 37, no. 3 (2011): 283–305.

Frankfurt, Harry G. *On Bullshit*. Princeton, NJ: Princeton University Press, 2005.

Franklin, Allan. "Experiment in Physics." *Stanford Encyclopedia of Philosophy*. 2019. https://plato.stanford.edu/entries/physics-experiment/.

Frost, Helen, Pauline Campbell, Margaret Maxwell, Ronan E. O'Carroll, Stephan U. Dombrowski, Brian Williams, Helen Cheyne, Emma Coles, and Alex Pollock. "Effectiveness of Motivational Interviewing on Adult Behaviour Change in Health and Social Care Settings: A Systematic Review of Reviews." *PLoS One* 13, no. 10 (2018): e0204890.

Galvani, Alison P., Alyssa S. Parpia, Abhishek Pandey, Pratha Sah, Kenneth Colon, Gerald Friedman, Travis Campbell, et al. "Universal Healthcare as Pandemic Preparedness: The Lives and Costs That Could Have Been Saved During the Covid-19 Pandemic." *Proceedings of the Natlional Academy of Sciences* 119, no. 25 (June 21 2022): e2200536119.

Garry, Amanda, Samantha Walther, Rukaya Rukaya, and Ayan Mohammed. "Qanon Conspiracy Theory: Examining Its Evolution and Mechanisms of Radicalization." *Journal for Deradicalization* 26, no. 3 (2021): 152–216.

Gilbert, Daniel T. "How Mental Systems Believe." *American Psychologist* 46, no. 2 (1991): 107–19.

Gilovich, Thomas. *How We Know What Isn't So*. New York: Free Press, 1991.

Gimpelson, Vladimir, and Daniel Treisman. "Misperceiving Inequality." *Economics & Politics* 30, no. 1 (2018): 27–54.

Glennan, Stuart. *The New Mechanical Philosophy*. Oxford: Oxford University Press, 2017.

Goldenberg, Maya J. *Vaccine Hesitancy: Public Trust, Expertise, and the War on Science*. Pittsburgh, PA: University of Pittsburgh Press, 2021.

Gordin, Michael D. *On the Fringe: Where Science Meets Pseudoscience*. Oxford: Oxford University Press, 2021.

Gottman, John M. *Principia Amoris: The New Science of Love*. New York: Routledge, 2015.

Graeber, David, and David Wengrow. *The Dawn of Everything: A New History of Humanity*. London: Allen Lane, 2021.

Graham, Loren R. *Science in Russia and the Soviet Union: A Short History*. Cambridge: Cambridge University Press, 1993.

Graham, Michael H. *Federal Rules of Evidence in a Nutshell*. St. Paul, MN: West Publishing, 1987.

Grant, Adam. *Think Again: The Power of Knowing What You Don't Know*. New York: Penguin, 2021.

Greenberg, Leslie S. *Emotion-Focused Therapy: Coaching Clients to Work Through Their Feelings*. Washington, DC: American Psychological Association, 2015.

Greene, Preston. "The Termination Risks of Simulation Science." *Erkenntnis* 85 (2018): 485–509.

Guay, Brian, and Christopher D. Johnston. "Ideological Asymmetries and the Determinants of Politically Motivated Reasoning." *American Journal of Political Science* 66, no. 2 (2022): 284–301.

Harris, Sam, Sameer A. Sheth, and Mark S. Cohen. "Functional Neuroimaging of Belief, Disbelief, and Uncertainty." *Annals of Neurology* 63 (2008): 141–47.

Hasen, Richard L. *Cheap Speech: How Disinformation Poisons Our Politics—and How to Cure It*. New Haven, CT: Yale University Press, 2022.

Hauser, Oliver P., and Michael I Norton. "(Mis)Perceptions of Inequality." *Current Opinion in Psychology* 18 (2017): 21–25.

Hensrud, Donald D. "Clinical Preventive Medicine in Primary Care: Background and Practice: 1. Rationale and Current Preventive Practices." *Mayo Clinic Proceedings* 75, no. 2 (2000): 165–72.

Herrnstein, Richard J., and Charles Murray. *The Bell Curve: Intelligence and Class Structure in American Life*. New York: Free Press, 1994.

Hill, Austin Bradford. "The Environment and Disease: Association or Causation?" *Proceedings of the Royal Society of Medicine* 58 (1965): 295–300.

Hines, Terence. *Pseudoscience and the Paranormal*. Amherst, NY: Prometheus, 2003.

Hobbes, Thomas. *Leviathan*. Oxford: Basil Blackwell, 1947.

Hoffman, Donald. *The Case Against Reality: Why Evolution Hid the Truth from Our Eyes*. New York: Norton, 2019.

Holland, John. H., Keith. J. Holyoak, Richard. E. Nisbett, and Paul R. Thagard. *Induction: Processes of Inference, Learning, and Discovery*. Cambridge, MA: MIT Press, 1986.

Holyoak, Keith J., and Paul Thagard. *Mental Leaps: Analogy in Creative Thought*. Cambridge, MA: MIT Press, 1995.

Homer-Dixon, Thomas, Manjana Milkoreit, Stehemn J. Mock, Tobias Schröder, and Paul Thagard. "The Conceptual Structure of Social Disputes: Cognitive-Affective Maps as a Tool for Conflict Analysis and Resolution." *SAGE Open* 4 (2014): 1–20.

Horne, Zachary, Derek Powell, John E. Hummel, and Keith J. Holyoak. "Countering Antivaccination Attitudes." *Proceedings of the National Academy of Sciences* 112, no. 33 (2015): 10321–24.

Hornsey, Matthew J. "Why Facts Are Not Enough: Understanding and Managing the Motivated Rejection of Science." *Current Directions in Psychological Science* 29, no. 6 (2020): 583–91.

Howard, Jeremy, Austin Huang, Zhiyuan Li, Zeynep Tufekci, Vladimir Zdimal, Helene-Mari van der Westhuizen, Arne von Delft, et al. "An Evidence Review of Face Masks Against Covid-19." *Proceedings of the National Academy of Sciences* 118, no. 4 (2021): e2014564118.

Howard, Philip N. *Lie Machines: How to Save Democracy from Troll Armies, Deceitful Robots, Junk News Operations, and Political Operatives*. New Haven, CT: Yale University Press, 2020.

Howick, Jeremy H. *The Philosophy of Evidence-Based Medicine*. Oxford: Wiley-Blackwell, 2011.

Hoynes, Hilary W., and Jesse Rothstein. "Universal Basic Income in the US and Advanced Countries." *Annual Review of Economics* 11 (2019): 929–58.

Hughes, Brent L., and Jennifer S. Beer. "Protecting the Self: The Effect of Social-Evaluative Threat on Neural Representations of Self." *Journal of Cognitive Neuroscience* 25, no. 4 (2013): 613–22.

Hughes, Brent L., and Jamil Zaki. "The Neuroscience of Motivated Cognition." *Trends in Cognitive Sciences* 19, no. 2 (2015): 62–64.

Hull, Sharon K. "A Larger Role for Preventive Medicine." *AMA Journal of Ethics* 10, no. 11 (2008): 724–29.

Iacoboni, Marco. *Mirroring People: The New Science of How We Connect with Others*. New York: Farrar, Straus and Giroux, 2008.

Jackson, Sam. *Oath Keepers: Patriots and the Edge of Violence in a Right-Wing Antigovernment Group*. New York: Columbia University Press, 2020.

Janis, Irving L. *Groupthink: Psychological Studies of Policy Decisions and Fiascoes.* 2nd ed. Boston: Houghton Mifflin, 1982.

Jefferson, Tom, Liz Dooley, Eliana Ferroni, Lubna A. Al-Ansary, Mieke L. van Driel, Ghada A. Bawazeer, Mark A. Jones, et al. "Physical Interventions to Interrupt or Reduce the Spread of Respiratory Viruses." *Cochrane Database Systematic Reviews* 1, no. 1 (January 30, 2023): CD006207.

Jefferson, Tom, Melanie Rudin, Suzanne Brodney Folse, and Frank Davidoff. "Editorial Peer Review for Improving the Quality of Reports of Biomedical Studies." *Cochrane Database of Systematic Reviews*, no. 2 (2007): 1–35.

Jones, Matt, and Bradley C. Love. "Bayesian Fundamentalism or Enlightenment: On the Explanatory Status and Theoretical Contributions of Bayesian Models of Cognition." *Behavioral and Brain Sciences* 34 (2011): 169–231.

Jost, John T. *A Theory of System Justification.* Cambridge, MA: Harvard University Press, 2020.

Kahneman, Daniel, Olivier Sibony, and Cass R. Sunstein. *Noise: A Flaw in Human Judgment.* New York: Little, Brown Spark, 2021.

Kahneman, Daniel, and Amos Tversky, eds. *Choices, Values, and Frames.* Cambridge: Cambridge University Press, 2000.

Kajić, Ivana, Tobias C. Schröder, Terrence C. Stewart, and Paul Thagard. "The Semantic Pointer Theory of Emotions: Integrating Physiology, Appraisal, and Construction." *Cognitive Systems Research* 58 (2019): 35–53.

Kendeou, Panayiota, Daniel H. Robinson, and Matthew T. McCrudden. *Misinformation and Fake News in Education.* Charlotte, NC: Information Age Publishing, 2019.

Kim, James S., Mary A. Burkhauser, Laura M. Mesite, Catherine A. Asher, Jackie Eunjung Relyea, Jill Fitzgerald, and Jeff Elmore. "Improving Reading Comprehension, Science Domain Knowledge, and Reading Engagement Through a First-Grade Content Literacy Intervention." *Journal of Educational Psychology* 113, no. 1 (2021): 3–26.

Koivisto, Mika, Henry Railo, Antti Revonsuo, Simo Vanni, and Niina Salminen-Vaparanta. "Recurrent Processing in V1/V2 Contributes to Categorization of Natural Scenes." *Journal of Neuroscience* 31, no. 7 (2011): 2488–92.

Kozyreva, Anastasia, Sam Wineburg, Stephan Lewandowsky, and Ralph Hertwig. "Critical Ignoring as a Core Competence for Digital Citizens." *Current Directions in Psychological Science* (2022): 09637214221121570.

Kraus, Michael W., Sa-kiera T. J. Hudson, and Jennifer A. Richeson. "Framing, Context, and the Misperception of Black–White Wealth Inequality." *Social Psychological and Personality Science* 13, no. 1 (2022): 4–13.

Krekó, Péter. "Conspiracy Theory as Collective Motivated Cognition." In *The Psychology of Conspiracy*, ed. Michal Bilewicz, Aleksandra Cichocka, and Wiktor Soral, 62–77. London: Routledge, 2015.

Kruglanski, Arie W., Jocelyn J. Belanger, Xiaoyan Chen, Catalina Kopetz, Antonio Pierro, and Lucia Mannetti. "The Energetics of Motivated Cognition: A Force-Field Analysis." *Psychological Review* 119, no. 1 (January 2012): 1–20.

Krugman, Paul. *Arguing with Zombies: Economics, Politics, and the Fight for a Better Future.* New York: Norton, 2020.

Kunda, Ziva. "The Case for Motivated Reasoning." *Psychological Bulletin* 108 (1990): 480–98.

——. "Motivated Inference: Self-Serving Generation and Evaluation of Evidence." *Journal of Personality and Social Psychology* 53 (1987): 636–47.

——. *Social Cognition: Making Sense of People*. Cambridge, MA: MIT Press, 1999.

Kunda, Ziva, and Paul Thagard. "Forming Impressions from Stereotypes, Traits, and Behaviors: A Parallel-Constraint-Satisfaction Theory." *Psychological Review* 103 (1996): 284–308.

Lakoff, George, and Mark Johnson. *Philosophy in the Flesh: The Embodied Mind and Its Challenge to Western Thought*. New York: Basic Books, 1999.

Latimer, Jon *Deception in War*. New York: Overlook Press, 2001.

Latour, Bruno, and Steve Woolgar. *Laboratory Life: The Construction of Scientific Facts*. Princeton, NJ: Princeton University Press., 1986.

Laudan, Larry. "A Confutation of Convergent Realism." *Philosophy of Science* 48 (1981): 19–49.

Lee, Junho, Emily F. Wong, and Patricia W. Cheng. "Promoting Climate Actions: A Cognitive-Constraints Approach." *Cognitive Psychology* 143 (Jun 2023): 101565.

Leong, Yuan Chang, Brent L. Hughes, Yiyu Wang, and Jamil Zaki. "Neurocomputational Mechanisms Underlying Motivated Seeing." *Nature Human Behavior* 3, no. 9 (September 2019): 962–73.

Letwin, William, ed. *Against Equality: Readings on Economic and Social Policy*. London: Macmillan, 1983.

Levy, Becca R., Martin D. Slade, Suzanne R. Kunkel, and Stanislav V. Kasl. "Longevity Increased by Positive Self-Perceptions of Aging." *Journal of Personality and Social Psychology* 83, no. 2 (2002): 261–70.

Lewandowsky, Stephan, and John Cook. *The Conspiracy Theory Handbook*. 2020. Available at http://sks.to/conspiracy.

Lewandowsky, Stephan, John Cook, Ullrich Ecker, Dolores Albarracin, Michelle Amazeen, Panayiota Kendou, Doug Lombardi, et al. *The Debunking Handbook 2020*. 2020. https://skepticalscience.com/docs/DebunkingHandbook2020.pdf.

Lewandowsky, Stephan, and Sander Van Der Linden. "Countering Misinformation and Fake News Through Inoculation and Prebunking." *European Review of Social Psychology* 32, no. 2 (2021): 348–84.

Lewin, Kurt. *Field Theory in Social Science*. New York: Harper and Row, 1951.

Lipton, Peter. *Inference to the Best Explanation*. 2nd ed. London: Routledge, 2004.

Locke, John. *John Locke: Two Treatises of Government*. Cambridge: Cambridge University Press, 1967.

Loftus, Elizabeith F. *Eyewitness Testimony*. Cambridge, MA: Harvard University Press, 1996.

Lutzke, Lauren, Caitlin Drummond, Paul Slovic, and Joseph Árvai. "Priming Critical Thinking: Simple Interventions Limit the Influence of Fake News About Climate Change on Facebook." *Global Environmental Change* 58 (2019): 101964.

MacMillan, Margaret. *The War That Ended Peace: The Road to 1914*. New York: Penguin, 2013.

Marlow, Thomas, Sean Miller, and J. Timmons Roberts. "Bots and Online Climate Discourses: Twitter Discourse on President Trump's Announcement of US Withdrawal from the Paris Agreement." *Climate Policy* 21, no. 6 (2021): 765–77.

McClelland, James L., Daniel Mirman, Donald J. Bolger, and Pranav Khaitan. "Interactive Activation and Mutual Constraint Satisfaction in Perception and Cognition." *Cognitive Science* 38, no. 6 (2014): 1139–89.

McIntyre, Lee. *How to Talk to a Science Denier: Conversations with Flat Earthers, Climate Deniers, and Others Who Defy Reason*. Cambridge, MA: MIT Press, 2021.

——. *Post-Truth*. Cambridge, MA: MIT Press, 2018.

Mehra, Mandeep R., Sapan S. Desai, Frank Ruschitzka, and Amit N. Patel. "Retracted: Hydroxychloroquine or Chloroquine with or Without a Macrolide for Treatment of Covid-19: A Multinational Registry Analysis." *The Lancet* (2020). https://doi.org/10.1016/s0140-6736(20)31180-6.

Meichenbaum, Donald. *A Clinical Handbook/Practical Therapist Manual for Assessing and Treating Adults with Post-Traumatic Stress Disorder (PTSD)*. Waterloo, Ontario: Institute Press, 1994.

Merrill, Roberto, Catarina Neves, and Bru Laín. *Basic Income Experiments: A Critical Examination of Their Goals, Contexts, and Methods*. London: Palgrave Macmillan, 2021.

Meyerson, Denise. *False Consciousness*. Oxford: Clarendon Press, 1991.

Milanovic, Branko. *The Haves and the Have-Nots: A Brief and Idiosyncratic History of Global Inequality*. New York: Basic Books, 2010.

Milkoreit, Manjana. *Mindmade Politics: The Cognitive Roots of International Climate Governance*. Cambridge, MA: MIT Press, 2017.

Mill, John Stuart. *A System of Logic*. 8th ed. London: Longman, 1970.

Miller, Dale T., and Michael Ross. "Self-Serving Biases in Attribution of Causality: Fact or Fiction?" *Psychological Bulletin* 82 (1975): 213–25.

Miller, Joanne M., Kyle L. Saunders, and Christina E. Farhart. "Conspiracy Endorsement as Motivated Reasoning: The Moderating Roles of Political Knowledge and Trust." *American Journal of Political Science* 60, no. 4 (2016): 824–44.

Miller, William R., and Stephen Rollnick. *Motivational Interviewing: Helping People Change*. 3rd ed. New York: Guilford, 2012.

Mills, Charles W. *The Racial Contract*. Ithaca, NY: Cornell University Press, 2014.

Montoya, R. M., R. S. Horton, J. L. Vevea, M. Citkowicz, and E. A. Lauber. "A Re-Examination of the Mere Exposure Effect: The Influence of Repeated Exposure on Recognition, Familiarity, and Liking." *Psychological Bulletin* 143, no. 5 (May 2017): 459–98. https://doi.org/10.1037/bul0000085.

Muirhead, Russell, and Nancy L Rosenblum. *A Lot of People Are Saying: The New Conspiracism and the Assault on Democracy*. Princeton, NJ: Princeton University Press, 2020.

Mullainathan, Sendhil, and Eldar Shafir. *Scarcity: Why Having Too Little Means So Much*. New York: Macmillan, 2013.

Muradoglu, Melis, Zachary Horne, Matthew D. Hammond, Sarah-Jane Leslie, and Andrei Cimpian. "Women—Particularly Underrepresented Minority Women—and Early-Career Academics Feel Like Impostors in Fields That Value Brilliance." *Journal of Educational Psychology* 114 (2021): 1086–1100.

Murphy, Gregory L. *The Big Book of Concepts*. Cambridge, MA: MIT Press, 2002.

Murray, Sandra. L., and John. G. Holmes. *Motivated Cognition in Relationships: The Pursuit of Belonging*. New York: Routledge, 2017.

Murthy, Vivek H. "Confronting Health Misinformation: The U.S. Surgeon General's Advisory on Building a Health Information Environment." 2021. Accessed October 4, 2021. https://www.hhs.gov/sites/default/files/surgeon-general-misinformation-advisory.pdf.

Nickerson, Raymond S. "Confirmation Bias: A Ubiquitous Phenomenon in Many Guises." *Review of General Psychology* 2, no. 2 (1998): 175–220.

Nisbett, Richard E. *Intelligence and How We Get It*. New York: Norton, 2009.

——. *Mindware: Tools for Smart Thinking*. New York: Farrar, Straus and Giroux, 2015.

Nisbett, Richard E., Joshua Aronson, Clancy Blair, William Dickens, James Flynn, Diane F. Halpern, and Eric Turkheimer. "Intelligence: New Findings and Theoretical Developments." *American Psychologist* 67, no. 2 (February–March 2012): 130–59.

Nisbett, Richard E., and Lee Ross. *Human Inference: Strategies and Shortcomings of Social Judgement*. Englewood Cliffs, NJ: Prentice Hall, 1980.

Nozick, Robert. *Anarchy, State, and Utopia*. New York: Basic Books, 1974.

O'Connor, Daryl B., John P. Aggleton, Bhismadev Chakrabarti, Cary L. Cooper, Cathy Creswell, Sandra Dunsmuir, Susan T. Fiske, et al. "Research Priorities for the Covid-19 Pandemic and Beyond: A Call to Action for Psychological Science." *British Journal of Psychology* 111, no. 4 (2020): 603–29.

O'Connell, Jeffrey, and Michael Ruse. *Social Darwinism*. Cambridge: Cambridge University Press, 2021.

Oettingen, Gabriele. *Rethinking Positive Thinking: Inside the New Science of Motivation*. New York: Current, 2015.

Oreskes, Naomi, and Erik M. Conway. *Merchants of Doubt: How a Handful of Scientists Obscured the Truth on Issues from Tobacco Smoke to Global Warming*. New York: Bloomsbury Publishing, 2011.

Orwell, George. *1984*. New York: Harper, 2014.

Osberg, Lars. *The Age of Increasing Inequality: The Astonishing Rise of Canada's 1 percent*. Toronto: Jame Lorimer, 2018.

Oyserman, Daphna, and Andrew Dawson. "Your Fake News, Our Facts: Identity-Based Motivation Shapes What We Believe, Share, and Accept." In *The Psychology of Fake News: Accepting, Sharing, and Correctig Misinformation*, ed. Rainer Greifeneder, Mariela E. Jaffé, Eryn J. Newman, and Norbert Schwarz, 173–95. London: Routledge, 2020.

Park, SoHyun, Jeewon Choi, Sungwoo Lee, Changhoon Oh, Changdai Kim, Soohyun La, Joonhwan Lee, and Bongwon Suh. "Designing a Chatbot for a Brief Motivational Interview on Stress Management: Qualitative Case Study." *Journal of Medical Internet Research* 21, no. 4 (2019): e12231.

Pashler, Harold, Mark McDaniel, Doug Rohrer, and Robert Bjork. "Learning Styles: Concepts and Evidence." *Psychological Science in the Public Interest* 9, no. 3 (2008): 105–19.

Pateman, Carole. *The Sexual Contract*. Cambridge: Polity Press, 1988.

Pears, David. *Motivated Irrationality*. Oxford: Oxford University Press, 1984.

Pekar, Jonathan E., et al. " The Molecular Epidemiology of Multiple Zoonotic Origins of Sars-Cov-2." *Science* 377 (2022): 960–66.

Pennycook, Gordon, Jonathon McPhetres, Yunhao Zhang, Jackson G. Lu, and David G. Rand. "Fighting Covid-19 Misinformation on Social Media: Experimental Evidence for a Scalable Accuracy-Nudge Intervention." *Psychological Science* 31, no. 7 (2020): 770–80.

Pennycook, Gordon, and David G. Rand. "Lazy, Not Biased: Susceptibility to Partisan Fake News Is Better Explained by Lack of Reasoning Than by Motivated Reasoning." *Cognition* 188 (2019): 39–50.

Peterson, Jordan B. *12 Rules for Life: An Antidote to Chaos*. Toronto: Random House Canada, 2018.

Pickett, Kate E., and Richard G. Wilkinson. "Income Inequality and Health: A Causal Review." *Social Science & Medicine* 128 (2015): 316–26.

Piketty, Thomas. *Capital in the Twenty-First Century*. Cambridge, MA: Harvard University Press, 2014.

Pinker, Steven. *Rationality: What It Is, Why It Seems Scarce, and Why It Matters*. New York: Viking, 2021.

Plato. *The Collected Dialogues*. Ed. E. Hamilton and E. Cairns. Princeton, NJ: Princeton University Press, 1961.

Popp, Maria, Miriam Stegemann, Maria-Inti Metzendorf, Susan Gould, Peter Kranke, Patrick Meybohm, Nicole Skoetz, and Stephanie Weibel. "Ivermectin for Preventing and Treating Covid-19." *Cochrane Database of Systematic Reviews* 7 (July 28, 2021): CD015017.

Poupis, Lauren Mayor. "Wishful Hearing: The Effect of Chronic Dieting on Auditory Perceptual Biases and Eating Behavior." *Appetite* 130 (November 1, 2018): 219–27.

Prakash, Chetan, Kyle D Stephens, Donald D. Hoffman, Manish Singh, and Chris Fields. "Fitness Beats Truth in the Evolution of Perception." *Acta Biotheoretica* 69, no. 3 (2021): 319–41.

Priniski, J. Hunter, and Keith J. Holyoak. "A Darkening Spring: How Preexisting Distrust Shaped Covid-19 Skepticism." *PLoS One* 17, no. 1 (2022): e0263191.

Railton, Peter. *Facts, Values, and Norms: Essays Toward a Morality of Consequence*. Cambridge: Cambridge University Press, 2003.

Ranney, Michael A., and Dav Clark. "Climate Change Conceptual Change: Scientific Information Can Transform Attitudes." *Topics in Cognitive Science* 8, no. 1 (January 2016): 49–75.

Ranney, Michael A., and Leela Velautham. "Climate Change Cognition and Education: Given No Silver Bullet for Denial, Diverse Information-Hunks Increase Global Warming Acceptance." *Current Opinion in Behavioral Sciences* 42 (2021): 139–46.

Rawls, John. *A Theory of Justice*. Cambridge, MA: Harvard University Press, 1971.

Razak, Fahad, Saeha Shin, C. David Naylor, and Arthur S Slutsky. "Canada's Response to the Initial 2 Years of the Covid-19 Pandemic: A Comparison with Peer Countries." *Canadian Medical Asociation Journal* 194, no. 25 (June 27, 2022): E870–E77.

RECOVERY Collaborative Group. "Dexamethasone in Hospitalized Patients with Covid-19—Preliminary Report." *The New England Journal of Medicine* 384 (2021): 693–704.

Reis, Gilmar, Eduardo Augusto dos Santos Moreira-Silva, Daniela Carla Medeiros Silva, Lehana Thabane, Aline Cruz Milagres, Thiago Santiago Ferreira, Castilho Vitor Quirino dos Santos, et al. "Effect of Early Treatment with Fluvoxamine on Risk of Emergency Care and Hospitalisation Among Patients with Covid-19: The Together Randomised, Platform Clinical Trial." *The Lancet Global Health* 10 (2022): E42–E51.

Reis, Gilmar, Eduardo A. S. M. Silva, Daniela C. M. Silva, Lehana Thabane, Aline C. Milagres, Thiago S. Ferreira, Castilho V. Q. Dos Santos, et al. "Effect of Early Treatment with Ivermectin Among Patients with Covid-19." *New England Journal of Medicine* 386, no. 18 (2022): 1721–31.

Rid, Thomas. *Active Measures: The Secret History of Disinformation and Political Warfare*. New York: Farrar, Straus and Giroux, 2020.

Rippon, Gina. *Gender and Our Brains: How New Neuroscience Explodes the Myths of the Male and Female Minds*. New York: Vintage, 2020.

Ritchie, Stuart. *Science Fictions: How Fraud, Bias, Negligence, and Hype Undermine the Search for Truth*. New York: Metropolitan Books, 2020.

Rittle-Johnson, Bethany, Jon R. Star, and Kelley Durkin. "How Can Cognitive-Science Research Help Improve Education? The Case of Comparing Multiple Strategies to Improve

Mathematics Learning and Teaching." *Current Directions in Psychological Science* 29, no. 6 (2020): 599–609.

Rode, Jacob B., Amy L. Dent, Caitlin N. Benedict, Daniel B. Brosnahan, Ramona L. Martinez, and Peter H. Ditto. "Influencing Climate Change Attitudes in the United States: A Systematic Review and Meta-Analysis." *Journal of Environmental Psychology* 76 (2021). https://doi.org/10.1016/j.jenvp.2021.101623.

Rogers, Timothy T., and James L. McClelland. *Semantic Cognition: A Parallel Distributed Processing Approach.* Cambridge, MA: MIT Press, 2004.

Roozenbeek, Jon, Sander Van Der Linden, Beth Goldberg, Steve Rathje, and Stephan Lewandowsky. "Psychological Inoculation Improves Resilience Against Misinformation on Social Media." *Science Advances* 8, no. 34 (2022): eabo6254.

Rosenbaum, Paul. *Observation and Experiment.* Cambridge, MA: Harvard University Press, 2018.

Rousseau, Jean-Jacques. *Discourse on the Origin of Inequality.* Oxford: Oxford University Press, 1999.

——. *The Social Contract and Discourses.* London: J. M. Dent, 1973.

Ruse, Michael. "Creationism." *Stanford Encyclopedia of Philosophy.* 2018. https://plato.stanford.edu/entries/creationism/.

Rushton, J. Philippe, and Arthur R. Jensen. "Thirty Years of Research on Race Differences in Cognitive Ability." *Psychology, Public Policy, and Law* 11, no. 2 (2005): 235–94.

Russo, J. Edward, and Paul J. H. Schoemaker. *Decision Traps.* New York: Simon and Schuster, 1989.

Ryan, Richard M., and Edward L. Deci. *Self-Determination Theory: Basic Psychological Needs in Motivation, Development, and Wellness.* New York: Guilford, 2017.

Saag, Michael S. "Misguided Use of Hydroxychloroquine for Covid-19: The Infusion of Politics into Science." *Journal of the American Academy of Medicine* 324, no. 21 (December 1, 2020): 2161–62.

Sahdra, Baljinder, and Paul Thagard. "Self-Deception and Emotional Coherence." *Minds and Machines* 15 (2003): 213–31.

Schacter, Daniel L. *The Seven Sins of Memory: How the Mind Forgets and Remembers.* Boston: Houghton Mifflin, 2002.

Scheuering, Rachel White. *Shapers of the Great Debate on Conservation: A Biographical Dictionary.* Westport, CT: Greenwood Publishing Group, 2004.

Schick, Theodore, and Lewis Vaughn. *How to Think About Weird Things.* 8th ed. New York: McGraw Hill, 2020.

Schmid, Philipp, and Cornelia Betsch. "Effective Strategies for Rebutting Science Denialism in Public Discussions." *Nature Human Behavior* 3, no. 9 (September 2019): 931–39.

Schröder, Tobias, Terrence C. Stewart, and Paul Thagard. "Intention, Emotion, and Action: A Neural Theory Based on Semantic Pointers." *Cognitive Science* 38 (2014): 851–80.

Schwarz, Norbert. "Feelings-as-Information Theory." In *Handbook of Theories of Social Psychology*, ed. Paul A. M. Van Lange, Arie W. Kruglanski and E. Tory Higgins, 289–308. Thousand Oaks, CA: Sage, 2011.

Senay, Emily, Mona Sarfaty, and Mary B. Rice. "Strategies for Clinical Discussions About Climate Change." *Annals of Internal Medicine* 174, no. 3 (March 2021): 417–18.

Sennett, Richard. *Respect in a World of Inequality.* New York: Norton, 2003.

Seth, Anil. *Being You: A New Science of Consciousness*. New York: Dutton 2021.

Seto, Michael C. "Pedophilia." *Annual Review of Clinical Psychology* 5 (2009): 391–407.

Shaw, Amy, Ou Lydia Liu, Lin Gu, Elena Kardonova, Igor Chirikov, Guirong Li, Shangfeng Hu, et al. "Thinking Critically About Critical Thinking: Validating the Russian Heighten® Critical Thinking Assessment." *Studies in Higher Education* 45, no. 9 (2019): 1933–48.

Shermer, Michael. *Conspiracy: Why the Rational Believe the Irrational*. Baltimore, MD: Johns Hopkins University Press, 2022.

Shoemaker, Pamela J., and Timothy Vos. *Gatekeeping Theory*. London: Routledge, 2009.

Shrout, Patrick E., and Joseph L. Rodgers. "Psychology, Science, and Knowledge Construction: Broadening Perspectives from the Replication Crisis." *Annual Review of Psychology* 69 (2018): 487–510.

Simon, Dan, Douglas M. Stenstrom, and Stephen. J. Read. "The Coherence Effect: Blending Cold and Hot Cognitions." *Journal of Personality and Social Psychology* 109 (2015): 369–94.

Sinclair, Lisa, and Ziva Kunda. "Reactions to a Black Professional: Motivated Inhibition and Activation of Conflicting Stereotypes." *Journal of Personality and Social Psychology* 77 (1999): 885–904.

Singer, Peter Warren, and Emerson T. Brooking. *Likewar: The Weaponization of Social Media*. Boston: Houghton Mifflin Harcourt, 2018.

Stanley, Jason. *How Fascism Works: The Politics of Us and Them*. New York: Random House, 2020.

——. *How Propaganda Works*. Princeton, NJ: Princeton University Press, 2015.

Stanovich, Keith E. *The Bias That Divides Us: The Science and Politics of Myside Thinking*. Cambridge, MA: MIT Press, 2021.

Sterelny, Kim. *The Pleistocene Social Contract: Culture and Cooperation in Human Evolution*. Oxford: Oxford University Press, 2021.

Stiglitz, Joseph E. *The Price of Inequality: How Today's Divided Society Endangers Our Future*. New York: Norton, 2013.

Strauss, Barry. *The Death of Caesar: The Story of History's Most Famous Assassination*. New York: Simon and Schuster, 2015.

Sunstein, Cass Robert, and Cornelius Adrian Vermeule. "Conspiracy Theories: Causes and Cures." *Journal of Political Philosophy* 17 (2009): 202–27.

Taber, Charles S., and Milton Lodge. "Motivated Skepticism in the Evaluation of Political Beliefs." *American Journal of Political Science* 50, no. 3 (2006): 755–69.

Tay, Li Qian, Mark J. Hurlstone, Tim Kurz, and Ullrich K. H. Ecker. "A Comparison of Prebunking and Debunking Interventions for Implied Versus Explicit Misinformation." *British Journal of Psychology* 113 (2021): 591–607.

Taylor, Shelley E., and Jonathon D. Brown. "Illusion and Well-Being: A Social Psychological Perspective on Mental Health." *Psychological Bulletin* 103, no. 2 (March 1988): 193–210.

Tegmark, M. *Our Mathematical Universe*. New York: Knopf, 2014.

Thagard, Paul. "Adversarial Problem Solving: Modelling an Opponent Using Explanatory Coherence." *Cognitive Science* 16 (1992): 123–49.

——. *Balance: How It Works and What It Means*. New York: Columbia University Press, 2022.

——. *Bots and Beasts: What Makes Machines, Animals, and People Smart?* Cambridge, MA: MIT Press, 2021.

——. *The Brain and the Meaning of Life*. Princeton, NJ: Princeton University Press, 2010.

——. *Brain-Mind: From Neurons to Consciousness and Creativity*. New York: Oxford University Press, 2019.
——. "The Cognitive Science of Covid-19: Acceptance, Denial, and Belief Change." *Methods* 195 (2021): 92–102.
——. *The Cognitive Science of Science: Explanation, Discovery, and Conceptual Change*. Cambridge, MA: MIT Press, 2012.
——. *Coherence in Thought and Action*. Cambridge, MA: MIT Press, 2000.
——. "Collaborative Knowledge." *NoÛs* 31 (1997): 242–61.
——. *Computational Philosophy of Science*. Cambridge, MA: MIT Press, 1988.
——. *Conceptual Revolutions*. Princeton, NJ: Princeton University Press, 1992.
——. "Critical Thinking and Informal Logic: Neuropsychological Perspectives." *Informal Logic* 31 (2011): 152–70.
——. "The Emotional Coherence of Religion." *Journal of Cognition and Culture* 5 (2005): 58–74.
——. "Energy Requirements Undermine Substrate Independence and Mind-Body Functionalism." *Philosophy of Science* 89 (2022): 70–88.
——. "Evolution, Creation, and the Philosophy of Science." In *Evolution, Epistemology, and Science Education*, ed. R. Taylor and M. Ferrari, 20–37. Milton Park, UK: Routledge, 2010.
——. "Explanatory Coherence." *Behavioral and Brain Sciences* 12 (1989): 435–67.
——. *Hot Thought: Mechanisms and Applications of Emotional Cognition*. Cambridge, MA: MIT Press, 2006.
——. "How Rationality Is Bounded by the Brain." In *Routledge Handbook of Bounded Rationality*, ed. Riccardo Viale, 398–406. London: Routledge, 2021.
——. *How Scientists Explain Disease*. Princeton, NJ: Princeton University Press, 1999.
——. "How to Collaborate: Procedural Knowledge in the Cooperative Development of Science." *Southern Journal of Philosophy* 44, (2006): 177–96.
——. *Mind-Society: From Brains to Social Sciences and Professions* New York: Oxford University Press, 2019.
——. "The Moral Psychology of Conflicts of Interest: Insights from Affective Neuroscience." *Journal of Applied Philosophy* 24 (2007): 367–80.
——. *Natural Philosophy: From Social Brains to Knowledge, Reality, Morality, and Beauty*. New York: Oxford University Press, 2019.
——. "The Self as a System of Multilevel Interacting Mechanisms." *Philosophical Psychology* 27 (2014): 145–63.
——. "Social Equality: Cognitive Modeling Based on Emotional Coherence Explains Attitude Change." *Policy Insights from Behavioral and Brain Sciences* 5, no. 2 (2018): 247–56.
——. "Testimony, Credibility, and Explanatory Coherence." *Erkenntnis* 63 (2005): 295–316.
——. "Thought Experiments Considered Harmful." *Perspectives on Science* 22 (2014): 288–305.
——. "Why Astrology Is a Pseudoscience." In *Psa 1978*, ed. P. Aquith and I. Hacking, 223–34. East Lansing MI: Philosophy of Science Association, 1978.
——. "Why Wasn't O. J. Convicted? Emotional Coherence in Legal Inference." *Cognition and Emotion* 17 (2003): 361–83.
Thagard, Paul, and Scott D. Findlay. "Changing Minds About Climate Change: Belief Revision, Coherence, and Emotion." In *Belief Revision Meets Philosophy of Science*, ed. E. J. Olsson and S. Enqvist, 329–45. Berlin: Springer, 2011.

Thagard, Paul, and Scott Findlay. "Getting to Darwin: Obstacles to Accepting Evolution by Natural Selection." *Science & Education* 19 (2010): 625–36.

Thagard, Paul, and Ziva Kunda. "Hot Cognition: Mechanisms of Motivated Inference.". In *Proceedings of the Ninth Annual Conference of the Cognitive Science Society.*, ed. E. Hunt, 753–63. Hillsdale, NJ: Erlbaum, 1987.

Thagard, Paul, Laurette Larocque, and Ivana Kajić. "Emotional Change: Neural Mechanisms Based on Semantic Pointers." *Emotion* 23 (2023): 182-93

Thagard, Paul, and Elijah Millgram. "Inference to the Best Plan: A Coherence Theory of Decision." In *Goal-Driven Learning*, ed. A. Ram and D. B. Leake, 439–54. Cambridge, MA: MIT Press, 1995.

Thagard, Paul, and Karsten Verbeurgt. "Coherence as Constraint Satisfaction." *Cognitive Science* 22 (1998): 1–24.

Thagard, Paul, and Joanne V. Wood. "Eighty Phenomena About the Self: Representation, Evaluation, Regulation, and Change." *Frontiers in Psychology* 6 (March 27, 2015): 34.

Thomson, Oliver. *Easily Led: A History of Propaganda.* Phoenix Mill, UK: Sutton, 1999.

Treen, Kathie M. d'I, Hywel T. P. Williams, and Saffron J. O'Neill. "Online Misinformation About Climate Change." *WIREs Climate Change* 11, no. 5 (2020): e665.

Tucker, William H. *The Funding of Scientific Racism: Wickliffe Draper and the Pioneer Fund.* Champaign: University of Illinois Press, 2002.

Uscinski, Joseph E., ed. *Conspiracy Theories and the People Who Believe Them.* New York: Oxford University Press, 2018.

van der Linden, Sander. *Foolproof: Why Misinformation Infects Our Minds and How to Build Immunity.* New York: Norton, 2023.

Van Der Zee, Sophie, Ronald Poppe, Alice Havrileck, and Aurelien Baillon. "A Personal Model of Trumpery: Linguistic Deception Detection in a Real-World High-Stakes Setting." *Psychological Science* 33, no. 1 (January 2022): 3–17.

van Fraassen, B. *The Scientific Image.* Oxford: Clarendon Press, 1980.

van Huijstee, Dian, Ivar Vermeulen, Peter Kerkhof, and Ellen Droog. "Continued Influence of Misinformation in Times of Covid-19." *International Journal of Psychology* 57, no. 1 (2022): 136–45.

Van Parijs, Phillippe, and Yannick Vanderborght. *Basic Income: A Radical Proposal for a Free Society and a Sane Economy.* Cambridge, MA: Harvard University Press, 2017.

van Prooijen, Jan-Willem. "Injustice Without Evidence: The Unique Role of Conspiracy Theories in Social Justice Research." *Social Justice Research* (September 28, 2021): 1–19.

——. *The Psychology of Conspiracy Theories.* London: Routledge, 2018.

Vansteenkiste, Maarten, and Kennon M. Sheldon. "There's Nothing More Practical Than a Good Theory: Integrating Motivational Interviewing and Self-Determination Theory." *British Journal of Clinical Psychology* 45 (2006): 63–82.

Van Zee, Art. "The Promotion and Marketing of Oxycontin: Commercial Triumph, Public Health Tragedy." *American Journal of Public Health* 99 (2009): 221–27.

Vegetti, Federico, and Moreno Mancosu. "The Impact of Political Sophistication and Motivated Reasoning on Misinformation." *Political Communication* 37, no. 5 (2020): 678–95.

Venner, Kamilla L., and Steven P. Verney. "Motivational Interviewing: Reduce Student Reluctance and Increase Engagement in Learning Multicultural Concepts." *Professional Psychology: Research and Practice* 46, no. 2 (2015): 116.

Vervoort, Dominique, Xiya Ma, and Mark G Shrime. "Money Down the Drain: Predatory Publishing in the Covid-19 Era." *Canadian Journal of Public Health* 111, no. 5 (2020): 665–66.

Wagar, Brandon M., and Paul Thagard. "Spiking Phineas Gage: A Neurocomputational Theory of Cognitive-Affective Integration in Decision Making." *Psychological Review* 111 (2004): 67–79.

Walzer, Michael. *Spheres of Justice: A Defense of Pluralism and Equality*. New York: Basic Books, 1983.

Wang, Dang, Hongyun Liu, and Kit-Tai Hau. "Automated and Interactive Game-Based Assessment of Critical Thinking." *Education and Information Technologies* 27, no. 4 (2022): 4553–75.

Ward, Stephen J. A. *Disrupting Journalism Ethics: Radical Change on the Frontier of Digital Media*. London: Routledge, 2019.

Watson, James D. *The Double Helix: A Personal Account of the Discovery of the Structure of DNA*. New York: New American Library, 1968.

Watson, Oliver J., Gregory Barnsley, Jaspreet Toor, Alexandra B. Hogan, Peter Winskill, and Azra C. Ghani. "Global Impact of the First Year of Covid-19 Vaccination: A Mathematical Modelling Study." *The Lancet Infectious Diseases* 22 (2022): P1293–302.

Weatherald, Jason, Kevin Solverson, Danny J. Zuege, Nicole Loroff, Kirsten M. Fiest, and Ken Kuljit S. Parhar. "Awake Prone Positioning for Covid-19 Hypoxemic Respiratory Failure: A Rapid Review." *Journal of Critical Care* 61 (2021): 63–70.

Westen, Drew, Pavel S. Blagov, Keith Harenski, Clint Kilts, and Stephan Hamann. "Neural Bases of Motivated Reasoning: An fMRI Study of Emotional Constraints on Partisan Political Judgment in the 2004 U.S. Presidential Election." *Journal of Cognitive Neuroscience* 18 (2006): 1947–58.

Wilkinson, Richard G., and Kate Pickett. *The Inner Level: How More Equal Societies Reduce Stress, Restore Sanity and Improve Everyone's Well-Being*. New York: Penguin Press, 2020.

Wilkinson, Richard G., and Kate Pickett. *The Spirit Level: Why Greater Equality Makes Societies Stronger*. New York: Penguin, 2010.

Williams, Daniel. "Motivated Ignorance, Rationality, and Democratic Politics." *Synthese* 198, no. 8 (2020): 7807–27.

Worobey, Michael, Joshua I. Levy, Lorena Malpica Serrano, Alexander Crits-Christoph, Jonathan E. Pekar, Stephen A. Goldstein, Angela L. Rasmussen, et al. "The Huanan Seafood Wholesale Market in Wuhan Was the Early Epicenter of the Covid-19 Pandemic." *Science* 377, no. 6609 (2022): 951–59.

Yasmin, Seema. *Viral BS: Medical Myths and Why We Fall for Them*. Baltimore, MD: Johns Hopkins University Press, 2021.

Yeo, Sara K., and Meaghan McKasy. "Emotion and Humor as Misinformation Antidotes." *Proceedings of the National Academy of Sciences* 118, no. 15 (2021).

Zhu, Na, Dingyu Zhang, Wenling Wang, Xingwang Li, Bo Yang, Jingdong Song, Xiang Zhao, et al. "A Novel Coronavirus from Patients with Pneumonia in China, 2019." *New England Journal of Medicine* 382, no. 8 (February 20, 2020): 727–33.

Zins, Chaim. "Conceptual Approaches for Defining Data, Information, and Knowledge." *Journal of the American Society for Information Science and Technology* 58, no. 4 (2007): 479–93.

INDEX

Page numbers in *italics* indicate figures or tables.